PRINCIPLES OF CONTINUUM MECHANICS

PRINCIPLES OF CONTINUUM MECHANICS

David J Steigmann
Milad Shirani
University of California, Berkeley, USA

World Scientific
NEW JERSEY • LONDON • SINGAPORE • BEIJING • SHANGHAI • TAIPEI • CHENNAI

Published by

World Scientific Publishing Co. Pte. Ltd.
5 Toh Tuck Link, Singapore 596224
USA office: 27 Warren Street, Suite 401-402, Hackensack, NJ 07601
UK office: 57 Shelton Street, Covent Garden, London WC2H 9HE

Library of Congress Control Number: 2025001827

British Library Cataloguing-in-Publication Data
A catalogue record for this book is available from the British Library.

PRINCIPLES OF CONTINUUM MECHANICS

ISBN 978-981-98-0685-0 (hardcover)
ISBN 978-981-98-0754-3 (paperback)
ISBN 978-981-98-0686-7 (ebook for institutions)
ISBN 978-981-98-0687-4 (ebook for individuals)

For any available supplementary material, please visit
https://www.worldscientific.com/worldscibooks/10.1142/14150#t=suppl

Desk Editors: Selas Hamilton/Julio Hong

Typeset by Stallion Press
Email: enquiries@stallionpress.com

Preface

This book is an introduction to the basic principles that form the foundation of modern continuum mechanics. Included are treatments of the motions and deformations of material bodies; of their momenta, energy and entropy; and of the states of stress within bodies. These, in turn, are interrelated by constitutive equations that codify the manner in which idealized materials respond to various processes. The concepts developed here are common to all continua belonging to a class which is sufficiently broad to encompass the classical models of viscous and inviscid fluids, and of elastic solids. These particular models are emphasized because of their wide range of applicability to practical problems, and because they furnish a foundation upon which theories of more complex material behavior can be built.

The book is based on a long-running course delivered to advanced undergraduates and beginning graduate students at Berkeley. This is a gateway course for a graduate-level curriculum treating specialized topics in continuum mechanics, including fluid dynamics, the linear and nonlinear theories of elasticity, plasticity theory, thermodynamics, electromagnetic interactions in materials, and generalized continua.

Most books in this genre reflect the influence of the monumental treatise, *The Non-Linear Field Theories of Mechanics*, by C. Truesdell and W. Noll (3rd edn. (S.S. Antman, ed.) reprinted by Springer, Berlin, 2004), and the present book is no exception. The importance of that work to the advancement of the field since the mid-1960s can hardly be overstated. This is especially true of the textbook and monograph literatures that have emerged since that time. An important example, and the inspiration for the present work, is the well known and widely adopted text: *Continuum Mechanics: Concise Theory and Problems*, by P. Chadwick (Dover, N.Y., 1999). This strikes a particularly good balance between coverage of theoretical concepts and their use in the solution of illustrative problems, a model that we have endeavored to emulate.

Distinguishing features of the present book include detailed treatments of the concept of material symmetry and its use in the classification of different constitutive models, and also of the principle of material frame indifference, which posits that constitutive response is intrinsic to material behavior and hence insensitive to the observer tasked with characterizing such response. This has important consequences for the constitutive equations employed by any one observer. The notion of a frame of reference is also emphasized due to the central role played by the so-called inertial frames in formulating the equations of motion.

A chapter on linear elasticity theory, arguably the most important single subject in all of solid mechanics, is included. It contains a demonstration that this theory, properly formulated, is frame invariant. This resolves a long-standing controversy surrounding the status of linear elasticity theory. In view of the great importance of the theory in applications, this is a question whose resolution is long overdue.

The various chapters are mostly devoted to the development of theoretical concepts. These culminate in problems designed to exhibit the use of these concepts in the solution of illustrative examples. Extensive solutions to the problems are provided at the back of the book. These can be consulted on a chapter-by-chapter basis to reinforce understanding. Several appendices detailing technical points alluded to in the text are also provided.

About the Authors

David Steigmann is a professor of mechanical engineering at the University of California, Berkeley. He is the recipient of the SES Medal of the Society of Engineering Science, awarded for 'singular contributions to engineering science', in addition to the 'Tullio Levi-Civita' prize for the mathematical and physical sciences, awarded for '... the high quality and originality of the scientific research of eminent Italian or foreign scientists'. He is the author or co-author of research monographs on nonlinear elasticity theory, plasticity theory, and the theory of plates and shells. He is the editor-in-chief of the journal *Mathematics and Mechanics of Solids* and the solid mechanics editor for the journal *Zeitschrift fur angewandte Mathematik und Physik.*

Milad Shirani was born in Isfahan, Iran, and lived there before coming to the United States. He obtained his B.Sc. and M.Sc. in Mechanical Engineering from Isfahan University of Technology, Isfahan, Iran. Subsequently, he attended the University of California, Berkeley to do his Ph.D. under the guidance and supervision of Prof. David Steigmann in the Department of Mechanical Engineering. He was the recipient of the Robert F. Steidel Fellowship (UC Berkeley, 2021) and the Paul Naghdi Fellowship (UC Berkeley, 2020).

He has co-authored several research papers, chapters, and a book on the theory of plates and shells with Prof. David Steigmann.

Contents

Chapter 1

Mathematical Preliminaries

A working knowledge of the elements of vector and tensor algebra and analysis is a prerequisite to a modern course on continuum mechanics. In this chapter, we summarize the main results that we will need to proceed.

1.1 Vector and tensor theory

1.1.1 *Vector spaces*

The concept of a vector space, also known as a linear space, is fundamental to our development of continuum mechanics. For our purposes a vector space, denoted by V, is a set of elements $\{\boldsymbol{a}, \boldsymbol{b}, \boldsymbol{c}, \ldots\}$, called *vectors*, plus two operations — addition and scalar multiplication — such that the following hold for all $\alpha, \beta \in \mathbb{R}$, the set of real numbers:

(a) $\alpha\boldsymbol{a} + \beta\boldsymbol{b} \in V$ (any linear combination of vectors is a vector),
(b) $(\boldsymbol{a} + \boldsymbol{b}) + \boldsymbol{c} = \boldsymbol{a} + (\boldsymbol{b} + \boldsymbol{c})$ (associativity with respect to "+"),
(c) $\boldsymbol{a} + \boldsymbol{b} = \boldsymbol{b} + \boldsymbol{a}$ (commutativity with respect to "+"),
(d) There is $\mathbf{0} \in V$ such that $\boldsymbol{a} + \mathbf{0} = \boldsymbol{a}$ (existence of a null element),
(e) For each $\boldsymbol{a} \in V$ there is $\boldsymbol{b} \in V$ such that $\boldsymbol{a} + \boldsymbol{b} = \mathbf{0}$; we write $\boldsymbol{b} = -\boldsymbol{a}$ (existence of an additive inverse),
(f) $(\alpha\beta)\boldsymbol{v} = \alpha(\beta\boldsymbol{v})$ (associativity with respect to $\cdot$),
(g) $(\alpha + \beta)\boldsymbol{a} = \alpha\boldsymbol{a} + \beta\boldsymbol{b}$ (distributivity with respect to $\mathbb{R}$),
(h) $\alpha(\boldsymbol{a} + \boldsymbol{b}) = \alpha\boldsymbol{a} + \alpha\boldsymbol{b}$ (distributivity with respect to V),
(i) $1\boldsymbol{a} = \boldsymbol{a}$ (existence of identity),
(j) $0\boldsymbol{a} = \mathbf{0}$.

We always write vectors in bold face to distinguish them from real numbers, written in light face.

Let $\{\boldsymbol{v}_1, \boldsymbol{v}_2, \ldots, \boldsymbol{v}_p\}$ be a set of elements of V, and $\{\alpha_1, \alpha_2, \ldots, \alpha_p\}$ a set of real numbers. Suppose that

$$\alpha_1 \boldsymbol{v}_1 + \alpha_2 \boldsymbol{v}_2 + \cdots + \alpha_p \boldsymbol{v}_p = \mathbf{0}. \tag{1.1}$$

Clearly, this is satisfied if the alphas are all equal to zero. If the equality holds *only if all* the alphas are equal to zero, i.e., if (1.1) implies that $\alpha_1 = \alpha_2 = \cdots = \alpha_p = 0$, then the set $\{\boldsymbol{v}_1, \boldsymbol{v}_2, \ldots, \boldsymbol{v}_p\}$ is said to be *linearly independent.* This set is *linearly dependent* if equality (1.1) holds with at least one member of the set $\{\alpha_1, \alpha_2, \ldots, \alpha_p\}$ unequal to zero. For, in this case, we can express at least one of the vectors as a linear combination of the others.

The vector space V is *finite-dimensional* if there is an integer n such that all linearly independent sets contain at most n elements. Then, n is the *dimension* of the space, denoted henceforth by V^n.

Any set of n linearly independent vectors is a *basis* for V^n. Let $\{\boldsymbol{u}_1, \boldsymbol{u}_2, \ldots, \boldsymbol{u}_n\}$ be a basis. Then any $\boldsymbol{v} \in V^n$ can be expressed as a linear combination of the basis elements, i.e.,

$$\boldsymbol{v} = \alpha_1 \boldsymbol{u}_1 + \alpha_2 \boldsymbol{u}_2 + \cdots + \alpha_n \boldsymbol{u}_n. \tag{1.2}$$

To demonstrate this, observe that $\{\boldsymbol{u}_1, \boldsymbol{u}_2, \ldots, \boldsymbol{u}_n, \boldsymbol{v}\}$ is a set of $n+1$ vectors. These are linearly dependent in V^n, and therefore

$$\lambda \boldsymbol{v} + \sum_{i=1}^{n} \lambda_i \boldsymbol{u}_i = \mathbf{0}, \tag{1.3}$$

with at least one member of $\{\lambda_1, \lambda_2, \ldots, \lambda_n, \lambda\}$ unequal to zero. In fact, $\lambda \neq 0$ because, were it otherwise we would have that $\sum_{i=1}^{n} \lambda_i \boldsymbol{u}_i = \mathbf{0}$, and hence that $\lambda_1, \lambda_2, \ldots, \lambda_n$ are all equal to zero by virtue of the linear independence of $\{\boldsymbol{u}_1, \boldsymbol{u}_2, \ldots, \boldsymbol{u}_n\}$. We would thus conclude that $\lambda_1 = \lambda_2 = \cdots = \lambda_n = \lambda = 0$, and hence that $\{\boldsymbol{u}_1, \boldsymbol{u}_2, \ldots, \boldsymbol{u}_n, \boldsymbol{v}\}$ is a linearly independent set, contrary to the fact that V^n has dimension n. Accordingly we can express $\boldsymbol{v}$ in the form (1.2), with $\alpha_i = -\lambda_i/\lambda$.

The coefficients α_i are unique. To see this suppose that $\boldsymbol{v} = \sum_{i=1}^{n} \beta_i \boldsymbol{u}_i$. Then, from (1.2), $\sum_{i=1}^{n} (\beta_i - \alpha_i)\boldsymbol{u}_i = \mathbf{0}$, and the linear independence of the $\boldsymbol{u}_i$ yields $\beta_i = \alpha_i$; $i = 1, \ldots, n$.

Of particular relevance to our development is the *Euclidean* vector space E^n. This is V^n, endowed with the additional structure conferred by the *inner product*, or *dot product* operation $\boldsymbol{u} \cdot \boldsymbol{v} \in \mathbb{R}$ defined by

(a) $\boldsymbol{u} \cdot \boldsymbol{v} = \boldsymbol{v} \cdot \boldsymbol{u}$ (commutativity),
(b) $\boldsymbol{u} \cdot (\boldsymbol{v} + \boldsymbol{w}) = \boldsymbol{u} \cdot \boldsymbol{v} + \boldsymbol{u} \cdot \boldsymbol{w}$ (associativity),
(c) $(\alpha \boldsymbol{u}) \cdot \boldsymbol{v} = \boldsymbol{u} \cdot (\alpha \boldsymbol{v}) = \alpha(\boldsymbol{u} \cdot \boldsymbol{v})$,
(d) $\boldsymbol{u} \cdot \boldsymbol{u} \geq 0$ and $\boldsymbol{u} \cdot \boldsymbol{u} = 0$ if and only if $\boldsymbol{u} = \boldsymbol{0}$.

The *Euclidean norm* of $\boldsymbol{u}$ is $|\boldsymbol{u}| = \sqrt{\boldsymbol{u} \cdot \boldsymbol{u}}$, a non-negative real number by virtue of property (d). The vector $\boldsymbol{u}$ is a *unit vector* if $|\boldsymbol{u}| = 1$.

The norm satisfies the *Cauchy–Schwarz* inequality

$$|\boldsymbol{u} \cdot \boldsymbol{v}| \leq |\boldsymbol{u}|\,|\boldsymbol{v}|, \tag{1.4}$$

implying that there exists $\theta \in \mathbb{R}$, the *angle between* $\boldsymbol{u}$ *and* $\boldsymbol{v}$, such that

$$\cos\theta = \frac{\boldsymbol{u} \cdot \boldsymbol{v}}{|\boldsymbol{u}|\,|\boldsymbol{v}|}; \tag{1.5}$$

and the *Triangle* inequality

$$|\boldsymbol{u} + \boldsymbol{v}| \leq |\boldsymbol{u}| + |\boldsymbol{v}|. \tag{1.6}$$

The vectors $\boldsymbol{u}$, $\boldsymbol{v}$ are *orthogonal* if $\boldsymbol{u} \cdot \boldsymbol{v} = 0$, and we require that $\boldsymbol{0} \cdot \boldsymbol{v} = 0$ for all $\boldsymbol{v} \in E^n$.

The elements of $\{\boldsymbol{u}_1, \boldsymbol{u}_2, \ldots, \boldsymbol{u}_n\}$ are *orthonormal* if

$$\boldsymbol{u}_i \cdot \boldsymbol{u}_j = \delta_{ij}, \tag{1.7}$$

where δ_{ij}, the *Kronecker delta*, is defined by

$$\delta_{ij} = \begin{cases} 1, & i = j \\ 0, & i \neq j. \end{cases} \tag{1.8}$$

In E^n every set $\{\boldsymbol{e}_1, \boldsymbol{e}_2, \ldots, \boldsymbol{e}_k\}$ of orthonormal vectors, with $k \leq n$, is linearly independent, since, if

$$\alpha_1 \boldsymbol{e}_1 + \alpha_2 \boldsymbol{e}_2 + \cdots + \alpha_k \boldsymbol{e}_k = \boldsymbol{0}, \tag{1.9}$$

then, by forming the dot product with $\boldsymbol{e}_j$, with $j \in \{1, \ldots, k\}$ fixed, we get

$$0 = \boldsymbol{0} \cdot \boldsymbol{e}_j = \left(\sum_{i=1}^{k} \alpha_i \boldsymbol{e}_i\right) \cdot \boldsymbol{e}_j = \sum_{i=1}^{k} \alpha_i \boldsymbol{e}_i \cdot \boldsymbol{e}_j = \sum_{i=1}^{k} \alpha_i \delta_{ij} = \alpha_j. \tag{1.10}$$

Because $\{\boldsymbol{e}_1, \boldsymbol{e}_2, \ldots, \boldsymbol{e}_n\}$ is a basis for E^n, for any $\boldsymbol{v} \in E^n$ we have

$$\boldsymbol{v} = \sum_{i=1}^{n} v_i \boldsymbol{e}_i, \tag{1.11}$$

in which the *components* v_i are uniquely determined by $\boldsymbol{v}$ and the $\boldsymbol{e}_i$.

In E^n it is possible to construct an orthonormal basis $\{\boldsymbol{e}_1, \boldsymbol{e}_2, \ldots, \boldsymbol{e}_n\}$ from any basis $\{\boldsymbol{u}_1, \boldsymbol{u}_2, \ldots, \boldsymbol{u}_n\}$ via the Gram–Schmidt procedure [1]. Accordingly, there exist infinitely many orthonormal bases in E^n.

It is rather cumbersome, and visually distracting, to write out the summation sign in (1.11) explicitly. Instead, we write

$$\boldsymbol{v} = v_i \boldsymbol{e}_i, \tag{1.12}$$

on the understanding that we are to sum on the repeated index i over the range $1, \ldots, n$. This convention extends to all the *terms* in an equation; that is, to expressions separated by "+", "−", or "=" signs. This *summation convention*, due to Einstein, facilitates the conveying of information in a concise manner, with a minimum of visual clutter. A summed index (e.g., i in (1.12)) is called a *dummy index*, since it may be relabeled without affecting the meaning of an equation. For example, (1.12) is the same as $\boldsymbol{v} = v_j \boldsymbol{e}_j$, a fact that becomes immediately evident on carrying out the sum explicitly. As we shall see, the freedom to relabel repeated indices proves to be a great convenience. The summation convention applies only to indices that are repeated once in a given term, i.e., that appear exactly twice.

Note that

$$\boldsymbol{v} \cdot \boldsymbol{e}_j = v_i \boldsymbol{e}_i \cdot \boldsymbol{e}_j = v_i \delta_{ij} = v_j; \quad j = 1, \ldots, n, \tag{1.13}$$

where, in the final equality, we have summed over the repeated index i and invoked the properties of the Kronecker delta. Here, the index j is not a repeated index. It is called a *free index*, and must match in every term of an equation, as in this example. Crucially, the free index must have a different label than a dummy index if the intended meaning is to be preserved. This rule is best reinforced by recasting (1.13) as $\sum_{i=1}^{n} v_i \delta_{ij} = v_j$, at least until the summation convention becomes second nature.

We have shown, for any orthonormal basis $\{\boldsymbol{e}_i\}$; $i = 1, \ldots, n$ in E^n, that

$$\boldsymbol{v} = (\boldsymbol{v} \cdot \boldsymbol{e}_k)\boldsymbol{e}_k, \tag{1.14}$$

where, of course, k is summed from 1 to n. Thus, the component v_i, with $i \in \{1, \ldots, n\}$ fixed, is the projection of $\boldsymbol{v}$ onto the corresponding basis element $\boldsymbol{e}_i$.

Consider the dot product $\boldsymbol{u} \cdot \boldsymbol{v}$ of two vectors $\boldsymbol{u} = u_i\boldsymbol{e}_i$ and $\boldsymbol{v} = v_i\boldsymbol{e}_i$. We have

$$\boldsymbol{u} \cdot \boldsymbol{v} = \boldsymbol{u} \cdot (v_i\boldsymbol{e}_i) = \boldsymbol{u} \cdot \boldsymbol{e}_i v_i = u_i v_i (= u_1v_1 + u_2v_2 + \cdots + u_nv_n). \tag{1.15}$$

Thus, the dot predict of two vectors is computable from their components. Alternatively,

$$\boldsymbol{u} \cdot \boldsymbol{v} = u_i\boldsymbol{e}_i \cdot v_j\boldsymbol{e}_j = u_iv_j\boldsymbol{e}_i \cdot \boldsymbol{e}_j = u_iv_j\delta_{ij} = u_iv_i, \tag{1.16}$$

where, after the first, second and third equalities, to avoid conflict with the summation convention we have been careful *not* to use the same labels for the dummy indices. In these terms both indices i and j are repeated, and hence a double sum over i and j is implied, each from 1 to n, comprising n^2 terms altogether. We begin to appreciate the great economy afforded by Einstein's convention.

Example

Suppose $f(\boldsymbol{v})$ is a scalar valued function of vectors $\boldsymbol{v} \in E^n$; that is, the value of the function is a real number (e.g., $f(\boldsymbol{v}) = |\boldsymbol{v}|$). Suppose $f(\boldsymbol{v})$ is *linear*. This means that

$$f(\alpha_1\boldsymbol{v}_1 + \alpha_2\boldsymbol{v}_2) = \alpha_1 f(\boldsymbol{v}_1) + \alpha_2 f(\boldsymbol{v}_2) \tag{1.17}$$

for all $\boldsymbol{v}_1, \boldsymbol{v}_2 \in E^n$ and all $\alpha_1, \alpha_2 \in \mathbb{R}$. We seek the general form of such a function. To establish this form we combine (1.12) with (1.17) to obtain

$$f(\boldsymbol{v}) = f(v_i\boldsymbol{e}_i) = v_i f(\boldsymbol{e}_i). \tag{1.18}$$

Here, the $f(\boldsymbol{e}_i)$ are the values of the function acting on fixed input vectors. These are fixed constants, independent of the variable $\boldsymbol{v}$. We call them $a_i : a_i = f(\boldsymbol{e}_i) \in \mathbb{R}$; $i = 1, \ldots, n$. Thus, $f(\boldsymbol{v}) = a_iv_i = \boldsymbol{a} \cdot \boldsymbol{v}$, where $\boldsymbol{a} = a_k\boldsymbol{e}_k$ is a fixed vector that characterizes the particular linear function at hand. Thus, the most general *linear* scalar valued function defined on E^n is of the form

$$f(\boldsymbol{v}) = \boldsymbol{a} \cdot \boldsymbol{v}, \tag{1.19}$$

where $\boldsymbol{a} \in E^n$ is fixed. This is the finite-dimensional form of the *Riesz Representation Theorem* for linear functions.

1.1.2 *Change-of-basis formulas*

Consider an orthonormal basis $\{\boldsymbol{e}'_i\}$ for E^n. Using (1.14), we can write

$$\boldsymbol{e}_i = (\boldsymbol{e}_i \cdot \boldsymbol{e}'_j)\boldsymbol{e}'_j. \tag{1.20}$$

Consider a vector $\boldsymbol{v} \in E^n$ with components $v_i = \boldsymbol{v} \cdot \boldsymbol{e}_i$ relative to the basis $\{\boldsymbol{e}_i\}$, and $v'_i = \boldsymbol{v} \cdot \boldsymbol{e}'_i$ relative to $\{\boldsymbol{e}'_i\}$. We have

$$v'_j\boldsymbol{e}'_j = \boldsymbol{v} = v_i\boldsymbol{e}_i = v_i(\boldsymbol{e}_i \cdot \boldsymbol{e}'_j)\boldsymbol{e}'_j, \quad \text{or } [v'_j - v_i(\boldsymbol{e}_i \cdot \boldsymbol{e}'_j)]\boldsymbol{e}'_j = \boldsymbol{0}, \tag{1.21}$$

and the linear independence of the basis elements furnishes the *transformation formula*

$$v'_j = v_i a_{ij}, \quad \text{where } a_{ij} = \boldsymbol{e}_i \cdot \boldsymbol{e}'_j. \tag{1.22}$$

Note that the components of the vector change when the basis is changed, while the vector itself remains invariant. The matrix (a_{ij}) is called the matrix of *direction cosines* relating the two sets of basis elements.

It is simple and instructive to repeat this exercise, with the $\{\boldsymbol{e}'_i\}$ and $\{\boldsymbol{e}_i\}$ interchanged, to obtain the inverse of (1.22).

1.1.3 *Vector products in E^3*

A further useful operation, for vectors in E^3, is the *vector product*, or *cross product*. This is equivalent, in E^3, to the *exterior product*, or *wedge product*, defined for vectors in E^n with n arbitrary [1, 2]. For vectors $\boldsymbol{u}, \boldsymbol{v} \in E^3$, the vector product $\boldsymbol{u} \times \boldsymbol{v} \in E^3$ is defined by the rules:

(a) $\boldsymbol{u} \times \boldsymbol{v} = -\boldsymbol{v} \times \boldsymbol{u}$.
(b) $|\boldsymbol{u} \times \boldsymbol{v}| = |\boldsymbol{u}|\,|\boldsymbol{v}| \sin\theta$ (the same θ as in (1.5)).

(c) $(\alpha \boldsymbol{u} + \beta \boldsymbol{v}) \times \boldsymbol{w} = \alpha \boldsymbol{u} \times \boldsymbol{w} + \beta \boldsymbol{v} \times \boldsymbol{w}$.

(d) $\boldsymbol{u} \cdot \boldsymbol{v} \times \boldsymbol{w} = \boldsymbol{v} \cdot \boldsymbol{w} \times \boldsymbol{u} = \boldsymbol{w} \cdot \boldsymbol{u} \times \boldsymbol{v} = [\boldsymbol{u}, \boldsymbol{v}, \boldsymbol{w}]$ (the scalar triple product, or *box product*).

From (a) and (d) we have $\boldsymbol{u} \times \boldsymbol{u} = \boldsymbol{0}$ and $[\boldsymbol{u}, \boldsymbol{u}, \boldsymbol{v}] = [\boldsymbol{v}, \boldsymbol{u}, \boldsymbol{v}] = \boldsymbol{0}$, so that $\boldsymbol{u} \times \boldsymbol{v}$ is orthogonal to $\boldsymbol{u}$ and $\boldsymbol{v}$. Further, the box product is linear in each argument. To demonstrate this in respect of the first argument, we form the product $(\alpha \boldsymbol{a} + \beta \boldsymbol{b}) \cdot \boldsymbol{e}$, with $\boldsymbol{e} = \boldsymbol{c} \times \boldsymbol{d}$. Thus,

$$[\alpha \boldsymbol{a} + \beta \boldsymbol{b}, \boldsymbol{c}, \boldsymbol{d}] = (\alpha \boldsymbol{a} + \beta \boldsymbol{b}) \cdot \boldsymbol{e} = \alpha [\boldsymbol{a}, \boldsymbol{c}, \boldsymbol{d}] + \beta [\boldsymbol{b}, \boldsymbol{c}, \boldsymbol{d}], \tag{1.23}$$

and similarly for the second and third arguments. In addition, $[\boldsymbol{u}, \boldsymbol{v}, \boldsymbol{w}] = \boldsymbol{u} \cdot \boldsymbol{v} \times \boldsymbol{w} = -\boldsymbol{u} \cdot \boldsymbol{w} \times \boldsymbol{v} = -[\boldsymbol{u}, \boldsymbol{w}, \boldsymbol{v}]$, etc. Thus, the box product reverses sign whenever two of its arguments are interchanged.

Suppose $\{\boldsymbol{e}_1, \boldsymbol{e}_2, \boldsymbol{e}_3\}$ is a *right-handed* orthonormal basis for E^3. This means that $\boldsymbol{e}_1 \times \boldsymbol{e}_2 = \boldsymbol{e}_3$, $\boldsymbol{e}_2 \times \boldsymbol{e}_3 = \boldsymbol{e}_1$ and $\boldsymbol{e}_3 \times \boldsymbol{e}_1 = \boldsymbol{e}_2$. Let e_{ijk} be the three-index object defined by

$$e_{ijk} = \boldsymbol{e}_i \cdot \boldsymbol{e}_j \times \boldsymbol{e}_k = [\boldsymbol{e}_i, \boldsymbol{e}_j, \boldsymbol{e}_k], \tag{1.24}$$

so that $e_{123} = 1$ in a right-handed system. From the definition, we have that

$$e_{ijk} = \begin{cases} +1, & \text{if } (i, j, k) \text{ is an even permutation of } (1, 2, 3) \\ -1, & \text{if } (i, j, k) \text{ is an odd permutation of } (1, 2, 3) \\ 0, & \text{if any two indices are equal.} \end{cases} \tag{1.25}$$

Accordingly, e_{ijk} is called the *permutation symbol*, or the *unit alternator*. It follows, from (1.14) and (1.24), that

$$\boldsymbol{e}_j \times \boldsymbol{e}_k = (\boldsymbol{e}_i \cdot \boldsymbol{e}_j \times \boldsymbol{e}_k) \boldsymbol{e}_i = e_{ijk} \boldsymbol{e}_i = e_{kij} \boldsymbol{e}_i = e_{jki} \boldsymbol{e}_i. \tag{1.26}$$

Note that the free indices j and k match in each term and, of course, the repeated index i is summed from 1 to 3. Thus, we have nine equations, each with three terms on the right-hand side.

Let $\boldsymbol{c} = \boldsymbol{a} \times \boldsymbol{b}$. We wish to compute the components of $\boldsymbol{c}$ in terms of those of $\boldsymbol{a}$ and $\boldsymbol{b}$. Using property (c) above and taking care not to misuse the summation convention, we have

$$\boldsymbol{c} = a_i \boldsymbol{e}_i \times b_j \boldsymbol{e}_j = a_i b_j \boldsymbol{e}_i \times \boldsymbol{e}_j = a_i b_j e_{ijk} \boldsymbol{e}_k, \tag{1.27}$$

where, on the right-hand side, we sum on the repeated indices i, j, k, each from 1 to 3, for a total of 27 terms. Writing $\boldsymbol{c} = c_k \boldsymbol{e}_k$, we then have

$$(c_k - e_{ijk}a_ib_j)\boldsymbol{e}_k = \boldsymbol{0}, \tag{1.28}$$

and the linear independence of the basis elements yields the result

$$c_k = e_{ijk}a_ib_j, \tag{1.29}$$

in which, here and in the parenthesis in (1.28), a double sum on i and j is implied for each fixed $k \in \{1, 2, 3\}$. This represents three equations, each with nine terms on the right-hand side.

With this result in hand we may compute the box product $[\boldsymbol{a}, \boldsymbol{b}, \boldsymbol{d}]$ in terms of components:

$$[\boldsymbol{a}, \boldsymbol{b}, \boldsymbol{d}] = \boldsymbol{a} \times \boldsymbol{b} \cdot \boldsymbol{d} = \boldsymbol{c} \cdot \boldsymbol{d} = c_kd_k = e_{ijk}a_ib_jd_k, \tag{1.30}$$

which, when written out in full, contains 27 terms on the right-hand side.

The permutation symbol furnishes compact expressions for the determinants of 3×3 matrices. Let (b_{ij}) be such a matrix, and let $B = \det(b_{ij})$. Then,

$$B = \begin{vmatrix} b_{11} & b_{12} & b_{13} \\ b_{21} & b_{22} & b_{23} \\ b_{31} & b_{32} & b_{33} \end{vmatrix} = e_{ijk}b_{i1}b_{j2}b_{k3}. \tag{1.31}$$

It is a simple matter to verify that $B = e_{ijk}b_{1i}b_{2j}b_{3k} = e_{ijk}b^t_{i1}b^t_{j2}b^t_{k3}$, where the superscript ^{t} is used to identify the matrix transpose. Thus, the matrix and its transpose have the same determinant. Equation (1.31) is the same as

$$e_{123}B = e_{ijk}b_{i1}b_{j2}b_{k3}, \tag{1.32}$$

which is a special case of the easily confirmed relation

$$e_{mnp}B = e_{ijk}b_{im}b_{jn}b_{kp} = \begin{vmatrix} b_{1m} & b_{1n} & b_{1p} \\ b_{2m} & b_{2n} & b_{2p} \\ b_{3m} & b_{3n} & b_{3p} \end{vmatrix}. \tag{1.33}$$

Of course the latter equation is preserved if we pre-multiply the left-hand side by $e_{123}(= 1)$. Doing so makes it easier to observe that it is a special case of the general relation

$$e_{ijk}e_{mnp}B = \begin{vmatrix} b_{im} & b_{in} & b_{ip} \\ b_{jm} & b_{jn} & b_{jp} \\ b_{km} & b_{kn} & b_{kp} \end{vmatrix}. \tag{1.34}$$

Consider the matrix with entries $b_{ij} = \delta_{ij}$, the Kronecker delta. This is the unit matrix, with determinant $B = 1$. Expanding the determinant by the third row, we obtain

$$\begin{aligned} e_{ijk}e_{mnp} &= \begin{vmatrix} \delta_{im} & \delta_{in} & \delta_{ip} \\ \delta_{jm} & \delta_{jn} & \delta_{jp} \\ \delta_{km} & \delta_{kn} & \delta_{kp} \end{vmatrix} \\ &= \delta_{km}(\delta_{in}\delta_{jp} - \delta_{jn}\delta_{ip}) - \delta_{kn}(\delta_{im}\delta_{jp} - \delta_{jm}\delta_{ip}) \\ &\quad + \delta_{kp}(\delta_{im}\delta_{jn} - \delta_{jm}\delta_{in}). \end{aligned} \tag{1.35}$$

Setting $p = k$, summing on k, and making use of the properties of the Kronecker delta yields the important $e - \delta$ *identity*

$$e_{ijk}e_{mnk} = \delta_{im}\delta_{jn} - \delta_{in}\delta_{jm}. \tag{1.36}$$

Putting $n = j$ and summing again, we obtain

$$e_{ijk}e_{mjk} = 3\delta_{im} - \delta_{ij}\delta_{jm} = 2\delta_{im}, \tag{1.37}$$

and therefore

$$e_{ijk}e_{ijk} = 2\delta_{ii} = 6, \tag{1.38}$$

which combines with $(1.33)_1$ to give

$$B = \frac{1}{6}e_{ijk}e_{mnp}b_{im}b_{jn}b_{kp}. \tag{1.39}$$

1.1.4 *Tensors*

Consider a vector-valued function $\boldsymbol{f}(\boldsymbol{v})$: $E^3 \to E^3$, i.e., $\boldsymbol{f}(\cdot)$ maps $\boldsymbol{v} \in E^3$ to $\boldsymbol{u} = \boldsymbol{f}(\boldsymbol{v}) \in E^3$. A *linear* function of this kind is called a *tensor*. To emphasize the linearity, we write $\boldsymbol{f}(\boldsymbol{v}) = \boldsymbol{A}\boldsymbol{v}$ in which the fixed object $\boldsymbol{A}$ identifies the particular linear function at hand. Because $\boldsymbol{f}(\cdot)$ is fully specified by $\boldsymbol{A}$, we regard $\boldsymbol{A}$ itself as the tensor.

There is nothing special about E^3; the tensor concept applies equally to E^n with n arbitrary. However, to avoid undue abstraction, we shall confine attention to the case $n = 3$. As before, linearity means

$$\boldsymbol{f}(\alpha_1 \boldsymbol{v}_1 + \alpha_2 \boldsymbol{v}_2) = \alpha_1 \boldsymbol{f}(\boldsymbol{v}_1) + \alpha_2 \boldsymbol{f}(\boldsymbol{v}_2), \tag{1.40}$$

for all $\boldsymbol{v}_1, \boldsymbol{v}_2 \in E^3$ and all $\alpha_1, \alpha_2 \in \mathbb{R}$.

Tensors $\boldsymbol{A}$ and $\boldsymbol{B}$ are said to be *equal*, written $\boldsymbol{A} = \boldsymbol{B}$, provided that

$$\boldsymbol{A}\boldsymbol{v} = \boldsymbol{B}\boldsymbol{v} \quad \text{for all } \boldsymbol{v} \in E^3. \tag{1.41}$$

For $\alpha \in \mathbb{R}$, we define the product $\alpha \boldsymbol{A}$ by

$$(\alpha \boldsymbol{A})\boldsymbol{v} = \alpha(\boldsymbol{A}\boldsymbol{v}), \tag{1.42}$$

and the sum $\boldsymbol{A} + \boldsymbol{B}$ by

$$(\boldsymbol{A} + \boldsymbol{B})\boldsymbol{v} = \boldsymbol{A}\boldsymbol{v} + \boldsymbol{B}\boldsymbol{v}. \tag{1.43}$$

These imply that $\alpha \boldsymbol{A} + \beta \boldsymbol{B}$ is a tensor for all $\alpha, \beta \in \mathbb{R}$ and hence that the set of tensors is a *linear space*. Its *zero* element, $\boldsymbol{O}$, is such that $\boldsymbol{O}\boldsymbol{v} = \boldsymbol{0}$ for all $\boldsymbol{v} \in E^3$, yielding $\boldsymbol{A} + \boldsymbol{O} = \boldsymbol{A}$ and hence the existence of an additive inverse $-\boldsymbol{A} : \boldsymbol{A} + (-\boldsymbol{A}) = \boldsymbol{O}$. The *identity* element, $\boldsymbol{I}$, is such that $\boldsymbol{I}\boldsymbol{v} = \boldsymbol{v}$ for all $\boldsymbol{v} \in E^3$.

Given fixed vectors $\boldsymbol{a}, \boldsymbol{b} \in E^3$, we define a particular tensor $\boldsymbol{a} \otimes \boldsymbol{b}$, called the *tensor product* of $\boldsymbol{a}$ and $\boldsymbol{b}$, by

$$(\boldsymbol{a} \otimes \boldsymbol{b})\boldsymbol{v} = (\boldsymbol{b} \cdot \boldsymbol{v})\boldsymbol{a} \quad \text{for all} \quad \boldsymbol{v} \in E^3. \tag{1.44}$$

This is linear and vector valued, and hence a tensor. Its significance lies in the fact that a basis for the linear space of tensors can be constructed from such tensor products. To see this we combine (1.12) and (1.40) to write

$$\boldsymbol{f}(\boldsymbol{v}) = \boldsymbol{f}(v_i \boldsymbol{e}_i) = v_i \boldsymbol{f}(\boldsymbol{e}_i), \tag{1.45}$$

and invoke (1.12) in the form $\boldsymbol{f}(\boldsymbol{e}_i) = f_j(\boldsymbol{e}_i)\boldsymbol{e}_j$, where $f_j(\boldsymbol{e}_i) = \boldsymbol{e}_j \cdot \boldsymbol{f}(\boldsymbol{e}_i)$. Writing A_{ji} for the numbers $\boldsymbol{e}_j \cdot \boldsymbol{f}(\boldsymbol{e}_i)$, we have $\boldsymbol{f}(\boldsymbol{e}_i) = A_{ji}\boldsymbol{e}_j$ and

$$\boldsymbol{A}\boldsymbol{v} = \boldsymbol{f}(\boldsymbol{v}) = v_i \boldsymbol{f}(\boldsymbol{e}_i) = A_{ji} v_i \boldsymbol{e}_j = A_{ji} \boldsymbol{e}_j (\boldsymbol{e}_i \cdot \boldsymbol{v}) = (A_{ji} \boldsymbol{e}_j \otimes \boldsymbol{e}_i)\boldsymbol{v}, \tag{1.46}$$

and (1.41) yields

$$\boldsymbol{A} = A_{ij}\boldsymbol{e}_i \otimes \boldsymbol{e}_j, \quad \text{where } A_{ij} = \boldsymbol{e}_i \cdot \boldsymbol{A}\boldsymbol{e}_j \tag{1.47}$$

are the components of the tensor $\boldsymbol{A}$ relative to the basis $\{\boldsymbol{e}_i \otimes \boldsymbol{e}_j\}$, consisting of nine elements. Similarly to (1.14), we have

$$\boldsymbol{A} = (\boldsymbol{e}_i \cdot \boldsymbol{A}\boldsymbol{e}_j)\boldsymbol{e}_i \otimes \boldsymbol{e}_j, \tag{1.48}$$

in which the parenthetical terms are the projections of the tensor onto the corresponding basis elements.

Another way to arrive at (1.47) is to first show that $\{\boldsymbol{e}_i \otimes \boldsymbol{e}_j\}$ is a linearly independent set. To see that this is so, we form the equation $\alpha_{ij}\boldsymbol{e}_i \otimes \boldsymbol{e}_j = \boldsymbol{O}$, from which it follows that $(\alpha_{ij}\boldsymbol{e}_i \otimes \boldsymbol{e}_j)\boldsymbol{e}_k = \boldsymbol{0}$. Using (1.8) and (1.44) we reduce this to $\alpha_{ik}\boldsymbol{e}_i = \boldsymbol{0}$, and then invoke the linear independence of $\{\boldsymbol{e}_i\}$ to conclude that all the α_{ij} vanish. We can then repeat the argument leading to (1.12) to arrive at (1.47).

The components A_{ij} can be arranged in a 3×3 matrix (A_{ij}). It is instructive to work out the matrices of these components in a few illustrative examples.

Examples

1. Consider the tensor product $\boldsymbol{a} \otimes \boldsymbol{b}$ defined in (1.44). We have $(\boldsymbol{a}\otimes\boldsymbol{b})\boldsymbol{e}_j = (\boldsymbol{b}\cdot\boldsymbol{e}_j)\boldsymbol{a} = b_j\boldsymbol{a}$, and thus $\boldsymbol{e}_i\cdot(\boldsymbol{a}\otimes\boldsymbol{b})\boldsymbol{e}_j = b_j\boldsymbol{e}_i\cdot\boldsymbol{a} = a_ib_j$. Equation (1.48) then gives

$$\boldsymbol{a} \otimes \boldsymbol{b} = a_ib_j\boldsymbol{e}_i \otimes \boldsymbol{e}_j, \tag{1.49}$$

and the associated matrix is

$$(a_ib_j) = \begin{pmatrix} a_1b_1 & a_1b_2 & a_1b_3 \\ a_2b_1 & a_2b_2 & a_2b_3 \\ a_3b_1 & a_3b_2 & a_3b_3 \end{pmatrix}. \tag{1.50}$$

This is the same as the matrix product of the column matrix $\{a_i\}$ and the row matrix $\lfloor b_j \rfloor$, often called the *outer product* to distinguish it from the *inner product* of $\lfloor a_i \rfloor$ with $\{b_j\}$, the latter being the same as $a_ib_i = \boldsymbol{a} \cdot \boldsymbol{b}$.

2. Consider the linear function $\boldsymbol{f}(\boldsymbol{v})$ such that $\boldsymbol{f}(\boldsymbol{e}_1) = -\boldsymbol{e}_1$, $\boldsymbol{f}(\boldsymbol{e}_2) = \boldsymbol{e}_2$ and $\boldsymbol{f}(\boldsymbol{e}_3) = \boldsymbol{e}_3$. We have $\boldsymbol{f}(\boldsymbol{v}) = \boldsymbol{A}\boldsymbol{v}$, where $\boldsymbol{A} = (A_{ij}\boldsymbol{e}_i) \otimes \boldsymbol{e}_j$ with $A_{ij} = \boldsymbol{e}_i \cdot \boldsymbol{A}\boldsymbol{e}_j = \boldsymbol{e}_i \cdot \boldsymbol{f}(\boldsymbol{e}_j)$. Thus, fixing $j \in \{1, 2, 3\}$ inside the parenthesis,

$$\begin{aligned} A_{i1} &= \boldsymbol{e}_i \cdot \boldsymbol{A}\boldsymbol{e}_1 = -\boldsymbol{e}_i \cdot \boldsymbol{e}_1 = -\delta_{i1}; \quad A_{i1}\boldsymbol{e}_i = -\delta_{i1}\boldsymbol{e}_i = -\boldsymbol{e}_1, \\ A_{i2} &= \boldsymbol{e}_i \cdot \boldsymbol{A}\boldsymbol{e}_2 = \boldsymbol{e}_i \cdot \boldsymbol{e}_2 = \delta_{i2}; \quad A_{i2}\boldsymbol{e}_i = \delta_{i2}\boldsymbol{e}_i = \boldsymbol{e}_2, \quad \text{and} \\ A_{i3} &= \boldsymbol{e}_i \cdot \boldsymbol{A}\boldsymbol{e}_3 = \boldsymbol{e}_i \cdot \boldsymbol{e}_3 = \delta_{i3}; \quad A_{i3}\boldsymbol{e}_i = \delta_{i3}\boldsymbol{e}_i = \boldsymbol{e}_3, \end{aligned} \tag{1.51}$$

yielding

$$\boldsymbol{A} = (A_{ij}\boldsymbol{e}_i) \otimes \boldsymbol{e}_j = -\boldsymbol{e}_1 \otimes \boldsymbol{e}_1 + \boldsymbol{e}_2 \otimes \boldsymbol{e}_2 + \boldsymbol{e}_3 \otimes \boldsymbol{e}_3, \tag{1.52}$$

with matrix

$$(A_{ij}) = \begin{pmatrix} -1 & 0 & 0 \\ 0 & 1 & 0 \\ 0 & 0 & 1 \end{pmatrix}. \tag{1.53}$$

3. Recall that the identity $\boldsymbol{I}$ is defined by $\boldsymbol{I}\boldsymbol{v} = \boldsymbol{v}$ for all $\boldsymbol{v}$. Then,

$$\begin{aligned} \boldsymbol{I} &= I_{ij}\boldsymbol{e}_i \otimes \boldsymbol{e}_j = (\boldsymbol{e}_i \cdot \boldsymbol{I}\boldsymbol{e}_j)\boldsymbol{e}_i \otimes \boldsymbol{e}_j = (\boldsymbol{e}_i \cdot \boldsymbol{e}_j)\boldsymbol{e}_i \otimes \boldsymbol{e}_j \\ &= \delta_{ij}\boldsymbol{e}_i \otimes \boldsymbol{e}_j = \boldsymbol{e}_i \otimes \boldsymbol{e}_i, \end{aligned} \tag{1.54}$$

with $(I_{ij}) = (\delta_{ij}) = diag(1, 1, 1)$, the unit matrix.

4. The zero tensor $\boldsymbol{O}$ is such that $\boldsymbol{O}\boldsymbol{v} = \boldsymbol{0}$ for all $\boldsymbol{v}$. Thus, $\boldsymbol{O} = O_{ij}\boldsymbol{e}_i \otimes \boldsymbol{e}_j$ with $O_{ij} = \boldsymbol{e}_i \cdot \boldsymbol{O}\boldsymbol{e}_j = \boldsymbol{e}_i \cdot \boldsymbol{0} = 0$; the matrix (O_{ij}) is simply the matrix with every entry equal to zero.

We can derive the transformation formula for tensor components in much the same way as for ordinary vector components. Thus, let A_{ij} be the components of tensor $\boldsymbol{A}$ relative to $\{\boldsymbol{e}_i \otimes \boldsymbol{e}_j\}$, and A'_{ij} those of the same tensor relative to $\{\boldsymbol{e}'_i \otimes \boldsymbol{e}'_j\}$:

$$A'_{ij}\boldsymbol{e}'_i \otimes \boldsymbol{e}'_j = \boldsymbol{A} = A_{kl}\boldsymbol{e}_k \otimes \boldsymbol{e}_l. \tag{1.55}$$

Then,

$$\boldsymbol{A}\boldsymbol{e}'_n = (A_{kl}\boldsymbol{e}_k \otimes \boldsymbol{e}_l)\boldsymbol{e}'_n = A_{kl}\boldsymbol{e}_k(\boldsymbol{e}_l \cdot \boldsymbol{e}'_n) = A_{kl}a_{ln}\boldsymbol{e}_k, \tag{1.56}$$

and

$$A'_{jn} = \boldsymbol{e}'_j \cdot \boldsymbol{A}\boldsymbol{e}'_n = A_{kl}a_{ln}\boldsymbol{e}'_j \cdot \boldsymbol{e}_k = A_{kl}a_{kj}a_{ln}. \tag{1.57}$$

Thus,

$$A'_{ij} = a_{ki}a_{lj}A_{kl}, \tag{1.58}$$

which may be compared to (1.22). This is the transformation law for the components of *second-order* tensors, involving the matrix of direction cosines *twice*. Accordingly, the object $\boldsymbol{A}$ is a second-order tensor. Thus, vectors are first-order tensors. The transformation law for a scalar φ, say, is $\varphi' = \varphi$, and so scalars may be regarded as zeroth-order tensors. It is straightforward to define tensors of arbitrary finite order, but most of our development will be concerned with tensors of order less than or equal to two.

1.1.5 *Algebraic operations with second-order tensors*

Several algebraic operations with tensors will prove to be indispensable. These are defined here, and the consequences of the definitions are explored.

(a) The *transpose* $\boldsymbol{A}^t$ of tensor $\boldsymbol{A}$ is defined by

$$\boldsymbol{u} \cdot \boldsymbol{A}\boldsymbol{v} = \boldsymbol{v} \cdot \boldsymbol{A}^t\boldsymbol{u} \quad \text{for all } \boldsymbol{u}, \boldsymbol{v} \in E^3. \tag{1.59}$$

To obtain the matrix of $\boldsymbol{A}^t$ relative to an orthonormal basis, let $\boldsymbol{u} = \boldsymbol{e}_i$ and $\boldsymbol{v} = \boldsymbol{e}_j$ for some *fixed* $i, j \in \{1, 2, 3\}$. Then,

$$A^t_{ji} = \boldsymbol{e}_j \cdot \boldsymbol{A}^t\boldsymbol{e}_i = \boldsymbol{e}_i \cdot \boldsymbol{A}\boldsymbol{e}_j = A_{ij}. \tag{1.60}$$

Thus, the matrix of $\boldsymbol{A}^t$ is simply the usual matrix transpose of the matrix of $\boldsymbol{A}$. Accordingly,

$$\boldsymbol{A}^t = A^t_{ji}\boldsymbol{e}_j \otimes \boldsymbol{e}_i = A_{ij}\boldsymbol{e}_j \otimes \boldsymbol{e}_i = A_{ji}\boldsymbol{e}_i \otimes \boldsymbol{e}_j. \tag{1.61}$$

$\boldsymbol{A}$ is *symmetric* if $\boldsymbol{A}^t = \boldsymbol{A}$, i.e., if $(A_{ji} - A_{ij})\boldsymbol{e}_i \otimes \boldsymbol{e}_j = \boldsymbol{O}$. The linear independence of $\{\boldsymbol{e}_i \otimes \boldsymbol{e}_j\}$ then yields the necessary and sufficient conditions $A_{ji} = A_{ij}$. $\boldsymbol{A}$ is *skew* if $\boldsymbol{A}^t = -\boldsymbol{A}$; that is, if and only if $A_{ji} = -A_{ij}$. In this case the matrix of $\boldsymbol{A}$ consists of only three independent components:

$$(A_{ij}) = \begin{pmatrix} 0 & A_{12} & A_{13} \\ -A_{12} & 0 & A_{23} \\ -A_{13} & -A_{23} & 0 \end{pmatrix}, \tag{1.62}$$

and is therefore equivalent to the components of a vector.

(b) The *product* of tensors $\boldsymbol{A}$ and $\boldsymbol{B}$, written $\boldsymbol{AB}$, is the *tensor* defined by

$$(\boldsymbol{AB})\boldsymbol{v} = \boldsymbol{A}(\boldsymbol{B}\boldsymbol{v}), \quad \text{for all } \boldsymbol{v} \in E^3. \tag{1.63}$$

The corresponding matrix follows from

$$\begin{aligned}(\boldsymbol{AB})\boldsymbol{e}_j &= \boldsymbol{A}(\boldsymbol{B}\boldsymbol{e}_j) = \boldsymbol{A}[(B_{kl}\boldsymbol{e}_k \otimes \boldsymbol{e}_l)\boldsymbol{e}_j] \\ &= \boldsymbol{A}(B_{kj}\boldsymbol{e}_k) = B_{kj}\boldsymbol{A}\boldsymbol{e}_k \quad \text{(by linearity).}\end{aligned} \tag{1.64}$$

Thus,

$$\boldsymbol{AB} = [\boldsymbol{e}_i \cdot (\boldsymbol{AB})\boldsymbol{e}_j]\boldsymbol{e}_i \otimes \boldsymbol{e}_j = A_{ik}B_{kj}\boldsymbol{e}_i \otimes \boldsymbol{e}_j, \tag{1.65}$$

and hence, denoting $\boldsymbol{AB}$ by $\boldsymbol{C} = C_{ij}\boldsymbol{e}_i \otimes \boldsymbol{e}_j$,

$$C_{ij} = A_{ik}B_{kj}. \tag{1.66}$$

This is simply the standard formula for matrix multiplication. Note that, in general, $A_{ik}B_{kj} \neq B_{ik}A_{kj}$, i.e., that $\boldsymbol{AB} \neq \boldsymbol{BA}$. Thus, the product does not commute.

An alternative derivation proceeds directly from (1.63):

$$\begin{aligned}(\boldsymbol{AB})\boldsymbol{v} &= \boldsymbol{A}[\boldsymbol{B}(v_i\boldsymbol{e}_i)] = \boldsymbol{A}(v_i\boldsymbol{B}\boldsymbol{e}_i) \\ &= v_i\boldsymbol{A}(B_{ji}\boldsymbol{e}_j) = v_iB_{ji}\boldsymbol{A}\boldsymbol{e}_j \\ &= A_{kj}B_{ji}\boldsymbol{e}_k(\boldsymbol{e}_i \cdot \boldsymbol{v}) = (A_{kj}B_{ji}\boldsymbol{e}_k \otimes \boldsymbol{e}_i)\boldsymbol{v},\end{aligned} \tag{1.67}$$

where repeated use has been made of linearity, and (1.65) then follows from the arbitrariness of $\boldsymbol{v}$.

As an application of (1.63), consider

$$(\boldsymbol{AI})\boldsymbol{v} = \boldsymbol{A}(\boldsymbol{I}\boldsymbol{v}) = \boldsymbol{A}\boldsymbol{v} = \boldsymbol{I}(\boldsymbol{A}\boldsymbol{v}) = (\boldsymbol{IA})\boldsymbol{v}. \tag{1.68}$$

Thus, $\boldsymbol{AI} = \boldsymbol{A} = \boldsymbol{IA}$. Every tensor commutes with the identity.

(c) The *inverse* $\boldsymbol{A}^{-1}$ of $\boldsymbol{A}$ exists provided that $\boldsymbol{A}$ is *invertible*; that is, if the equation $\boldsymbol{Au} = \boldsymbol{v}$ has a unique solution $\boldsymbol{u}$ for any $\boldsymbol{v}$. This is equivalent to the requirement that the function $\boldsymbol{f}(\boldsymbol{u}) = \boldsymbol{Au}$

be one-to-one, i.e., that if $\boldsymbol{A}\boldsymbol{u}_1 = \boldsymbol{v}_1$ and $\boldsymbol{A}\boldsymbol{u}_2 = \boldsymbol{v}_2$, then $\boldsymbol{v}_1 = \boldsymbol{v}_2$ if and only if $\boldsymbol{u}_1 = \boldsymbol{u}_2$. We then write $\boldsymbol{u} = \boldsymbol{A}^{-1}\boldsymbol{v}$. Thus,

$$(\boldsymbol{A}^{-1}\boldsymbol{A})\boldsymbol{u} = \boldsymbol{A}^{-1}(\boldsymbol{A}\boldsymbol{u}) = \boldsymbol{A}^{-1}\boldsymbol{v} = \boldsymbol{u} = \boldsymbol{I}\boldsymbol{u}, \tag{1.69}$$

so that

$$\boldsymbol{A}^{-1}\boldsymbol{A} = \boldsymbol{I}; \tag{1.70}$$

and,

$$(\boldsymbol{A}\boldsymbol{A}^{-1})\boldsymbol{v} = \boldsymbol{A}(\boldsymbol{A}^{-1}\boldsymbol{v}) = \boldsymbol{A}\boldsymbol{u} = \boldsymbol{v} = \boldsymbol{I}\boldsymbol{v}, \tag{1.71}$$

so that

$$\boldsymbol{A}\boldsymbol{A}^{-1} = \boldsymbol{I}. \tag{1.72}$$

Every invertible tensor commutes with its inverse.

(d) A tensor $\boldsymbol{A}$ is *orthogonal* if it preserves dot products:

$$\boldsymbol{A}\boldsymbol{u} \cdot \boldsymbol{A}\boldsymbol{v} = \boldsymbol{u} \cdot \boldsymbol{v} \quad \text{for all } \boldsymbol{u}, \boldsymbol{v} \in E^3. \tag{1.73}$$

Thus orthogonal tensors preserve the norms of vectors (set $\boldsymbol{u} = \boldsymbol{v}$) and the angles between them (see (1.5)). To explore the implications of orthogonality, observe that

$$\boldsymbol{A}\boldsymbol{u} \cdot \boldsymbol{A}\boldsymbol{v} = \boldsymbol{u} \cdot \boldsymbol{A}^t(\boldsymbol{A}\boldsymbol{v}) = \boldsymbol{u} \cdot (\boldsymbol{A}^t\boldsymbol{A})\boldsymbol{v}. \tag{1.74}$$

Thus, $\boldsymbol{u} \cdot (\boldsymbol{A}^t\boldsymbol{A})\boldsymbol{v} = \boldsymbol{u} \cdot \boldsymbol{v}$, or

$$\boldsymbol{w} \cdot \boldsymbol{u} = 0 \quad \text{for all } \boldsymbol{u}, \boldsymbol{v} \in E^3, \quad \text{where } \boldsymbol{w} = (\boldsymbol{A}^t\boldsymbol{A})\boldsymbol{v} - \boldsymbol{I}\boldsymbol{v}. \tag{1.75}$$

Fixing $\boldsymbol{v}$ (hence $\boldsymbol{w}$) and choosing $\boldsymbol{u} = \boldsymbol{w}$, we conclude that $|\boldsymbol{w}|^2 = 0$ and hence that $\boldsymbol{w} = \boldsymbol{0}$, i.e., that

$$(\boldsymbol{A}^t\boldsymbol{A})\boldsymbol{v} = \boldsymbol{I}\boldsymbol{v} \quad \text{for all } \boldsymbol{v} \in E^3, \tag{1.76}$$

and therefore that $\boldsymbol{A}^t\boldsymbol{A} = \boldsymbol{I}$. Orthogonal tensors are invertible, with inverse

$$\boldsymbol{A}^{-1} = \boldsymbol{A}^t, \tag{1.77}$$

and (1.72) implies that $\boldsymbol{A}\boldsymbol{A}^t = \boldsymbol{I}$. Thus, orthogonal tensors commute with their transposes.

(e) The *principal invariants* of a tensor $\boldsymbol{A}$ are the scalars $I_1(\boldsymbol{A}), I_2(\boldsymbol{A})$ and $I_3(\boldsymbol{A})$ defined, respectively, by

$$I_1(\boldsymbol{A})[\boldsymbol{a},\boldsymbol{b},\boldsymbol{c}] = [\boldsymbol{A}\boldsymbol{a},\boldsymbol{b},\boldsymbol{c}] + [\boldsymbol{a},\boldsymbol{A}\boldsymbol{b},\boldsymbol{c}] + [\boldsymbol{a},\boldsymbol{b},\boldsymbol{A}\boldsymbol{c}],$$
$$I_2(\boldsymbol{A})[\boldsymbol{a},\boldsymbol{b},\boldsymbol{c}] = [\boldsymbol{A}\boldsymbol{a},\boldsymbol{A}\boldsymbol{b},\boldsymbol{c}] + [\boldsymbol{A}\boldsymbol{a},\boldsymbol{b},\boldsymbol{A}\boldsymbol{c}] + [\boldsymbol{a},\boldsymbol{A}\boldsymbol{b},\boldsymbol{A}\boldsymbol{c}], \quad \text{and}$$
$$I_3(\boldsymbol{A})[\boldsymbol{a},\boldsymbol{b},\boldsymbol{c}] = [\boldsymbol{A}\boldsymbol{a},\boldsymbol{A}\boldsymbol{b},\boldsymbol{A}\boldsymbol{c}] \qquad \text{for all } \boldsymbol{a},\boldsymbol{b},\boldsymbol{c} \in E^3. \tag{1.78}$$

Reference may be made to Appendix A for proofs of various statements made here concerning these functions.

The set $\{\boldsymbol{a},\boldsymbol{b},\boldsymbol{c}\}$ is linearly independent if and only if $[\boldsymbol{a},\boldsymbol{b},\boldsymbol{c}] \neq 0$, in which case these relations may be used to compute $I_{1,2,3}(\boldsymbol{A})$ explicitly (see Appendix A). Remarkably, these functions are independent of the particular set $\{\boldsymbol{a},\boldsymbol{b},\boldsymbol{c}\}$ chosen. The function $I_1(\boldsymbol{A})$ is also called the *trace* of $\boldsymbol{A}$, written $tr\boldsymbol{A}$, and $I_3(\boldsymbol{A}) = \det \boldsymbol{A}$, the determinant. An interesting interpretation of the determinant follows on observing that, if $[\boldsymbol{a},\boldsymbol{b},\boldsymbol{c}]$ is positive, then it is the volume of the parallelepiped formed by the vectors $\boldsymbol{a},\boldsymbol{b},\boldsymbol{c}$ (Figure 1.1). Similarly, if $[\boldsymbol{A}\boldsymbol{a},\boldsymbol{A}\boldsymbol{b},\boldsymbol{A}\boldsymbol{c}]$ is positive, then it is the volume of the parallelepiped formed by $\boldsymbol{A}\boldsymbol{a},\boldsymbol{A}\boldsymbol{b},\boldsymbol{A}\boldsymbol{c}$. Thus, $\det \boldsymbol{A}$ is the ratio of the two volumes.

It is easy to show that $\det \boldsymbol{A} = \det(A_{ij})$, where $A_{ij} = \boldsymbol{e}_i \cdot \boldsymbol{A}\boldsymbol{e}_j$ are the components of $\boldsymbol{A}$ relative to a basis $\{\boldsymbol{e}_i \otimes \boldsymbol{e}_j\}$ constructed from orthonormal vectors $\boldsymbol{e}_i$. It is also true that $\boldsymbol{A}$ is invertible if and only if $\det \boldsymbol{A} \neq 0$.

Another fact of considerable importance is that I_1, the trace, is a linear function. To see this we use $(1.78)_1$ to obtain

$$\begin{aligned} I_1(\alpha\boldsymbol{A} + \beta\boldsymbol{B})[\boldsymbol{a},\boldsymbol{b},\boldsymbol{c}] &= \alpha[\boldsymbol{A}\boldsymbol{a},\boldsymbol{b},\boldsymbol{c}] + \beta[\boldsymbol{B}\boldsymbol{a},\boldsymbol{b},\boldsymbol{c}] \\ &\quad + \alpha[\boldsymbol{a},\boldsymbol{A}\boldsymbol{b},\boldsymbol{c}] + \beta[\boldsymbol{a},\boldsymbol{B}\boldsymbol{b},\boldsymbol{c}] \\ &\quad + \alpha[\boldsymbol{a},\boldsymbol{b},\boldsymbol{A}\boldsymbol{c}] + \beta[\boldsymbol{a},\boldsymbol{b},\boldsymbol{B}\boldsymbol{c}] \\ &= [\boldsymbol{a},\boldsymbol{b},\boldsymbol{c}]\{\alpha I_1(\boldsymbol{A}) + \beta I_1(\boldsymbol{B})\}, \end{aligned} \tag{1.79}$$

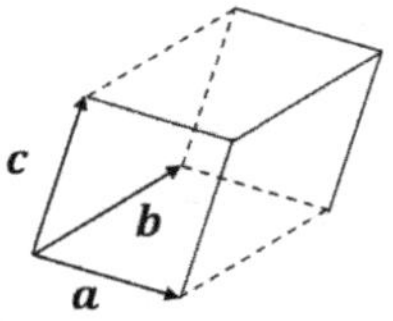

Figure 1.1. A parallelepiped formed by the vectors $\boldsymbol{a},\boldsymbol{b},\boldsymbol{c}$.

and conclude that

$$tr(\alpha \boldsymbol{A} + \beta \boldsymbol{B}) = \alpha tr(\boldsymbol{A}) + \beta tr(\boldsymbol{B}). \tag{1.80}$$

For example, with $[\boldsymbol{e}_1, \boldsymbol{e}_2, \boldsymbol{e}_3] = 1$, and for fixed i and j, we have

$$\begin{aligned} tr(\boldsymbol{e}_i \otimes \boldsymbol{e}_j) &= [(\boldsymbol{e}_i \otimes \boldsymbol{e}_j)\boldsymbol{e}_1, \boldsymbol{e}_2, \boldsymbol{e}_3] + [\boldsymbol{e}_1, (\boldsymbol{e}_i \otimes \boldsymbol{e}_j)\boldsymbol{e}_2, \boldsymbol{e}_3] \\ &\quad + [\boldsymbol{e}_1, \boldsymbol{e}_2, (\boldsymbol{e}_i \otimes \boldsymbol{e}_j)\boldsymbol{e}_3] \\ &= \delta_{j1}[\boldsymbol{e}_i, \boldsymbol{e}_2, \boldsymbol{e}_3] + \delta_{j2}[\boldsymbol{e}_1, \boldsymbol{e}_i, \boldsymbol{e}_3] + \delta_{j3}[\boldsymbol{e}_1, \boldsymbol{e}_2, \boldsymbol{e}_i] \\ &= \delta_{j1}e_{i23} + \delta_{j2}e_{1i3} + \delta_{j3}e_{12i}. \end{aligned} \tag{1.81}$$

For $i = 1$, the right-hand side reduces to δ_{j1}; for $i = 2$, to δ_{j2}; and, for $i = 3$, to δ_{j3}. Thus,

$$tr(\boldsymbol{e}_i \otimes \boldsymbol{e}_j) = \delta_{ji} = \delta_{ij}, \tag{1.82}$$

and the linearity of the trace operation furnishes

$$tr\boldsymbol{A} = tr(A_{ij}\boldsymbol{e}_i \otimes \boldsymbol{e}_j) = A_{ij}tr(\boldsymbol{e}_i \otimes \boldsymbol{e}_j) = A_{ij}\delta_{ij} = A_{ii}, \tag{1.83}$$

the trace of the matrix (A_{ij}). In particular, the trace converts tensor products into dot products:

$$tr(\boldsymbol{u} \otimes \boldsymbol{v}) = u_i v_i = \boldsymbol{u} \cdot \boldsymbol{v}. \tag{1.84}$$

The function $I_2(\boldsymbol{A})$ is not so easily interpreted. However, it is possible to show that

$$I_2(\boldsymbol{A}) = I_1(\boldsymbol{A}^*) = tr\boldsymbol{A}^*, \tag{1.85}$$

where $\boldsymbol{A}^*$ is the *cofactor* of $\boldsymbol{A}$, defined by

$$\boldsymbol{A}^*(\boldsymbol{a} \times \boldsymbol{b}) = \boldsymbol{A}\boldsymbol{a} \times \boldsymbol{A}\boldsymbol{b}, \quad \text{for all } \boldsymbol{a}, \boldsymbol{b} \in E^3. \tag{1.86}$$

The functions $I_2(\boldsymbol{A})$ and $I_3(\boldsymbol{A})$ are *nonlinear*; it follows from $(1.78)_{2,3}$ that they are quadratic and cubic functions, respectively.

(f) The *dot product* (or *inner product*), $\boldsymbol{A} \cdot \boldsymbol{B}$, of tensors $\boldsymbol{A}$ and $\boldsymbol{B}$ is defined by

$$\boldsymbol{A} \cdot \boldsymbol{B} = tr(\boldsymbol{A}\boldsymbol{B}^t). \tag{1.87}$$

Resolving on $\{\boldsymbol{e}_i \otimes \boldsymbol{e}_j\}$, we have

$$\begin{aligned}\boldsymbol{A} \cdot \boldsymbol{B} &= tr(A_{kj}B_{ij}\boldsymbol{e}_k \otimes \boldsymbol{e}_i)\\ &= A_{kj}B_{ij}tr(\boldsymbol{e}_k \otimes \boldsymbol{e}_i) \quad \text{(by linearity)}\\ &= A_{kj}B_{ij}\delta_{ki}\\ &= A_{ij}B_{ij} = B_{ij}A_{ij} = \boldsymbol{B} \cdot \boldsymbol{A}.\end{aligned} \tag{1.88}$$

The induced norm is

$$|\boldsymbol{A}| = \sqrt{\boldsymbol{A} \cdot \boldsymbol{A}} = \sqrt{A_{ij}A_{ij}}, \tag{1.89}$$

in which the second radicand is the sum of squares and hence non-negative. Thus, the norm is well defined. It is identical in form to the norm of ordinary vectors, and furnishes post facto motivation for the definition (1.89). In fact, from the structure introduced thus far, it is clear that the set of tensors $Span\{\boldsymbol{e}_i \otimes \boldsymbol{e}_j\}$ is simply the nine-dimensional Euclidean space E^9. Accordingly, we have the Cauchy–Schwarz inequality

$$|\boldsymbol{A} \cdot \boldsymbol{B}| \leq |\boldsymbol{A}|\,|\boldsymbol{B}| \tag{1.90}$$

and the triangle inequality

$$|\boldsymbol{A} + \boldsymbol{B}| \leq |\boldsymbol{A}| + |\boldsymbol{B}|. \tag{1.91}$$

Consider the dot product $\boldsymbol{A} \cdot \boldsymbol{u} \otimes \boldsymbol{v}$. From (1.84) and (1.87),

$$\boldsymbol{A} \cdot \boldsymbol{u} \otimes \boldsymbol{v} = tr[\boldsymbol{A}(\boldsymbol{v} \otimes \boldsymbol{u})] = tr[(\boldsymbol{A}\boldsymbol{v}) \otimes \boldsymbol{u}] = \boldsymbol{A}\boldsymbol{v} \cdot \boldsymbol{u} = \boldsymbol{u} \cdot \boldsymbol{A}\boldsymbol{v}, \tag{1.92}$$

where we have used $\boldsymbol{A}(\boldsymbol{a} \otimes \boldsymbol{b}) = \boldsymbol{A}\boldsymbol{a} \otimes \boldsymbol{b}$, which follows from $[\boldsymbol{A}(\boldsymbol{a} \otimes \boldsymbol{b})]\boldsymbol{v} = \boldsymbol{A}[(\boldsymbol{a} \otimes \boldsymbol{b})\boldsymbol{v}] = (\boldsymbol{b} \cdot \boldsymbol{v})\boldsymbol{A}\boldsymbol{a} = (\boldsymbol{A}\boldsymbol{a} \otimes \boldsymbol{b})\boldsymbol{v}$. Alternatively, in terms of components,

$$\boldsymbol{A} \cdot \boldsymbol{u} \otimes \boldsymbol{v} = A_{ij}u_iv_j = u_i(A_{ij}v_j) = u_i\boldsymbol{e}_i \cdot \boldsymbol{A}\boldsymbol{v} = \boldsymbol{u} \cdot \boldsymbol{A}\boldsymbol{v}. \tag{1.93}$$

Thus,

$$A_{ij} = \boldsymbol{e}_i \cdot \boldsymbol{A}\boldsymbol{e}_j = \boldsymbol{A} \cdot \boldsymbol{e}_i \otimes \boldsymbol{e}_j, \tag{1.94}$$

a result that, on comparison with (1.14), reinforces the interpretation of A_{ij} as the projection of $\boldsymbol{A}$ onto the basis element $\boldsymbol{e}_i \otimes \boldsymbol{e}_j$.

1.2 Scalar, vector and tensor fields; gradient, divergence and curl; integral formulae

1.2.1 *Euclidean point spaces*

We associate *positions* $\boldsymbol{x}, \boldsymbol{y}$ respectively with *points* x, y in a three-dimensional space $\mathcal{E}^3$. These are not elements of E^3 because their specification requires an origin O (Figure 1.2), which plays no role in the definition of a vector space. Thus, we say that $\mathcal{E}^3$ is a *point space.* However, the difference $\boldsymbol{y}-\boldsymbol{x}$, constructed by the usual parallelogram law, as in the figure, is independent of an origin and is regarded as a vector in the usual sense; that is, as an element of E^3. Accordingly, E^3 is called the *translation space* of $\mathcal{E}^3$. It is the set of all the position differences.

To reconcile the notion of point space with the conventional notion of a position *vector*, we assign the position $\boldsymbol{x}$ to a point x and the position $\boldsymbol{o}$ to O, and form the position vector $\boldsymbol{x} - \boldsymbol{o}$. At the risk of causing confusion, this is normally written simply as $\boldsymbol{x}$, i.e.,

$$\boldsymbol{x}(= \boldsymbol{x} - \boldsymbol{o}) = x_i \boldsymbol{e}_i, \tag{1.95}$$

where, on the right-hand side, we have decomposed the vector into components relative to an orthonormal basis. The components x_i are

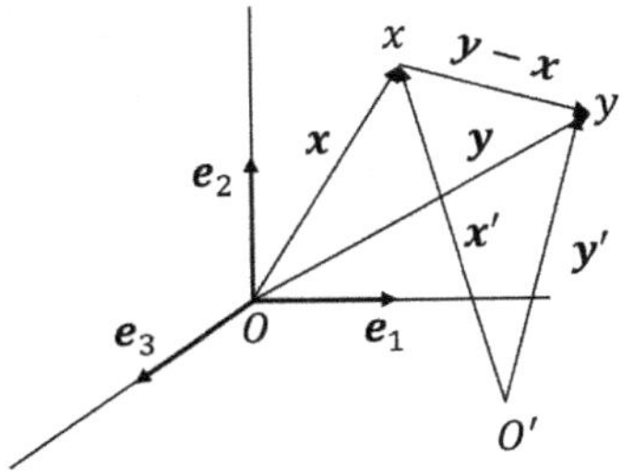

Figure 1.2. Positions in Euclidean point space.

the *Cartesian coordinates* of the point x relative to this basis and the selected origin.

We are concerned with *fields*; that is, with scalar-, vector-, and tensor-valued functions of position (or vector) $\boldsymbol{x}$.

1.2.2 *Gradients*

A scalar-valued function, or *field*, $\phi(\boldsymbol{x}) : \mathcal{E}^3 \to \mathbb{R}$, is *differentiable* at $\boldsymbol{x}_0 \in \mathcal{E}^3$ if it approximated by a linear function for $\boldsymbol{x}$ sufficiently close to $\boldsymbol{x}_0$; that is, if there is a *linear* scalar-valued function $f_0(\boldsymbol{x} - \boldsymbol{x}_0) : E^3 \to \mathbb{R}$ such that

$$\phi(\boldsymbol{x}) = \phi(\boldsymbol{x}_0) + f_0(\boldsymbol{x} - \boldsymbol{x}_0) + o(|\boldsymbol{x} - \boldsymbol{x}_0|), \tag{1.96}$$

where the *Landau symbol* $o(\epsilon)$ means

$$o(\epsilon)/\epsilon \to 0 \quad \text{as} \quad \epsilon \to 0. \tag{1.97}$$

The one-dimensional version of this condition is depicted in Figure 1.3.

From the representation formula (1.19) for linear functions, there is a unique $\boldsymbol{c} \in E^3$, independent of $\boldsymbol{x} - \boldsymbol{x}_0$, such that

$$f_0(\boldsymbol{x} - \boldsymbol{x}_0) = \boldsymbol{c} \cdot (\boldsymbol{x} - \boldsymbol{x}_0). \tag{1.98}$$

We adopt the conventional notation $\boldsymbol{c} = \nabla\phi(\boldsymbol{x})_{|\boldsymbol{x}_0}$, where $\nabla\phi(\boldsymbol{x}) \in E^3$ is the *gradient* of ϕ with respect to $\boldsymbol{x}$. Thus,

$$\phi(\boldsymbol{x}) = \phi(\boldsymbol{x}_0) + \nabla\phi(\boldsymbol{x})_{|\boldsymbol{x}_0} \cdot (\boldsymbol{x} - \boldsymbol{x}_0) + o(|\boldsymbol{x} - \boldsymbol{x}_0|). \tag{1.99}$$

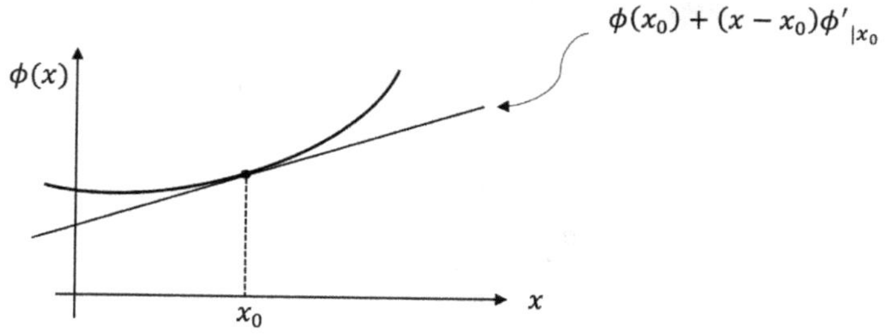

Figure 1.3. A differentiable function $\phi(x)$.

1.2.3 *Directional derivatives*

Let $\boldsymbol{v} = \boldsymbol{x} - \boldsymbol{x}_0$ and let $s = |\boldsymbol{v}|$. Then, $\boldsymbol{v} = s\boldsymbol{u}$ with $|\boldsymbol{u}| = 1$, and $\boldsymbol{x} = \boldsymbol{x}_0 + s\boldsymbol{u}$. The derivative of ϕ in the direction of $\boldsymbol{u}$ is the coefficient of s in the expression

$$\phi(\boldsymbol{x}) = \phi(\boldsymbol{x}_0) + s\nabla\phi(\boldsymbol{x})_{|\boldsymbol{x}_0} \cdot \boldsymbol{u} + o(s). \tag{1.100}$$

Thus,

$$\nabla\phi(\boldsymbol{x})_{|\boldsymbol{x}_0} \cdot \boldsymbol{u} = \frac{1}{s}[\phi(\boldsymbol{x}_0 + s\boldsymbol{u}) - \phi(\boldsymbol{x}_0)] + o(s)/s, \tag{1.101}$$

and passage to the limit $s \to 0$ yields the directional derivative

$$\left(\frac{df}{ds}\right)_{|s=0} = \boldsymbol{u} \cdot \nabla\phi(\boldsymbol{x})_{|\boldsymbol{x}_0}, \tag{1.102}$$

evaluated at $\boldsymbol{x}_0$, where $f(s) = \phi(\boldsymbol{x}_0 + s\boldsymbol{u})$ in which $\boldsymbol{x}_0$ and $\boldsymbol{u}$ are fixed.

For example, if we define $\hat{\phi}(x_1, x_2, x_3) = \phi(x_i\boldsymbol{e}_i)$ in which the $\boldsymbol{e}_i$ are fixed, then

$$\begin{aligned} \boldsymbol{e}_1 \cdot \nabla\phi(\boldsymbol{x})_{|\boldsymbol{x}_0} &= \lim_{s\to 0} \frac{1}{s}[\hat{\phi}(x_1^0 + s, x_2^0, x_3^0) - \hat{\phi}(x_1^0, x_2^0, x_3^0)] \\ &= \left(\frac{\partial\hat{\phi}}{\partial x_1}\right)_{|(x_1^0, x_2^0, x_3^0)}, \end{aligned} \tag{1.103}$$

where x_i^0 are the Cartesian coordinates of the point with position $\boldsymbol{x}_0$. In the same way, we have

$$\boldsymbol{e}_i \cdot \nabla\phi(\boldsymbol{x})_{|\boldsymbol{x}_0} = \left(\frac{\partial\hat{\phi}}{\partial x_i}\right)_{|(x_1^0, x_2^0, x_3^0)}, \tag{1.104}$$

or, more simply, because $\boldsymbol{x}_0$ is arbitrary,

$$\boldsymbol{e}_i \cdot \nabla\phi(\boldsymbol{x}) = \hat{\phi}_{,i}, \tag{1.105}$$

where

$$\hat{\phi}_{,i} = \frac{\partial\hat{\phi}}{\partial x_i}. \tag{1.106}$$

This comma notation for partial derivatives with respect to the coordinates is quite convenient and will be used henceforth.

From (1.14) and (1.105) it follows that

$$\nabla\phi = \hat{\phi}_{,i}\boldsymbol{e}_i. \tag{1.107}$$

To avoid a proliferation of symbols, we typically drop the caret and write this as

$$\nabla\phi = \phi_{,i}\boldsymbol{e}_i, \tag{1.108}$$

on the understanding that $\hat{\phi}$ and ϕ, while having the same *values*, are different *functions*. This formula gives the representation of $\nabla\phi$ in terms of Cartesian coordinates.

1.2.4 *Parameterized paths and the chain rule*

The notion of directional derivative pertains to a straight-line path with a uniform orientation along its length. More generally, consider a smooth curve with parametric representation $\boldsymbol{x}(u)$: $u_0 \leq u \leq u_1$ having no points of self intersection. Let $\boldsymbol{x}_0 = \boldsymbol{x}(u_0)$ and $\boldsymbol{x}_1 = \boldsymbol{x}(u_1)$, and let $\psi(u) = \phi(\boldsymbol{x}(u))$. Then, from (1.99),

$$\psi(u_1) = \psi(u_0) + \nabla\phi(\boldsymbol{x})_{|\boldsymbol{x}_0} \cdot (\boldsymbol{x}_1 - \boldsymbol{x}_0) + o(|\boldsymbol{x}_1 - \boldsymbol{x}_0|). \tag{1.109}$$

We also have

$$\boldsymbol{x}_1 - \boldsymbol{x}_0 = \boldsymbol{x}'(u)_{|u_0}(u_1 - u_0) + o(u_1 - u_0), \tag{1.110}$$

where $(\cdot)'(u) = d(\cdot)/du$, implying that $|\boldsymbol{x}_1 - \boldsymbol{x}_0| = O(u_1 - u_0)$. Combining these gives

$$\frac{\psi(u_1) - \psi(u_0)}{u_1 - u_0} = \nabla\phi(\boldsymbol{x})_{|\boldsymbol{x}_0} \cdot \boldsymbol{x}'(u)_{|u_0} + \frac{o(u_1 - u_0)}{u_1 - u_0}, \tag{1.111}$$

and passing to the limit $u_1 \to u_0$ we obtain the *Chain Rule*

$$\psi'(u)_{|u_0} = \nabla\phi(\boldsymbol{x})_{|\boldsymbol{x}_0} \cdot \boldsymbol{x}'(u)_{|u_0}. \tag{1.112}$$

Because of the arbitrariness of $\boldsymbol{x}(u_0)$, and with a slight abuse of notation, we write this as

$$\frac{d\phi}{du} = \nabla\phi \cdot \frac{d\boldsymbol{x}}{du}, \tag{1.113}$$

or, more simply, as

$$d\phi = \nabla\phi \cdot d\boldsymbol{x}. \tag{1.114}$$

This exceedingly useful formula does not involve any coordinates. It can be used to generate an expression for $\nabla\phi$ in any coordinate system, if desired. We simply invoke (1.113) for arbitrary curves $\boldsymbol{x}(u)$ and hence for arbitrary $d\boldsymbol{x}$.

Examples

1. Using the conventional form of the chain rule, we have
$$d\hat{\phi}(x_1, x_2, x_3) = \hat{\phi}_{,i} dx_i = \hat{\phi}_{,i} \boldsymbol{e}_i \cdot d\boldsymbol{x}, \tag{1.115}$$
where (1.95) has been used with fixed $\boldsymbol{e}_i$, independent of position, i.e., $d\boldsymbol{x} = \boldsymbol{e}_i dx_i$. On comparison with (1.114), we have $(\nabla\phi - \hat{\phi}_{,i}\boldsymbol{e}_i) \cdot d\boldsymbol{x} = 0$ for all $d\boldsymbol{x}$. Choosing $d\boldsymbol{x}$ to be a constant multiple of the expression in parentheses, we have that the squared norm of this expression vanishes, and hence that it is equal to the zero vector. This yields the Cartesian formula (1.108) for the gradient.
2. Consider a system of cylindrical polar coordinates $\{r, \theta, z\}$, where $r = \sqrt{x_1^2 + x_2^2}$, $\tan\theta = x_2/x_1$ and $z = x_3$. We define unit-vector functions $\boldsymbol{e}_r(\theta)$ and $\boldsymbol{e}_\theta(\theta)$ by (Figure 1.4)
$$\boldsymbol{e}_r(\theta) = (\boldsymbol{e}_i \cdot \boldsymbol{e}_r)\boldsymbol{e}_i = \cos\theta\boldsymbol{e}_1 + \sin\theta\boldsymbol{e}_2 \quad \text{and}$$
$$\boldsymbol{e}_\theta(\theta) = \boldsymbol{k} \times \boldsymbol{e}_r(\theta) = -\sin\theta\boldsymbol{e}_1 + \cos\theta\boldsymbol{e}_2, \quad \text{where} \quad \boldsymbol{k} = \boldsymbol{e}_3. \tag{1.116}$$
These point in the directions of increasing r, θ and z, respectively. Given any point x with position vector $\boldsymbol{x}$ relative to a specified

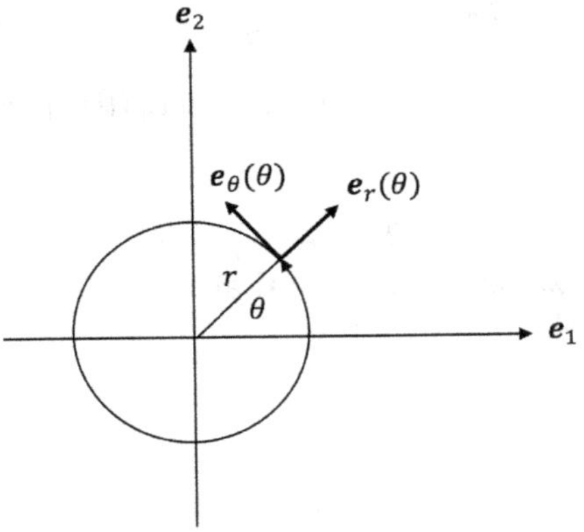

Figure 1.4. Basis vectors $\boldsymbol{e}_r(\theta)$ and $\boldsymbol{e}_\theta(\theta)$ in cylindrical polar coordinates.

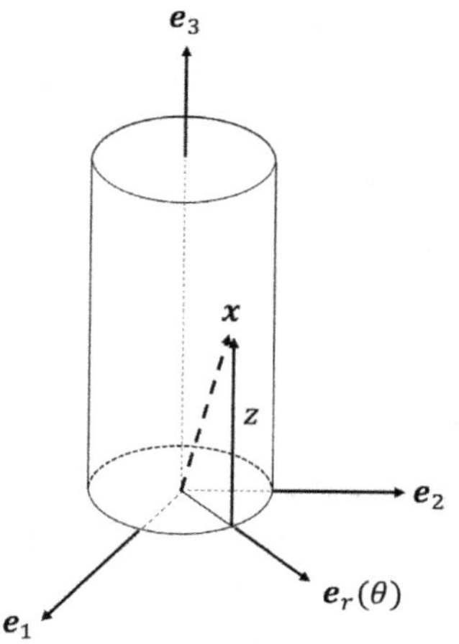

Figure 1.5. Position vector $\boldsymbol{x}$ in cylindrical polar coordinates.

origin, and given a fixed orthonormal set $\{\boldsymbol{e}_i\}$, we can construct a cylindrical surface containing x such that (Figure 1.5)

$$\boldsymbol{x} = r\boldsymbol{e}_r(\theta) + z\boldsymbol{k}. \tag{1.117}$$

Accordingly, any function $\phi(\boldsymbol{x})$ is also a function $\tilde{\phi}(r, \theta, z)$ of the coordinates. Assuming this to be differentiable, we have

$$d\tilde{\phi} = \tilde{\phi}_{,r}dr + \tilde{\phi}_{,\theta}d\theta + \tilde{\phi}_{,z}dz, \tag{1.118}$$

where $(\cdot)_{,r} = \partial(\cdot)/\partial r$, $(\cdot)_{,\theta} = \partial(\cdot)/\partial\theta$ and $(\cdot)_{,z} = \partial(\cdot)/\partial z$. To derive a formula for the gradient $\nabla\phi$, we need to express this as a linear function of $d\boldsymbol{x}$. To this end we observe, from (1.117), that

$$d\boldsymbol{x} = dr\boldsymbol{e}_r(\theta) + rd\boldsymbol{e}_r(\theta) + dz\boldsymbol{k} = dr\boldsymbol{e}_r(\theta) + rd\theta\boldsymbol{e}_\theta(\theta) + dz\boldsymbol{k}, \tag{1.119}$$

where we have used $d\boldsymbol{e}_r(\theta) = \boldsymbol{e}_r'(\theta)d\theta$ with $\boldsymbol{e}_r'(\theta) = \boldsymbol{e}_\theta(\theta)$. Because the set $\{\boldsymbol{e}_r, \boldsymbol{e}_\theta, \boldsymbol{k}\}$ is orthonormal at each θ, on scalar-multiplying by each of its elements we find that

$$dr = \boldsymbol{e}_r(\theta) \cdot d\boldsymbol{x}, \quad rd\theta = \boldsymbol{e}_\theta(\theta) \cdot d\boldsymbol{x} \quad \text{and} \quad dz = \boldsymbol{k} \cdot d\boldsymbol{x}. \tag{1.120}$$

Then,

$$d\tilde{\phi} = \tilde{\phi}_{,r}\boldsymbol{e}_r(\theta) \cdot d\boldsymbol{x} + \frac{1}{r}\tilde{\phi}_{,\theta}\boldsymbol{e}_\theta(\theta) \cdot d\boldsymbol{x} + \tilde{\phi}_{,z}\boldsymbol{k} \cdot d\boldsymbol{x}. \tag{1.121}$$

Comparing with (1.114), invoking the arbitrariness of $d\boldsymbol{x}$ and dropping the tilde, we finally obtain

$$\nabla\phi = \phi_{,r}\boldsymbol{e}_r(\theta) + \frac{1}{r}\phi_{,\theta}\boldsymbol{e}_\theta(\theta) + \phi_{,z}\boldsymbol{k}. \tag{1.122}$$

The structure of this formula differs fundamentally from that of the Cartesian formula (1.108). This is due to the fact that the basis $\{\boldsymbol{e}_i\}$ consists of fixed vectors whereas $\{\boldsymbol{e}_r, \boldsymbol{e}_\theta, \boldsymbol{k}\}$ is variable.

3. Consider the distance $r = |\boldsymbol{x}|$ of x from a specified origin. We wish to obtain a formula for ∇r. We pursue two approaches:

   (a) Using the Cartesian formula (1.95), we write $r^2 = \boldsymbol{x}\cdot\boldsymbol{x} = x_j x_j$. Then,

   $$2rr_{,i} = (r^2)_{,i} = (x_j x_j)_{,i} = x_{j,i}x_j + x_j x_{j,i} = 2\delta_{ji}x_j = 2x_i, \tag{1.123}$$

   where we have invoked the usual product rule for derivatives and the fact that $x_{j,i} = \partial x_j/\partial x_i = \delta_{ji}$. Thus, $r_{,i} = r^{-1}x_i$ and (1.108) yields

   $$\nabla r = r^{-1}\boldsymbol{x}. \tag{1.124}$$

   (b) Alternatively, we have

   $$2rdr = d(r^2) = d(\boldsymbol{x}\cdot\boldsymbol{x}) = d\boldsymbol{x}\cdot\boldsymbol{x} + \boldsymbol{x}\cdot d\boldsymbol{x} = 2\boldsymbol{x}\cdot d\boldsymbol{x}, \tag{1.125}$$

   where we have used the product rule in the form $d(\boldsymbol{a}\cdot\boldsymbol{b}) = d\boldsymbol{a}\cdot\boldsymbol{b} + \boldsymbol{a}\cdot d\boldsymbol{b}$, which is readily proved via (1.16) and the conventional product rule. Thus, $dr = r^{-1}\boldsymbol{x}\cdot d\boldsymbol{x}$ and (1.124) follows from (1.114).

1.2.5 *Gradients of vector fields*

Following on our discussion of scalar fields, a vector field $\boldsymbol{v}(\boldsymbol{x})$ is said to differentiable at $\boldsymbol{x}_0$ provided that there exists a linear vector-valued function $\boldsymbol{f}_0(\cdot)$: $E^3 \to E^3$ such that

$$\boldsymbol{v}(\boldsymbol{x}) = \boldsymbol{v}(\boldsymbol{x}_0) + \boldsymbol{f}_0(\boldsymbol{x}-\boldsymbol{x}_0) + \boldsymbol{o}(\boldsymbol{x}-\boldsymbol{x}_0),$$
$$\text{where } |\boldsymbol{o}(\boldsymbol{x}-\boldsymbol{x}_0)| = o(|\boldsymbol{x}-\boldsymbol{x}_0|); \tag{1.126}$$

that is, if there exists a fixed tensor $\boldsymbol{L}_0$ such that

$$\boldsymbol{v}(\boldsymbol{x}) = \boldsymbol{v}(\boldsymbol{x}_0) + \boldsymbol{L}_0(\boldsymbol{x}-\boldsymbol{x}_0) + \boldsymbol{o}(\boldsymbol{x}-\boldsymbol{x}_0). \tag{1.127}$$

We write

$$\boldsymbol{L}_0 = \nabla\boldsymbol{v}(\boldsymbol{x})_{|\boldsymbol{x}_0}, \tag{1.128}$$

where $\nabla\boldsymbol{v}$ is the tensor-valued *gradient* of $\boldsymbol{v}(\boldsymbol{x})$.

Let $\boldsymbol{x} - \boldsymbol{x}_0 = s\boldsymbol{u}$ with $|\boldsymbol{u}| = 1$. Then,

$$\boldsymbol{L}_0\boldsymbol{u} = \lim_{s\to 0}\frac{1}{s}[\boldsymbol{v}(\boldsymbol{x}_0 + s\boldsymbol{u}) - \boldsymbol{v}(\boldsymbol{x}_0)]. \tag{1.129}$$

Writing $\boldsymbol{v}(\boldsymbol{x}) = \boldsymbol{v}(x_i\boldsymbol{e}_i) = \hat{\boldsymbol{v}}(x_1, x_2, x_3) = \hat{v}_i(x_1, x_2, x_3)\boldsymbol{e}_i$ with $\{\boldsymbol{e}_i\}$ fixed, we obtain

$$\begin{aligned}\boldsymbol{L}_0\boldsymbol{e}_1 &= \boldsymbol{e}_i \lim_{s\to 0}\frac{1}{s}[\hat{v}_i(x_1^0 + s, x_2^0, x_3^0) - \hat{v}_i(x_1^0, x_2^0, x_3^0)]\\ &= \boldsymbol{e}_i\left(\frac{\partial \hat{v}_i}{\partial x_1}\right)_{|(x_1^0, x_2^0, x_3^0)}.\end{aligned} \tag{1.130}$$

Dropping the subscript $_0$ and the superposed caret, it follows in similar fashion that

$$\boldsymbol{L}\boldsymbol{e}_j = v_{k,j}\boldsymbol{e}_k, \quad \text{where } v_{k,j} = \partial v_k/\partial x_j. \tag{1.131}$$

Thus, $L_{ij} = \boldsymbol{e}_i \cdot \boldsymbol{L}\boldsymbol{e}_j = v_{k,j}\delta_{ik} = v_{i,j}$ and we have the Cartesian-coordinate representation

$$\nabla\boldsymbol{v} = v_{i,j}\boldsymbol{e}_i \otimes \boldsymbol{e}_j. \tag{1.132}$$

A more useful result, not involving any coordinate system, is the chain-rule formula

$$d\boldsymbol{v} = (\nabla\boldsymbol{v})d\boldsymbol{x}. \tag{1.133}$$

This is derived by repeating the discussion of Section 1.2.4 essentially verbatim. For example, from $d\hat{v}_i = \hat{v}_{i,j}dx_j = \hat{v}_{i,j}\boldsymbol{e}_j \cdot d\boldsymbol{x}$ we have

$$d\boldsymbol{v} = d\hat{v}_i\boldsymbol{e}_i = \hat{v}_{i,j}\boldsymbol{e}_i(\boldsymbol{e}_j \cdot d\boldsymbol{x}) = (\hat{v}_{i,j}\boldsymbol{e}_i \otimes \boldsymbol{e}_j)d\boldsymbol{x}, \tag{1.134}$$

and comparison with (1.133), with $d\boldsymbol{x}$ arbitrary, yields (1.132). However, (1.133) may be used directly to generate a representation of $\nabla\boldsymbol{v}$ in any coordinate system.

Examples

1. Consider the function

$$\boldsymbol{u}(\boldsymbol{x}) = \mu\boldsymbol{x} + \nu(\boldsymbol{x} \otimes \boldsymbol{x})\boldsymbol{a}, \tag{1.135}$$

where $\boldsymbol{a}$ is a fixed vector and μ, ν are constants. To obtain $\nabla\boldsymbol{u}$, we use:

(a) $u_i = \mu x_i + \nu x_i x_j a_j$. Then,

$$\begin{aligned} u_{i,k} &= \mu x_{i,k} + \nu(x_{i,k}x_j + x_i x_{j,k})a_j \\ &= \mu\delta_{ik} + \nu(\delta_{ik}x_j + x_i\delta_{jk})a_j \\ &= [\mu + \nu(\boldsymbol{x}\cdot\boldsymbol{a})]\delta_{ik} + \nu x_i a_k, \end{aligned} \tag{1.136}$$

and (1.132) gives

$$\nabla\boldsymbol{u} = [\mu + \nu(\boldsymbol{x}\cdot\boldsymbol{a})]\boldsymbol{I} + \nu\boldsymbol{x}\otimes\boldsymbol{a}, \tag{1.137}$$

or,

(b)

$$\begin{aligned} d\boldsymbol{u} &= \mu d\boldsymbol{x} + \nu(d\boldsymbol{x}\otimes\boldsymbol{x} + \boldsymbol{x}\otimes d\boldsymbol{x})\boldsymbol{a} \\ &= \mu d\boldsymbol{x} + \nu(\boldsymbol{x}\cdot\boldsymbol{a})d\boldsymbol{x} + \nu\boldsymbol{x}(d\boldsymbol{x}\cdot\boldsymbol{a}) \\ &= \{[\mu + \nu(\boldsymbol{x}\cdot\boldsymbol{a})]\boldsymbol{I} + \nu\boldsymbol{x}\otimes\boldsymbol{a}\}d\boldsymbol{x}, \end{aligned} \tag{1.138}$$

and comparison with (1.133) yields (1.137). Here, we have used the product rule in the form $d(\boldsymbol{b}\otimes\boldsymbol{c}) = d\boldsymbol{b}\otimes\boldsymbol{c} + \boldsymbol{b}\otimes d\boldsymbol{c}$, which is easily proved by using Cartesian representations.

2. Consider

$$\boldsymbol{u}(\boldsymbol{x}) = \boldsymbol{x} + (\boldsymbol{b}\cdot\boldsymbol{x})\boldsymbol{a}, \tag{1.139}$$

where $\boldsymbol{a}, \boldsymbol{b}$ are fixed vectors. Then,

$$d\boldsymbol{u} = d\boldsymbol{x} + \boldsymbol{a}(\boldsymbol{b}\cdot d\boldsymbol{x}) = (\boldsymbol{I} + \boldsymbol{a}\otimes\boldsymbol{b})d\boldsymbol{x}, \tag{1.140}$$

yielding

$$\nabla\boldsymbol{u} = \boldsymbol{I} + \boldsymbol{a}\otimes\boldsymbol{b}. \tag{1.141}$$

Note that this gradient is uniform, i.e., independent of $\boldsymbol{x}$. That this is to be expected follows from (1.139), written in the form $\boldsymbol{u}(\boldsymbol{x}) = (\boldsymbol{I} + \boldsymbol{a}\otimes\boldsymbol{b})\boldsymbol{x}$. Thus,

$$\boldsymbol{u}(\boldsymbol{x}) = \boldsymbol{u}(\boldsymbol{x}_0) + (\boldsymbol{I} + \boldsymbol{a}\otimes\boldsymbol{b})(\boldsymbol{x} - \boldsymbol{x}_0), \tag{1.142}$$

and comparison with (1.127) yields (1.141) directly as there is no remainder term.

3. Consider a vector field

$$\boldsymbol{u}(\boldsymbol{x}) = u_r(r,\theta)\boldsymbol{e}_r + u_\theta(r,\theta)\boldsymbol{e}_\theta + u_z(r,\theta)\boldsymbol{k}, \tag{1.143}$$

expressed in terms of polar coordinates. For the sake of simplicity we confine attention to the case in which the components u_r, u_θ, u_z are independent of the axial coordinate z. To obtain $\nabla\boldsymbol{u}$, we use the chain rule directly:

$$(\nabla\boldsymbol{u})d\boldsymbol{x} = d\boldsymbol{u} = u_{r,r}\boldsymbol{e}_r dr + u_{r,\theta}\boldsymbol{e}_r d\theta + u_r\boldsymbol{e}_\theta d\theta + \cdots. \tag{1.144}$$

Combining this with (1.120), after some effort, we arrive at

$$\begin{aligned}\nabla\boldsymbol{u} &= u_{r,r}\boldsymbol{e}_r\otimes\boldsymbol{e}_r + \frac{1}{r}(u_{r,\theta} - u_\theta)\boldsymbol{e}_r\otimes\boldsymbol{e}_\theta + u_{\theta,r}\boldsymbol{e}_\theta\otimes\boldsymbol{e}_r \\ &\quad + \frac{1}{r}(u_{\theta,\theta} + u_r)\boldsymbol{e}_\theta\otimes\boldsymbol{e}_\theta + u_{z,r}\boldsymbol{k}\otimes\boldsymbol{e}_r + \frac{1}{r}u_{z,\theta}\boldsymbol{k}\otimes\boldsymbol{e}_\theta.\end{aligned} \tag{1.145}$$

It is a useful exercise to generalize this formula to accommodate components that depend on all three coordinates r, θ and z.

1.2.6 *Divergence and curl*

The *divergence* of a vector field $\boldsymbol{v}(\boldsymbol{x})$ is the *scalar* field, $div\boldsymbol{v}$, defined by

$$div\boldsymbol{v} = tr(\nabla\boldsymbol{v}). \tag{1.146}$$

This has nothing to do with coordinates. However, to obtain a Cartesian-coordinate representation we combine (1.132) with the linearity of the trace operator:

$$div(v_j\boldsymbol{e}_j) = tr(v_{i,j}\boldsymbol{e}_i\otimes\boldsymbol{e}_j) = v_{i,j}tr(\boldsymbol{e}_i\otimes\boldsymbol{e}_j) = v_{i,j}\delta_{ij} = v_{j,j}. \tag{1.147}$$

The divergence of a tensor field $\boldsymbol{A}(\boldsymbol{x})$ is the vector field, $div\boldsymbol{A}$, defined by

$$\boldsymbol{c}\cdot div\boldsymbol{A} = div(\boldsymbol{A}^t\boldsymbol{c}), \quad \text{for any } \textit{fixed} \text{ vector } \boldsymbol{c}. \tag{1.148}$$

Thus,

$$\begin{aligned}\boldsymbol{e}_i\cdot div\boldsymbol{A} &= div(\boldsymbol{A}^t\boldsymbol{e}_i) = div(A^t_{ji}\boldsymbol{e}_j) \\ &= div(A_{ij}\boldsymbol{e}_j) \quad (\text{for fixed } i\in\{1,2,3\}) \\ &= A_{ij,j} \quad (\text{from (1.147)}).\end{aligned} \tag{1.149}$$

Hence, the Cartesian-coordinate representation

$$div\boldsymbol{A} = A_{ij,j}\boldsymbol{e}_i. \tag{1.150}$$

The conventional product-rule formula

$$u_i A_{ij,j} = (A_{ij}u_i)_{,j} - A_{ij}u_{i,j} \tag{1.151}$$

is thus seen to be the Cartesian-coordinate representation of the coordinate-independent rule

$$\boldsymbol{u} \cdot div\boldsymbol{A} = div(\boldsymbol{A}^t\boldsymbol{u}) - tr[\boldsymbol{A}^t(\nabla\boldsymbol{u})]. \tag{1.152}$$

Example

We use this result to construct explicit expressions for the coefficients in the orthogonal decomposition

$$div\boldsymbol{A} = (\boldsymbol{e}_r \cdot div\boldsymbol{A})\boldsymbol{e}_r + (\boldsymbol{e}_\theta \cdot div\boldsymbol{A})\boldsymbol{e}_\theta + (\boldsymbol{k} \cdot div\boldsymbol{A})\boldsymbol{k}, \tag{1.153}$$

where

$$\boldsymbol{A} = A_{rr}\boldsymbol{e}_r \otimes \boldsymbol{e}_r + A_{r\theta}\boldsymbol{e}_r \otimes \boldsymbol{e}_\theta + \cdots + A_{zz}\boldsymbol{k} \otimes \boldsymbol{k}, \tag{1.154}$$

in which the components $A_{rr}, A_{r\theta}, \ldots$ are functions of r and θ in a polar coordinate system. The result can be easily extended to the case when these components also depend on the axial coordinate z.

From (1.152), we have

$$\begin{aligned} \boldsymbol{e}_r \cdot div\boldsymbol{A} &= div(\boldsymbol{A}^t\boldsymbol{e}_r) - tr[\boldsymbol{A}^t(\nabla\boldsymbol{e}_r)], \\ \boldsymbol{e}_\theta \cdot div\boldsymbol{A} &= div(\boldsymbol{A}^t\boldsymbol{e}_\theta) - tr[\boldsymbol{A}^t(\nabla\boldsymbol{e}_\theta)] \quad \text{and} \\ \boldsymbol{k} \cdot div\boldsymbol{A} &= div(\boldsymbol{A}^t\boldsymbol{k}), \end{aligned} \tag{1.155}$$

where, from (1.145),

$$\nabla\boldsymbol{e}_r = \frac{1}{r}\boldsymbol{e}_\theta \otimes \boldsymbol{e}_\theta \quad \text{and} \quad \nabla\boldsymbol{e}_\theta = -\frac{1}{r}\boldsymbol{e}_r \otimes \boldsymbol{e}_\theta, \tag{1.156}$$

and, also,

$$\begin{aligned} \boldsymbol{A}^t\boldsymbol{e}_r &= A_{rr}\boldsymbol{e}_r + A_{r\theta}\boldsymbol{e}_\theta + A_{rz}\boldsymbol{k}, \\ \boldsymbol{A}^t\boldsymbol{e}_\theta &= A_{\theta r}\boldsymbol{e}_r + A_{\theta\theta}\boldsymbol{e}_\theta + A_{\theta z}\boldsymbol{k}, \\ \boldsymbol{A}^t\boldsymbol{k} &= A_{zr}\boldsymbol{e}_r + A_{z\theta}\boldsymbol{e}_\theta + A_{zz}\boldsymbol{k}. \end{aligned} \tag{1.157}$$

Next, we use these results with

$$div\boldsymbol{u} = tr(\nabla\boldsymbol{u}) = u_{r,r} + \frac{1}{r}(u_{\theta,\theta} + u_r), \tag{1.158}$$

which follows from (1.145), to obtain

$$\begin{aligned} div(\boldsymbol{A}^t\boldsymbol{e}_r) &= A_{rr,r} + \frac{1}{r}(A_{rr} + A_{r\theta,\theta}), \\ div(\boldsymbol{A}^t\boldsymbol{e}_\theta) &= A_{\theta r,r} + \frac{1}{r}(A_{\theta r} + A_{\theta\theta,\theta}), \\ div(\boldsymbol{A}^t\boldsymbol{k}) &= A_{zr,r} + \frac{1}{r}(A_{zr} + A_{z\theta,\theta}). \end{aligned} \tag{1.159}$$

We also require

$$\boldsymbol{A}^t(\nabla\boldsymbol{e}_r) = \frac{1}{r}(\boldsymbol{A}^t\boldsymbol{e}_\theta)\otimes\boldsymbol{e}_\theta \quad \text{and} \quad \boldsymbol{A}^t(\nabla\boldsymbol{e}_\theta) = -\frac{1}{r}(\boldsymbol{A}^t\boldsymbol{e}_r)\otimes\boldsymbol{e}_\theta. \tag{1.160}$$

These imply that

$$\begin{aligned} tr[\boldsymbol{A}^t(\nabla\boldsymbol{e}_r)] &= \frac{1}{r}\boldsymbol{e}_\theta\cdot\boldsymbol{A}\boldsymbol{e}_\theta = \frac{1}{r}A_{\theta\theta} \quad \text{and} \quad tr[\boldsymbol{A}^t(\nabla\boldsymbol{e}_\theta)] \\ &= -\frac{1}{r}\boldsymbol{e}_r\cdot\boldsymbol{A}\boldsymbol{e}_\theta = -\frac{1}{r}A_{r\theta}. \end{aligned} \tag{1.161}$$

Combining (1.155), (1.159) and (1.161), we finally arrive at

$$\begin{aligned} \boldsymbol{e}_r\cdot div\boldsymbol{A} &= A_{rr,r} + \frac{1}{r}(A_{rr} - A_{\theta\theta} + A_{r\theta,\theta}), \\ \boldsymbol{e}_\theta\cdot div\boldsymbol{A} &= A_{\theta r,r} + \frac{1}{r}(A_{\theta r} + A_{r\theta} + A_{\theta\theta,\theta}) \quad \text{and} \\ \boldsymbol{k}\cdot div\boldsymbol{A} &= A_{zr,r} + \frac{1}{r}(A_{zr} + A_{z\theta,\theta}). \end{aligned} \tag{1.162}$$

The last of the main differential operators is the *curl*. For a vector field $\boldsymbol{v}(\boldsymbol{x})$ this is the vector, $curl\boldsymbol{v}$, defined by

$$\mathbf{c}\cdot curl\boldsymbol{v} = div(\boldsymbol{v}\times\mathbf{c}) \quad \text{for any } \textit{fixed} \text{ vector } \mathbf{c}. \tag{1.163}$$

Accordingly,

$$\begin{aligned} \boldsymbol{e}_j\cdot curl\boldsymbol{v} &= div(\boldsymbol{v}\times\boldsymbol{e}_j) = div(v_i\boldsymbol{e}_i\times\boldsymbol{e}_j) \\ &= div(v_ie_{ijk}\boldsymbol{e}_k) = (v_ie_{ijk})_{,k} \\ &= e_{kij}v_{i,k}, \end{aligned} \tag{1.164}$$

yielding the Cartesian-coordinate formula

$$curl\boldsymbol{v} = e_{kij}v_{i,k}\boldsymbol{e}_j. \tag{1.165}$$

This may be used, as in the passage from (1.148) to (1.152), to derive the formula for the curl in any coordinate system.

For a tensor field $\boldsymbol{A}(\boldsymbol{x})$, the curl is the tensor, $curl\boldsymbol{A}$, defined by

$$(curl\boldsymbol{A})\boldsymbol{c} = curl(\boldsymbol{A}^t\boldsymbol{c}), \tag{1.166}$$

again for any *fixed* vector $\boldsymbol{c}$. We leave the derivation of the Cartesian representation as an exercise.

1.2.7 *Integral formulae*

We will make frequent use of the divergence theorem in various forms. For a continuously differentiable (and hence continuous) vector field $\boldsymbol{v}(\boldsymbol{x})$ defined on a bounded region R with a piecewise smooth orientable boundary ∂R, this is

$$\int_R div\boldsymbol{v}dv = \int_{\partial R} \boldsymbol{v}\cdot\boldsymbol{n}da, \tag{1.167}$$

where $\boldsymbol{n}$ is the piecewise continuous exterior unit normal field on ∂R. Proofs of this classical formula abound in the literature (e.g., [1, 2]). We will have occasion later to make use of variants of this formula, and to extend it to piecewise continuous fields.

Although we will not make use of it in this book, Stokes' theorem [1, 2] is also of great importance to our subject. For differentiable vector fields $\boldsymbol{v}(\boldsymbol{x})$, this is

$$\int_\Omega curl\boldsymbol{v}\cdot\boldsymbol{n}da = \int_{\partial\Omega} \boldsymbol{v}\cdot d\boldsymbol{x}, \tag{1.168}$$

where Ω is a piecewise smooth surface bounded by a piecewise smooth closed curve $\partial\Omega$. Here, the unit-normal field $\boldsymbol{n}(\boldsymbol{x})$ is chosen such that $\boldsymbol{\nu} = \boldsymbol{t}\times\boldsymbol{n}_{|\partial\Omega}$ is the exterior unit normal to $\partial\Omega$, lying in the tangent plane to Ω at $\boldsymbol{x}\in\partial\Omega$. Here, $\boldsymbol{t} = \boldsymbol{x}'(s)$ at points where the derivative is defined, where $\boldsymbol{x}(s)$ is an arclength parametrization of $\partial\Omega$.

1.3 Problems

1. Prove the Cauchy–Schwarz inequality $|\boldsymbol{a}\cdot\boldsymbol{b}| \le |\boldsymbol{a}|\,|\boldsymbol{b}|$ for vectors $\boldsymbol{a},\boldsymbol{b}$. [*Hint*: Consider the vector $(\boldsymbol{a}\cdot\boldsymbol{a})\boldsymbol{b} - (\boldsymbol{a}\cdot\boldsymbol{b})\boldsymbol{a}$ and use the positivity of the norm]. Prove the triangle inequality $|\boldsymbol{a}+\boldsymbol{b}| \le |\boldsymbol{a}| + |\boldsymbol{b}|$. Prove the Pythagorean theorem for orthogonal $\boldsymbol{a},\boldsymbol{b}$.
2. Let $f(\boldsymbol{v}) = |\boldsymbol{v}|$ for any vector $\boldsymbol{v}$. Is f a linear function?
3. Let $\boldsymbol{m} = \frac{\sqrt{2}}{2}(\boldsymbol{e}_1+\boldsymbol{e}_2)$ where $\boldsymbol{e}_1$ and $\boldsymbol{e}_2$ are orthonormal vectors. Is there a tensor $\boldsymbol{A}$ such that $\boldsymbol{A}\boldsymbol{v} = \boldsymbol{m}$ for any vector $\boldsymbol{v}$?
4. Let $\boldsymbol{m} = \frac{\sqrt{2}}{2}(\boldsymbol{e}_1+\boldsymbol{e}_2)$ and $\boldsymbol{n} = \frac{\sqrt{2}}{2}(\boldsymbol{e}_3-\boldsymbol{e}_1)$, where $\{\boldsymbol{e}_i\}$ is an orthonormal basis. Find the tensor $\boldsymbol{A}$ such that $\boldsymbol{A}\boldsymbol{v} = (\boldsymbol{v}\cdot\boldsymbol{n})\boldsymbol{m}$ for every vector $\boldsymbol{v}$.
5. Show that $e_{ijk}A_{jk} = 0$ if and only if $\boldsymbol{A} = \boldsymbol{A}^t$ (i.e., if $\boldsymbol{A} = \boldsymbol{A}^t$, then $e_{ijk}A_{jk} = 0$, and conversely.).
6. Show:
   (a) If $\boldsymbol{A}^t = -\boldsymbol{A}$ then $\boldsymbol{A}\cdot\boldsymbol{a}\otimes\boldsymbol{a} = 0$ for any vector $\boldsymbol{a}$.
   (b) If $\boldsymbol{T}\cdot\boldsymbol{S} = 0$ for any tensor $\boldsymbol{S}$, then $\boldsymbol{T} = \boldsymbol{O}$.
   (c) If $\boldsymbol{T}\cdot\boldsymbol{S} = 0$ for any symmetric tensor $\boldsymbol{S}$, then $\boldsymbol{T}$ is skew.
   (d) If $\boldsymbol{T}\cdot\boldsymbol{S} = 0$ for any skew tensor $\boldsymbol{S}$, then $\boldsymbol{T}$ is symmetric.
   (e) Establish the converses of the statements in (a)–(d).
7. $\boldsymbol{u}$ and $\boldsymbol{v}$ are arbitrary vectors and $\boldsymbol{A}$ is an arbitrary tensor. Does $\boldsymbol{A}(\boldsymbol{u}\times\boldsymbol{v}) = \boldsymbol{A}\boldsymbol{u}\times\boldsymbol{v} + \boldsymbol{u}\times\boldsymbol{A}\boldsymbol{v}$?
8. If $\{\boldsymbol{e}_i\}$ and $\{\boldsymbol{e}'_i\}$ are orthonormal bases we can always write $\boldsymbol{e}_i = a_{ij}\boldsymbol{e}'_j$ and $\boldsymbol{e}'_i = a'_{ij}\boldsymbol{e}_j$, for some a_{ij} and a'_{ij}. Show:
   (a) $a_{ik}a'_{kj} = \delta_{ij}$ and
   (b) the matrices (a), (a') are mutual transposes.
9. Let A_{ij} and A'_{ij} be the components of a tensor $\boldsymbol{A}$ with respect to $\{\boldsymbol{e}_i\otimes\boldsymbol{e}_j\}$ and $\left\{\boldsymbol{e}'_i\otimes\boldsymbol{e}'_j\right\}$, respectively. Show:
   (a) $A_{ii} = A'_{ii}$, i.e., the trace of $\boldsymbol{A}$ does not depend on the basis. For this reason, the trace is said to be *invariant*.
   (b) Use this result to show that the dot product of two vectors is also invariant.
   (c) Let $A_{11} = 1$, $A_{12} = A_{21} = 5$, $A_{13} = A_{31} = -5$, $A_{33} = 1$, $A_{22} = A_{23} = A_{32} = 0$. Find the components A'_{ij} if $\boldsymbol{e}'_1 = (2\boldsymbol{e}_3 - \boldsymbol{e}_1)/\sqrt{5}$, $\boldsymbol{e}'_2 = \boldsymbol{e}_2$ and $\{\boldsymbol{e}'_i\}$ is right-handed ($\boldsymbol{e}'_1\cdot\boldsymbol{e}'_2\times\boldsymbol{e}'_3 = 1$).

10. (a) Consider a plane P with unit normal vector $\boldsymbol{n}$. Let $\boldsymbol{v}$ (in E^3) be an arbitrary vector, and let $\boldsymbol{v}_p$ be its projection onto the plane P. Use a figure to verify that $\boldsymbol{v} = \boldsymbol{v}_p + (\boldsymbol{v} \cdot \boldsymbol{n})\boldsymbol{n}$. Find the tensor $\mathbb{P}_{(\boldsymbol{n})}$ in the representation $\boldsymbol{v}_p = \mathbb{P}_{(\boldsymbol{n})}\boldsymbol{v}$.
    (b) In general, a tensor $\boldsymbol{A}$ is called a *projection* if it is symmetric and $\boldsymbol{A}^2 = \boldsymbol{A}$, where $\boldsymbol{A}^2 = \boldsymbol{A}\boldsymbol{A}$. Show that $\mathbb{P}_{(\boldsymbol{n})}$ is a projection. Show that $\boldsymbol{O}, \boldsymbol{I}$ and $\boldsymbol{n} \otimes \boldsymbol{n}$ are also projections.
    (c) Let $\mathbb{R}_{(\boldsymbol{n})}$ be the *reflection* tensor for the plane P, i.e., the tensor that maps any vector $\boldsymbol{v}$ (in E^3) to its mirror image with respect to the plane. Find a general expression for $\mathbb{R}_{(\boldsymbol{n})}$ and calculate its components relative to $\{\boldsymbol{e}_i \otimes \boldsymbol{e}_j\}$ for the two cases: (i) $\boldsymbol{n} = \boldsymbol{e}_1$, (ii) $\boldsymbol{n} = \frac{\sqrt{2}}{2}(\boldsymbol{e}_1 + \boldsymbol{e}_2)$.
11. (a) Recall the $e - \delta$ identities: (i) $e_{ijm}e_{klm} = \delta_{ik}\delta_{jl} - \delta_{il}\delta_{jk}$, (ii) $e_{ilm}e_{jlm} = 2\delta_{ij}$, (iii) $e_{ijk}e_{ijk} = 6$. Use these as needed to show that:

$$\boldsymbol{a} \times (\boldsymbol{b} \times \boldsymbol{c}) = (\boldsymbol{a} \cdot \boldsymbol{c})\boldsymbol{b} - (\boldsymbol{a} \cdot \boldsymbol{b})\boldsymbol{c},$$

    and

$$\boldsymbol{v} = (\boldsymbol{v} \cdot \boldsymbol{e})\boldsymbol{e} + \boldsymbol{e} \times (\boldsymbol{v} \times \boldsymbol{e})$$

    for any vector $\boldsymbol{v}$ and any *unit* vector $\boldsymbol{e}$.
    (b) Let $\boldsymbol{e}$ be a fixed unit vector and $\boldsymbol{v}$ an arbitrary vector. Is there a tensor $\boldsymbol{A}$ such that $\boldsymbol{f}(\boldsymbol{v}) = \boldsymbol{A}\boldsymbol{v}$ if $\boldsymbol{f}(\boldsymbol{v}) = \boldsymbol{e} \times (\boldsymbol{v} \times \boldsymbol{e})$? If so, find it.
    (c) Do the same for $\boldsymbol{f}(\boldsymbol{v}) = \boldsymbol{v} + \boldsymbol{a} \times \boldsymbol{v}$ where $\boldsymbol{a}$ is a fixed vector.
12. We used a single orthonormal basis $\{\boldsymbol{e}_i\}$ to show that a linear vector-valued function $\boldsymbol{f}(\boldsymbol{v})$ is expressible as $\boldsymbol{f}(\boldsymbol{v}) = \boldsymbol{A}\boldsymbol{v}$ with $\boldsymbol{A} = A_{ij}\boldsymbol{e}_i \otimes \boldsymbol{e}_j$ and $A_{ij} = \boldsymbol{e}_i \cdot \boldsymbol{A}\boldsymbol{e}_j$. This applies to tensors that map a vector space to itself. Later, we will encounter *two-point* tensors that map one vector space to another vector space. Let $\{\boldsymbol{e}_i\}$ and $\{\hat{\boldsymbol{E}}_A\}$ respectively be orthonormal bases for the vector spaces E^3 and $\hat{E}^3$. Show that if $\boldsymbol{A}$ is a two-point tensor, i.e., a linear map from $\hat{E}^3$ to E^3, then $\boldsymbol{A} = A_{iB}\boldsymbol{e}_i \otimes \hat{\boldsymbol{E}}_B$, where $A_{iB} = \boldsymbol{e}_i \cdot \boldsymbol{A}\hat{\boldsymbol{E}}_B$.
13. Recall that a linear scalar-valued function f of vectors can be expressed as $f(\boldsymbol{v}) = \boldsymbol{a} \cdot \boldsymbol{v}$ where $\boldsymbol{a}$ is a unique vector. Using a similar argument, show that a linear scalar-valued function g of tensors is expressible as $g(\boldsymbol{V}) = \boldsymbol{A} \cdot \boldsymbol{V}$ where $\boldsymbol{A}$ is a unique tensor.

14. Any tensor $\boldsymbol{A}$ that maps a vector space to itself can be represented as the sum of a symmetric part and a skew part. Thus,

$$\boldsymbol{A} = \boldsymbol{S} + \boldsymbol{W}; \quad \boldsymbol{S} = \frac{1}{2}(\boldsymbol{A} + \boldsymbol{A}^t), \quad \boldsymbol{W} = \frac{1}{2}(\boldsymbol{A} - \boldsymbol{A}^t).$$

(a) Show that this representation is unique; i.e., if $\boldsymbol{A} = \boldsymbol{S}' + \boldsymbol{W}'$ with $(\boldsymbol{S}')^t = \boldsymbol{S}'$ and $(\boldsymbol{W}')^t = -\boldsymbol{W}'$, then $\boldsymbol{S}' = \boldsymbol{S}$ and $\boldsymbol{W}' = \boldsymbol{W}$.
Remark: Thus, every such tensor may be written uniquely as the sum of an element of the set $Sym = \{\boldsymbol{S} : \boldsymbol{S} = \boldsymbol{S}^t\}$ and an element of the set $Skw = \{\boldsymbol{W} : \boldsymbol{W} = -\boldsymbol{W}^t\}$.

(b) Show that Sym and Skw are vector spaces and that they are orthogonal in the sense that $\boldsymbol{S} \cdot \boldsymbol{W} = 0$ for every $\boldsymbol{S} \in Sym$ and $\boldsymbol{W} \in Skw$.
Remark: Vector (sub-)spaces with this property are called *orthogonal complements* with respect to the 9-dimensional vector space E^9 of tensors. It is common to indicate this by writing $E^9 = Sym \oplus Skw$, where the operation $\oplus$ is called the *direct sum*. Thus, E^9 may be regarded as the direct sum of Sym and Skw. The decomposition of a given vector space into the direct sum of subspaces is not unique, however (see Problem 15).

15. Show that if a tensor $\boldsymbol{T}$ maps a vector space to itself, then it can be represented as $\boldsymbol{T} = \boldsymbol{A} + \alpha\boldsymbol{I}$ with $tr\boldsymbol{A} = 0$ and $3\alpha = tr\boldsymbol{T}$. The term $\alpha\boldsymbol{I}$ is called the *spherical* part of $\boldsymbol{T}$ and $\boldsymbol{A}$ is called the *deviatoric* part of $\boldsymbol{T}$ (so named because it represents the *deviation* from a spherical tensor).

(a) Show that this representation is unique, i.e., if $\boldsymbol{T} = \boldsymbol{B} + \beta\boldsymbol{I}$ with $tr\boldsymbol{B} = 0$, then $\boldsymbol{B} = \boldsymbol{A}$ and $\beta = \alpha$.

(b) Let $Sph = \{\boldsymbol{C} : \boldsymbol{C} = \lambda\boldsymbol{I} \text{ for some } \lambda \in \mathbb{R}\}$ and $Dev = \{\boldsymbol{D} : tr\boldsymbol{D} = 0\}$. Show that these are orthogonal vector spaces and thus that $E^9 = Sph \oplus Dev$.

16. In Problem 11(c) above you showed that if $\boldsymbol{\omega}$ is a fixed vector then the vector-valued function $\boldsymbol{f}(\boldsymbol{v}) = \boldsymbol{\omega} \times \boldsymbol{v}$ is linear. It follows from this that there is a tensor $\boldsymbol{\Omega}$ such that $\boldsymbol{f}(\boldsymbol{v}) = \boldsymbol{\Omega}\boldsymbol{v}$.

(a) Use bases $\{\boldsymbol{e}_i\}$ and $\{\boldsymbol{e}_i \otimes \boldsymbol{e}_j\}$ to obtain the formula $\Omega_{ij} = e_{jik}\omega_k$. Thus, $\boldsymbol{\Omega} \in Skw$.

(b) Show that $\omega_k = \frac{1}{2}\Omega_{ij}e_{jik}$. Thus, there is a one-to-one relationship between Skw and E^3 in the sense that any element of one corresponds to an element of the other.

17. Verify:

(a) $(\boldsymbol{a} \otimes \boldsymbol{b})^t = \boldsymbol{b} \otimes \boldsymbol{a}$,
(b) $\boldsymbol{A}(\boldsymbol{a} \otimes \boldsymbol{b}) = (\boldsymbol{A}\boldsymbol{a}) \otimes \boldsymbol{b}$,
(c) $(\boldsymbol{a} \otimes \boldsymbol{b})\boldsymbol{A} = \boldsymbol{a} \otimes (\boldsymbol{A}^t\boldsymbol{b})$,
(d) $\boldsymbol{A} = (\boldsymbol{A}\boldsymbol{e}_i) \otimes \boldsymbol{e}_i = \boldsymbol{e}_i \otimes (\boldsymbol{A}^t\boldsymbol{e}_i)$.

18. Show that $(\boldsymbol{A}^t)^{-1} = (\boldsymbol{A}^{-1})^t$. For this reason we usually write $\boldsymbol{A}^{-t}$.
19. $(\boldsymbol{B}\boldsymbol{A})^{-1} = \boldsymbol{A}^{-1}\boldsymbol{B}^{-1}$.
20. $(\boldsymbol{B}\boldsymbol{A})^t = \boldsymbol{A}^t\boldsymbol{B}^t$.
21. Use box products to verify that:

(a) $\det(\alpha\boldsymbol{A}) = \alpha^3 \det \boldsymbol{A}$,
(b) $\det(\boldsymbol{A}\boldsymbol{B}) = (\det \boldsymbol{A})(\det \boldsymbol{B})$,
(c) $\det \boldsymbol{I} = 1$,
(d) $\det(\boldsymbol{A}^{-1}) = (\det \boldsymbol{A})^{-1}$,
(e) $\det(\boldsymbol{a} \otimes \boldsymbol{b}) = 0$,
(f) $\det(\boldsymbol{I} + \boldsymbol{a} \otimes \boldsymbol{b}) = 1 + \boldsymbol{a} \cdot \boldsymbol{b}$,
(g) $\det(\boldsymbol{A} + \boldsymbol{a} \otimes \boldsymbol{b}) = (\det \boldsymbol{A})(1 + \boldsymbol{a} \cdot \boldsymbol{A}^{-t}\boldsymbol{b})$.

22. Let $\boldsymbol{A}^*$ be the cofactor of $\boldsymbol{A}$ (see Appendix A), defined by $\boldsymbol{A}^* (\boldsymbol{a} \times \boldsymbol{b}) = \boldsymbol{A}\boldsymbol{a} \times \boldsymbol{A}\boldsymbol{b}$.

(a) Show that $I_2(\boldsymbol{A}) = I_1(\boldsymbol{A}^*)$.
(b) Show that $I_2(\boldsymbol{A}) = \frac{1}{2}[(tr\boldsymbol{A})^2 - tr(\boldsymbol{A}^2)]$.
(c) Show that $\boldsymbol{A}^t\boldsymbol{A}^* = (\det \boldsymbol{A})\boldsymbol{I}$, and therefore that $\boldsymbol{A}^* = (\det \boldsymbol{A})\boldsymbol{A}^{-t}$ if $\boldsymbol{A}$ is invertible.
(d) Use the result in (c) to recover the formula $\det \boldsymbol{A} = \frac{1}{6}e_{ijk}e_{lmn}A_{il}A_{jm}A_{kn}$.

23. Show that for $\lambda \in \mathbb{R}$, $\det(\boldsymbol{A} - \lambda\boldsymbol{I}) = -\lambda^3 + I_1(\boldsymbol{A})\lambda^2 - I_2(\boldsymbol{A})\lambda + I_3(\boldsymbol{A})$. [*Hint*: Use the box-product definitions of the I_k and the linearity of the box product with respect to each argument individually.]
24. $\phi(\boldsymbol{x})$, $\psi(\boldsymbol{x})$ are scalar fields and $\boldsymbol{u}(\boldsymbol{x})$, $\boldsymbol{v}(\boldsymbol{x})$ are vector fields. Show the following:

(a) $\nabla(\phi + \psi) = \nabla\phi + \nabla\psi$,
(b) $div(\boldsymbol{u} + \boldsymbol{v}) = div\boldsymbol{u} + div\boldsymbol{v}$,

(c) $div(\phi \boldsymbol{v}) = \nabla\phi \cdot \boldsymbol{v} + \phi div\boldsymbol{v}$,
(d) $curl(\nabla\phi) = \boldsymbol{0}$,
(e) $div(curl\boldsymbol{v}) = 0$,
(f) $\nabla(\phi\boldsymbol{u}) = \boldsymbol{u} \otimes \nabla\phi + \phi\nabla\boldsymbol{u}$,
(g) $div(\boldsymbol{u} \otimes \boldsymbol{v}) = (\nabla\boldsymbol{u})\boldsymbol{v} + (div\boldsymbol{v})\boldsymbol{u}$.

In 24(d) assume that ϕ is twice continuously differentiable (i.e., its second partial derivatives are continuous functions of the coordinates). This implies that the second partials commute, i.e., $\phi_{,ij} \equiv (\phi_{,i})_{,j} = (\phi_{,j})_{,i} = \phi_{,ji}$. Use the same assumption for v_i in part 24(e).

25. Consider a scalar-valued function of spherical coordinates $f(r, \theta, \phi)$, where r is radius from the center of the sphere, θ is the azimuthal angle (i.e., the longitude) and ϕ is the elevation angle above the plane of the equator (the latitude). Thus, $\boldsymbol{x} = r\boldsymbol{e}_\rho(\theta, \phi)$, where $\boldsymbol{e}_\rho = \cos\phi \boldsymbol{e}_r(\theta) + \sin\phi \boldsymbol{e}_3$. Find the components of ∇f relative to the (variable) basis $\{\boldsymbol{e}_\rho, \boldsymbol{e}_\theta, \boldsymbol{e}_\phi\}$, where $\boldsymbol{e}_\phi = \boldsymbol{e}_\rho \times \boldsymbol{e}_\theta$.
26. Let f be a function of radius $r(= |\boldsymbol{x}|)$ *only* in a spherical coordinate system. Obtain and solve *Laplace's equation* $div(\nabla f) = 0$ in the region $(0 <)a \leq r \leq b$ bounded by two spheres. Assume boundary conditions $f(a) = f_a$, $f(b) = f_b$. [*Hint*: Use $df(r) = \nabla f \cdot d\boldsymbol{x}$ to get ∇f. Let $\boldsymbol{v} = \nabla f$ and use $div\boldsymbol{v} = tr(\nabla\boldsymbol{v})$, where $d\boldsymbol{v} = (\nabla\boldsymbol{v})d\boldsymbol{x}$.]
27. Establish that $\boldsymbol{u} \cdot curl\boldsymbol{v} = div(\boldsymbol{v} \times \boldsymbol{u}) + \boldsymbol{v} \cdot curl\boldsymbol{u}$ and derive a formula for the curl of the vector field (1.143).
28. Establish the Cartesian-coordinate formula $curl\boldsymbol{A} = e_{ilm}A_{jl,i}\, \boldsymbol{e}_m \otimes \boldsymbol{e}_j$.

Chapter 2

Bodies, Configurations, and Motions

In this chapter, we discuss the concept of a material body and introduce a general framework for the description of its motions through space.

2.1 Position, velocity and acceleration

A body B is a collection of material points, labelled p (Figure 2.1). In the older literature, these points are sometimes referred to as particles to convey the notion of a small indivisible object. While this has some heuristic value, it is not really what is intended when speaking of a material point, which is better regarded as a primitive concept. At a particular time t, the body occupies a configuration κ_t, a region of a Euclidean point space $\mathcal{E}^3$, in which $p \in B$ is located at position $\boldsymbol{x} \in \kappa_t$ (Figure 2.2). We perceive the body only in such configurations.

Further, we assume that there is one, and only one, $\boldsymbol{x} \in \kappa_t$ for a given $p \in B$, and, conversely, that there is a unique $p \in B$ for each $\boldsymbol{x} \in \kappa_t$. That is, we assume that different material points occupy different positions at any fixed time. Thus, we assume the existence of an invertible function $\boldsymbol{\chi}(p, t)$ such that

$$\boldsymbol{x} = \boldsymbol{\chi}(p, t) \quad \text{and} \quad p = \boldsymbol{\chi}^{-1}(\boldsymbol{x}, t). \tag{2.1}$$

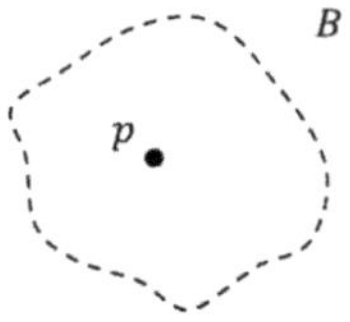

Figure 2.1. A material body.

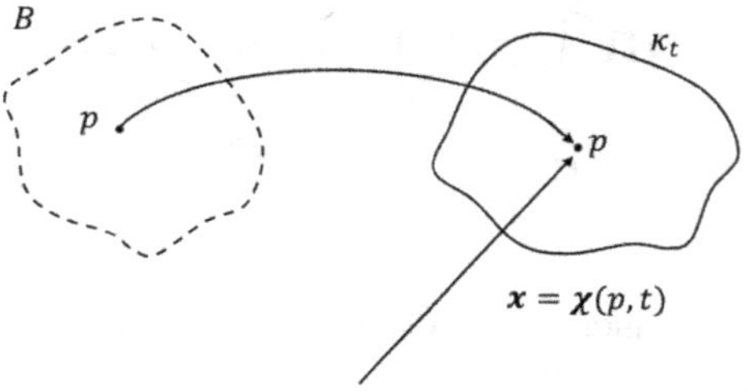

Figure 2.2. Configuration of the body at time t.

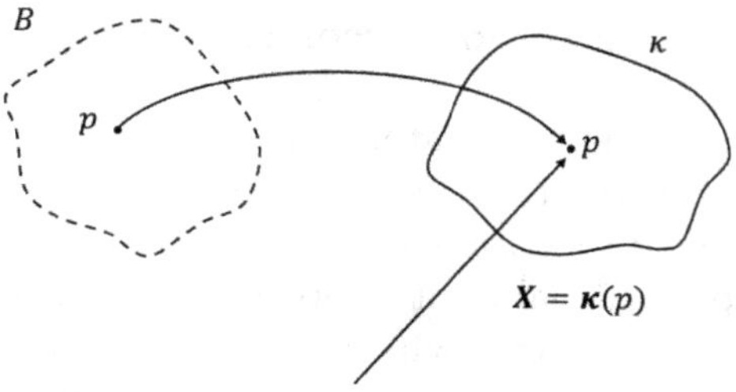

Figure 2.3. A reference configuration for the body.

At time t, $\boldsymbol{\chi}$ identifies the location occupied by a material point, and its inverse, $\boldsymbol{\chi}^{-1}$, the material point occupying that location. The function $\boldsymbol{\chi}(p, t)$ is called the *motion* of $p \in B$. The velocity $\boldsymbol{v}(p, t)$ and acceleration $\boldsymbol{a}(p, t)$ of p are given, respectively, by

$$\boldsymbol{v}(p, t) = \frac{\partial}{\partial t}\boldsymbol{\chi}(p, t) \quad \text{and} \quad \boldsymbol{a}(p, t) = \frac{\partial^2}{\partial t^2}\boldsymbol{\chi}(p, t). \tag{2.2}$$

To facilitate analysis, instead of working with the ethereal body B, we instead select a fixed *reference configuration* κ, positions $\boldsymbol{X}$ in which are in one-to-one correspondence with material points p (Figure 2.3). Thus, we assume the existence of an invertible function

$\boldsymbol{\kappa}$ such that

$$\boldsymbol{X} = \boldsymbol{\kappa}(p), \quad p = \boldsymbol{\kappa}^{-1}(\boldsymbol{X}). \tag{2.3}$$

In this way, we effectively label a material point using the fixed position that it occupies in κ. For example, analysts often prefer $\kappa = \kappa_{t_0}$, the actual configuration occupied at the instant t_0. In this case, we have

$$\boldsymbol{\kappa}(p) = \boldsymbol{\chi}(p, t_0). \tag{2.4}$$

However, this choice, while often convenient, is by no means necessary.

In the general case, we use (2.1) and (2.3) to write (Figure 2.4)

$$\boldsymbol{x} = \boldsymbol{\chi}_{\kappa}(\boldsymbol{X}, t), \tag{2.5}$$

where

$$\boldsymbol{\chi}_{\kappa}(\boldsymbol{X}, t) = \boldsymbol{\chi}(\boldsymbol{\kappa}^{-1}(\boldsymbol{X}), t). \tag{2.6}$$

This function obviously depends on the choice of κ, hence the subscript.

To each $p \in B$, there corresponds one, and only one, $\boldsymbol{x} \in \kappa_t$, and one, and only one, $\boldsymbol{X} \in \kappa$. Thus, $\boldsymbol{\chi}_{\kappa}$ is invertible at each t, and

$$\boldsymbol{X} = \boldsymbol{\chi}_{\kappa}^{-1}(\boldsymbol{x}, t). \tag{2.7}$$

This identifies a unique position in κ associated with a given position in κ_t. Said differently, there is only one referential position — equivalently, one material point — associated with any particular position in κ_t. Moreover, because B and κ are in fixed one-to-one

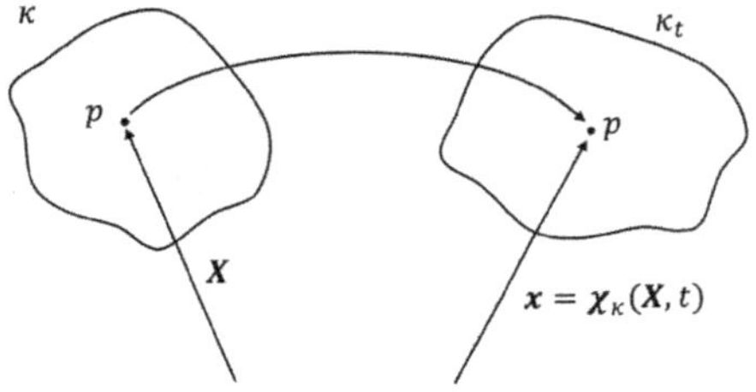

Figure 2.4. Mapping of reference configuration to current configuration.

correspondence, it follows that the use of the pair (p, t) as independent variables is tantamount to the use of $(\boldsymbol{X}, t)$ as independent variables. For example, the velocity of a material point p is given by the function

$$\hat{\boldsymbol{v}}(\boldsymbol{X}, t) = \boldsymbol{v}(\boldsymbol{\kappa}^{-1}(\boldsymbol{X}), t) = \frac{\partial}{\partial t}\boldsymbol{\chi}_{\kappa}(\boldsymbol{X}, t), \tag{2.8}$$

in which the partial derivative is evaluated at fixed $\boldsymbol{X}$, i.e., at fixed p. Alternatively, the material velocity is given by the function

$$\tilde{\boldsymbol{v}}(\boldsymbol{x}, t) = \boldsymbol{v}(\boldsymbol{\chi}^{-1}(\boldsymbol{x}, t), t). \tag{2.9}$$

We may relate $\tilde{\boldsymbol{v}}$ and $\hat{\boldsymbol{v}}$ directly using (2.5) and (2.7). Thus,

$$\hat{\boldsymbol{v}}(\boldsymbol{X}, t) = \tilde{\boldsymbol{v}}(\boldsymbol{\chi}_{\kappa}(\boldsymbol{X}, t), t) \quad \text{and} \quad \tilde{\boldsymbol{v}}(\boldsymbol{x}, t) = \hat{\boldsymbol{v}}(\boldsymbol{\chi}_{\kappa}^{-1}(\boldsymbol{x}, t), t). \tag{2.10}$$

2.2 Material, referential and spatial descriptions

To elaborate on the foregoing change-of-variable formulas for the velocity field, we explore various ways to describe a generic field f associated with a material point p. For example, we may write

$$f = \bar{f}(p, t) = \tilde{f}(\boldsymbol{x}, t) = \hat{f}(\boldsymbol{X}, t), \tag{2.11}$$

where

$$\tilde{f}(\boldsymbol{x}, t) = \bar{f}(\boldsymbol{\chi}^{-1}(\boldsymbol{x}, t), t) \quad \text{and} \quad \hat{f}(\boldsymbol{X}, t) = \bar{f}(\boldsymbol{\kappa}^{-1}(\boldsymbol{X}), t), \tag{2.12}$$

Of course, we can relate $\tilde{f}$ and $\hat{f}$ directly, as in (2.10):

$$\hat{f}(\boldsymbol{X}, t) = \tilde{f}(\boldsymbol{\chi}_{\kappa}(\boldsymbol{X}, t), t) \quad \text{and} \quad \tilde{f}(\boldsymbol{x}, t) = \hat{f}(\boldsymbol{\chi}_{\kappa}^{-1}(\boldsymbol{x}, t), t). \tag{2.13}$$

The functions $\bar{f}$, $\tilde{f}$ and $\hat{f}$ are the *material, spatial* and *referential* descriptions, respectively, of the field f. They are distinguished by the independent variables adopted. The material description is used mainly to formulate abstract concepts pertaining to material points. It is equivalent, as we have seen, to the referential description, which is useful for purposes of analysis.

In the spatial description, we regard position $\boldsymbol{x} \in \mathcal{E}^3$ and time t as independent variables. Fixing $\boldsymbol{x}$ means fixing the position, not

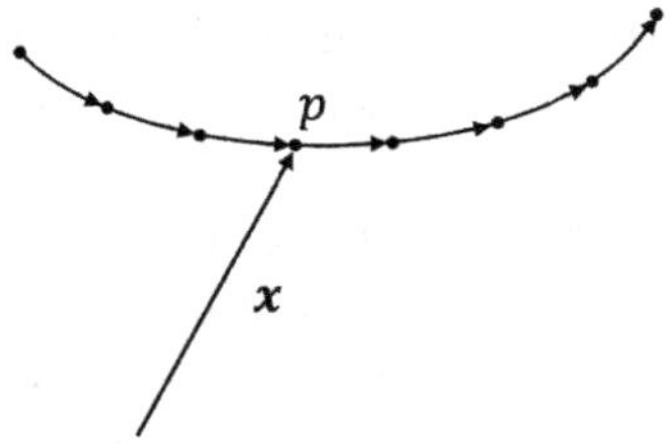

Figure 2.5. Different material points occupy position $\boldsymbol{x}$ at different times.

the material point. Accordingly, different material points occupy the fixed position $\boldsymbol{x}$ at different times (Figure 2.5).

The various fixed positions $\boldsymbol{x}$ identify points in a fixed region of space, or *control volume*, which generally contains different material at different times.

Combining the chain rule (1.114) with the spatial description, for example, we have

$$df = d\tilde{f}(\boldsymbol{x}, t) = \tilde{f}' dt + grad\, \tilde{f} \cdot d\boldsymbol{x}, \tag{2.14}$$

where $\tilde{f}' = \partial \tilde{f}/\partial t$ — the *spatial* time derivative — is the partial derivative with respect to t at fixed $\boldsymbol{x}$ and $grad\, \tilde{f}$, the *spatial* gradient, is the gradient of f with respect to $\boldsymbol{x}$ at fixed t. Alternatively, the referential description yields

$$df = d\hat{f}(\boldsymbol{X}, t) = (\hat{f})^{\cdot} dt + Grad\, \hat{f} \cdot d\boldsymbol{X}, \tag{2.15}$$

where $(\hat{f})^{\cdot} = \partial \hat{f}/\partial t$ — the *referential* time derivative — is the partial derivative with respect to t at fixed $\boldsymbol{X}$ and $Grad\, \hat{f}$ is the *referential* gradient, i.e., the gradient of f with respect to $\boldsymbol{X}$ at fixed t. As the preceding relations are simply two expressions for the same quantity, we have

$$\dot{f} dt + Grad\, f \cdot d\boldsymbol{X} = f' dt + grad\, f \cdot d\boldsymbol{x}, \tag{2.16}$$

where we have suppressed the superposed tildes and carets in an effort to minimize visual distraction. We shall do so routinely when converting between the referential and spatial descriptions, but it is important to keep in mind that these entail the use of different functions to describe the various fields involved. Since p and $\boldsymbol{X}$ are in fixed one-to-one correspondence, the referential time derivative $\dot{f}$ has the same meaning as the *material* derivative $\partial \bar{f}/\partial t$ at fixed p. Accordingly, we will henceforth refer to $\dot{f}$ as the material derivative.

Suppose we have knowledge of the function $\tilde{f}(\boldsymbol{x}, t)$. This function might describe the temperature distribution in a room, for example, as read off from thermometers located at various fixed positions $\boldsymbol{x}$. The t-dependence in this function accounts for the time variation of temperature at each thermometer. Using this information, we might wish to know how temperature varies at the material point that momentarily occupies position $\boldsymbol{x}$. Thus, we wish to compute $\dot{f}$ at $\boldsymbol{x}$, the lack of information about the functions $\hat{f}(\boldsymbol{X}, t)$ or $\bar{f}(p, t)$ notwithstanding. To this end, we fix p and hence $\boldsymbol{X}$, i.e., we set $d\boldsymbol{X} = \boldsymbol{0}$ in (2.16), in which $d\boldsymbol{x} = d\boldsymbol{\chi}_\kappa(\boldsymbol{X}, t) = \boldsymbol{v}dt$, which follows from (2.5), $(2.8)_1$ and (2.10), finally obtaining $\dot{f}dt = f'dt + grad f \cdot \boldsymbol{v}dt$. Thus,

$$\dot{f} = f' + grad f \cdot \boldsymbol{v}, \tag{2.17}$$

which is computable from the function $\tilde{f}(\boldsymbol{x}, t)$ and the material velocity field in the spatial description.

Similarly, for a vector field,

$$\boldsymbol{u} = \bar{\boldsymbol{u}}(p, t) = \tilde{\boldsymbol{u}}(\boldsymbol{x}, t) = \hat{\boldsymbol{u}}(\boldsymbol{X}, t), \tag{2.18}$$

we have

$$\dot{\boldsymbol{u}}dt + (Grad\boldsymbol{u})d\boldsymbol{X} = d\boldsymbol{u} = \boldsymbol{u}'dt + (grad\boldsymbol{u})d\boldsymbol{x}, \tag{2.19}$$

and hence,

$$\dot{\boldsymbol{u}} = \boldsymbol{u}' + (grad\boldsymbol{u})\boldsymbol{v}, \tag{2.20}$$

the computation of which requires knowledge of the function $\tilde{\boldsymbol{u}}(\boldsymbol{x}, t)$.

For example, the material acceleration $\boldsymbol{a}$, which figures prominently in the equations of motion for material bodies, is given in the spatial description by

$$\boldsymbol{a} = \dot{\boldsymbol{v}} = \boldsymbol{v}' + \boldsymbol{L}\boldsymbol{v}, \quad \text{where} \quad \boldsymbol{L} = grad\boldsymbol{v} \tag{2.21}$$

is the spatial velocity gradient.

2.3 Cartesian coordinates

To facilitate analysis and to promote understanding, we summarize our results in terms of components relative to orthonormal bases expressed in terms of associated Cartesian coordinates. For example, we may associate the position of p in κ_t with its position vector in terms of Cartesian coordinates x_i, as in (1.95):

$$\boldsymbol{x} = x_i \boldsymbol{e}_i; \quad i \in \{1,2,3\}, \tag{2.22}$$

where $\{\boldsymbol{e}_i\}$ is a fixed orthonormal basis. In the same way, we associate its position in κ with the position vector

$$\boldsymbol{X} = X_A \hat{\boldsymbol{E}}_A; \quad A \in \{1,2,3\}, \tag{2.23}$$

where X_A are the Cartesian coordinates of p in κ and $\{\hat{\boldsymbol{E}}_A\}$ is another fixed orthonormal basis (Figure 2.6).

For reasons that will become clear as we proceed, it is useful to maintain a conceptual distinction between the two bases. This entails a distinction between the vector spaces $E^3 = \text{Span}\{\boldsymbol{e}_i\}$ and $\hat{E}^3 = \text{Span}\{\hat{\boldsymbol{E}}_A\}$, these being the translation spaces of the point spaces $\mathcal{E}^3$ and $\hat{\mathcal{E}}^3$, respectively.

Consistency with (2.5) requires that $\boldsymbol{\chi}_\kappa$ also be regarded as a vector in E^3. Thus,

$$x_i \boldsymbol{e}_i = \boldsymbol{\chi}_\kappa(X_A \hat{\boldsymbol{E}}_A, t) = \chi_i(X_A, t)\boldsymbol{e}_i; \quad x_i = \chi_i(X_A, t), \tag{2.24}$$

where the subscript κ is suppressed in the components χ_i for clarity and we have invoked the fact that the $\hat{\boldsymbol{E}}_A$ are fixed. Thus, the various

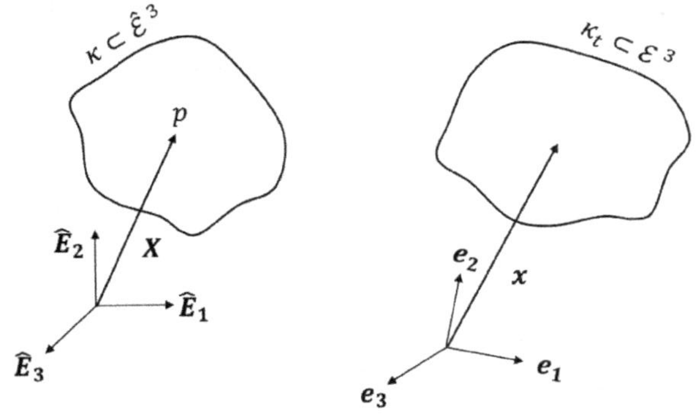

Figure 2.6. Position vectors in κ and κ_t.

material points are distinguished by their Cartesian coordinates X_A. The notation here is shorthand for $x_i = \chi_i(X_1, X_2, X_3, t)$; $i = 1, 2, 3$. The velocity field is then given, in the referential description, by

$$\boldsymbol{v} = \hat{\boldsymbol{v}}(X_A \hat{\boldsymbol{E}}_A, t) = \hat{v}_i(X_A, t)\boldsymbol{e}_i, \quad \text{with}$$
$$\hat{v}_i(X_A, t) = \dot{\chi}_i = \frac{\partial}{\partial t}\chi_i(X_A, t). \tag{2.25}$$

In detail, from (2.2), (2.5), (2.6) and (2.24), we have

$$\begin{aligned}\boldsymbol{v} &= \lim_{\alpha \to 0} \frac{1}{\alpha}[\boldsymbol{\chi}_\kappa(\boldsymbol{X}, t + \alpha) - \boldsymbol{\chi}_\kappa(\boldsymbol{X}, t)] \\ &= \boldsymbol{e}_i \lim_{\alpha \to 0} \frac{1}{\alpha}[\chi_i(X_A, t + \alpha) - \chi_i(X_A, t)] \\ &= \dot{\chi}_i \boldsymbol{e}_i.\end{aligned} \tag{2.26}$$

Similarly, the referential expression for the material acceleration is

$$\boldsymbol{a} = \hat{\boldsymbol{a}}(X_A \hat{\boldsymbol{E}}_A, t) = \hat{a}_i(X_A, t)\boldsymbol{e}_i, \quad \text{with}$$
$$\hat{a}_i(X_A, t) = \ddot{\chi}_i = \frac{\partial^2}{\partial t^2}\chi_i(X_A, t). \tag{2.27}$$

Of course, these vector fields may also be expressed in terms of the spatial description via (2.22). Thus,

$$\boldsymbol{v} = \tilde{\boldsymbol{v}}(x_i \boldsymbol{e}_i, t) = \tilde{v}_i(x_k, t)\boldsymbol{e}_i \quad \text{and} \quad \boldsymbol{a} = \tilde{\boldsymbol{a}}(x_i \boldsymbol{e}_i, t) = \tilde{a}_i(x_k, t)\boldsymbol{e}_i. \tag{2.28}$$

Equating the second of these to the decomposition of (2.21) in the basis $\{\boldsymbol{e}_i\}$ furnishes

$$\tilde{a}_i = \tilde{v}_i' + \tilde{v}_{i,j}\tilde{v}_j, \tag{2.29}$$

whereas application of (2.20) to the position vector $\boldsymbol{x}$, combined with $\boldsymbol{x}' = \boldsymbol{0}$, yields $\dot{x}_i = x_{i,j}\tilde{v}_j = \delta_{ij}\tilde{v}_j = \tilde{v}_i$, as expected.

Example

Consider the simple motion defined by

$$\begin{Bmatrix} x_1 \\ x_2 \\ x_3 \end{Bmatrix} = \begin{pmatrix} e^t & 0 & 0 \\ 0 & 1 & t \\ 0 & -t & 1 \end{pmatrix} \begin{Bmatrix} X_1 \\ X_2 \\ X_3 \end{Bmatrix}. \tag{2.30}$$

This motion is quite atypical in that $x_i = \chi_i(X_A, t)$ are linear functions of the X_A. However, it furnishes an explicit illustration of the process of conversion between the referential and spatial descriptions. For example, the material velocity is $\boldsymbol{v} = \hat{v}_i(X_A, t)\boldsymbol{e}_i$, with $\hat{v}_1 = X_1 e^t$, $\hat{v}_2 = X_3$ and $\hat{v}_3 = -X_2$. To express this in the spatial form $\boldsymbol{v} = \tilde{v}_i(x_k, t)\boldsymbol{e}_i$, we must first invert (2.30) to express X_A in terms of x_i and t. Thus,

$$\begin{Bmatrix} X_1 \\ X_2 \\ X_3 \end{Bmatrix} = \begin{pmatrix} e^{-t} & 0 & 0 \\ 0 & (1+t^2)^{-1} & -t(1+t^2)^{-1} \\ 0 & t(1+t^2)^{-1} & (1+t^2)^{-1} \end{pmatrix} \begin{Bmatrix} x_1 \\ x_2 \\ x_3 \end{Bmatrix}, \tag{2.31}$$

yielding $\tilde{v}_1 = x_1$, $\tilde{v}_2 = (1+t^2)^{-1}(tx_2 + x_3)$ and $\tilde{v}_3 = (1+t^2)^{-1}$ $(tx_3 - x_2)$.

The material acceleration is given by $\boldsymbol{a} = \hat{a}_i(X_A, t)\boldsymbol{e}_i = \tilde{a}_i(x_k, t)\boldsymbol{e}_i$, with $\hat{a}_1 = X_1 e^t$, $\hat{a}_2 = 0$ and $\hat{a}_3 = 0$, and with $\tilde{a}_1 = x_1$, $\tilde{a}_2 = 0$, and $\tilde{a}_3 = 0$, where we have again used (2.31). Alternatively, we may use (2.29) with

$$\tilde{v}_1' = 0, \quad \tilde{v}_2' = \frac{1-2t^2}{1+t^2}x_2 - \frac{2t}{1+t^2}x_3, \quad \tilde{v}_3' = \frac{1-2t^2}{1+t^2}x_3 - \frac{2t}{1+t^2}x_2, \tag{2.32}$$

together with

$$(\tilde{v}_{i,j}) = \begin{pmatrix} 1 & 0 & 0 \\ 0 & t(1+t^2)^{-1} & (1+t^2)^{-1} \\ 0 & -(1+t^2)^{-1} & t(1+t^2)^{-1} \end{pmatrix} \tag{2.33}$$

to derive $\tilde{a}_1 = \tilde{v}_1' + \tilde{v}_{1,j}\tilde{v}_j = \tilde{v}_{1,1}\tilde{v}_1 = x_1$, etc.

2.4 The deformation gradient

We assume the motion $\boldsymbol{\chi}_\kappa(\boldsymbol{X}, t)$ to be differentiable with respect to $\boldsymbol{X}$. Thus, holding t fixed,

$$d\boldsymbol{x} = \boldsymbol{F}d\boldsymbol{X}, \quad \text{where} \quad \boldsymbol{F} = Grad\boldsymbol{\chi}_\kappa \tag{2.34}$$

is the deformation gradient. It would be appropriate to append a subscript κ to $\boldsymbol{F}$, to identify the reference configuration adopted, but to avoid overburdening the notation, we shall refrain from doing so. The derivative with respect to $\boldsymbol{X}$ and t jointly is then given by

$$d\boldsymbol{x} = \boldsymbol{F}d\boldsymbol{X} + \boldsymbol{v}dt, \tag{2.35}$$

and combining this with (2.19), we obtain

$$\dot{\boldsymbol{u}}dt + (Grad\boldsymbol{u})d\boldsymbol{X} = [\boldsymbol{u}' + (grad\boldsymbol{u})\boldsymbol{v}]dt + (grad\boldsymbol{u})\boldsymbol{F}d\boldsymbol{X}. \tag{2.36}$$

It follows from this and (2.20) that $(Grad\boldsymbol{u})d\boldsymbol{X} = (grad\boldsymbol{u})\boldsymbol{F}d\boldsymbol{X}$ for all $d\boldsymbol{X}$ and hence that

$$Grad\boldsymbol{u} = (grad\boldsymbol{u})\boldsymbol{F}. \tag{2.37}$$

To obtain a representation for $\boldsymbol{F}$ in terms of Cartesian components, we write $d\boldsymbol{x} = d\chi_i\boldsymbol{e}_i$ with $d\chi_i = \chi_{i,A}dX_A = \chi_{i,A}\hat{\boldsymbol{E}}_A \cdot d\boldsymbol{X}$, where $\chi_{i,A} = \partial\chi_i/\partial X_A$ is evaluated at fixed t. Thus,

$$d\boldsymbol{x} = \chi_{i,A}\boldsymbol{e}_i(\hat{\boldsymbol{E}}_A \cdot d\boldsymbol{X}) = (\chi_{i,A}\boldsymbol{e}_i \otimes \hat{\boldsymbol{E}}_A)d\boldsymbol{X}, \tag{2.38}$$

and therefore, from (2.34),

$$\boldsymbol{F} = F_{iA}\boldsymbol{e}_i \otimes \hat{\boldsymbol{E}}_A, \quad \text{where} \quad F_{iA} = \chi_{i,A}. \tag{2.39}$$

This decomposition, in the mixed basis $\{\boldsymbol{e}_i \otimes \hat{\boldsymbol{E}}_A\}$, implies that the deformation gradient is a *two-point* tensor, i.e., a linear mapping from $\hat{E}^3$ to E^3. Note that

$$\begin{aligned} \boldsymbol{F}\hat{\boldsymbol{E}}_B &= (F_{iA}\boldsymbol{e}_i \otimes \hat{\boldsymbol{E}}_A)\hat{\boldsymbol{E}}_B = F_{iA}(\hat{\boldsymbol{E}}_A \cdot \hat{\boldsymbol{E}}_B)\boldsymbol{e}_i \\ &= F_{iA}\delta_{AB}\boldsymbol{e}_i = F_{iB}\boldsymbol{e}_i \end{aligned} \tag{2.40}$$

and hence that

$$F_{iA} = \boldsymbol{e}_i \cdot \boldsymbol{F}\hat{\boldsymbol{E}}_A. \tag{2.41}$$

For an arbitrary vector field $\boldsymbol{u} = \tilde{u}_i(x_k, t)\boldsymbol{e}_i = \hat{u}_i(X_A, t)\boldsymbol{e}_i$ we have, from (2.37), that

$$\begin{aligned} Grad\boldsymbol{u} &= (\tilde{u}_{i,k}\boldsymbol{e}_i \otimes \boldsymbol{e}_k)(F_{jA}\boldsymbol{e}_j \otimes \hat{\boldsymbol{E}}_A) \\ &= \tilde{u}_{i,k}F_{jA}(\boldsymbol{e}_k \cdot \boldsymbol{e}_j)\boldsymbol{e}_i \otimes \hat{\boldsymbol{E}}_A \\ &= \tilde{u}_{i,k}x_{j,A}\delta_{kj}\boldsymbol{e}_i \otimes \hat{\boldsymbol{E}}_A \\ &= \tilde{u}_{i,j}x_{j,A}\boldsymbol{e}_i \otimes \hat{\boldsymbol{E}}_A = \hat{u}_{i,A}\boldsymbol{e}_i \otimes \hat{\boldsymbol{E}}_A \end{aligned} \tag{2.42}$$

and hence the conclusion that $Grad\boldsymbol{u}$ is also a two-point tensor. In passing from the first line to the second, we have used the rule

$$(\boldsymbol{a} \otimes \boldsymbol{b})(\boldsymbol{c} \otimes \boldsymbol{d}) = (\boldsymbol{b} \cdot \boldsymbol{c})\boldsymbol{a} \otimes \boldsymbol{d}, \tag{2.43}$$

which follows from the fact that, for any vector $\boldsymbol{v}$,

$$\begin{aligned} [(\boldsymbol{a} \otimes \boldsymbol{b})(\boldsymbol{c} \otimes \boldsymbol{d})]\boldsymbol{v} &= (\boldsymbol{a} \otimes \boldsymbol{b})[(\boldsymbol{c} \otimes \boldsymbol{d})]\boldsymbol{v} \\ &= (\boldsymbol{d} \cdot \boldsymbol{v})(\boldsymbol{a} \otimes \boldsymbol{b})\boldsymbol{c} = (\boldsymbol{b} \cdot \boldsymbol{c})(\boldsymbol{d} \cdot \boldsymbol{v})\boldsymbol{a} \\ &= (\boldsymbol{b} \cdot \boldsymbol{c})(\boldsymbol{a} \otimes \boldsymbol{d})\boldsymbol{v} = [(\boldsymbol{b} \cdot \boldsymbol{c})\boldsymbol{a} \otimes \boldsymbol{d}]\boldsymbol{v}. \end{aligned} \tag{2.44}$$

Consider a material curve $C \subset \kappa$, i.e., a curve consisting of a fixed set of material points. Let $\boldsymbol{X}(S)$ be its parametric representation in terms of arclength parameter S (Figure 2.7).

In effect, S labels the material points on this curve. In the course of a motion, the material points constituting C are *convected* to the

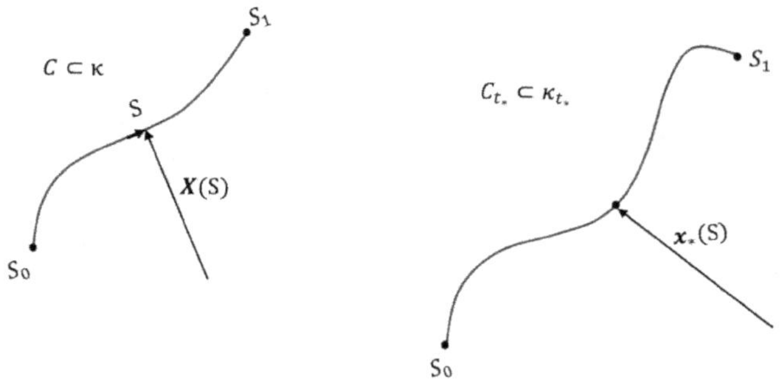

Figure 2.7. A material curve and its image in κ_{t_*}.

curve $C_{t_*} \subset \kappa_{t_*}$ at time t_*, say, with parametric representation

$$\boldsymbol{x}_*(S) = \chi_\kappa(\boldsymbol{X}(S), t_*). \tag{2.45}$$

If S_0 and S_1 are the values of S associated with two material points on these curves, then their relative position at time t_* is

$$\begin{aligned} \boldsymbol{x}_*(S_1) - \boldsymbol{x}_*(S_0) &= \boldsymbol{e}_i \int_{S_0}^{S_1} x'_{*i}(S)dS \\ &= \boldsymbol{e}_i \int_{S_0}^{S_1} F_{iA}(X_B(S), t_*)X'_A(S)dS, \end{aligned} \tag{2.46}$$

where $X_A(S)$ are the parametric equations of C. Thus, knowledge of the deformation gradient allows us to compute the relative position of any two material points, provided there exists a curve connecting them.

Our assumption that the motion $\chi_\kappa(\boldsymbol{X}, t)$ is invertible at each fixed t implies, by the inverse function theorem [2], that its gradient $\boldsymbol{F}$ is invertible at each $\boldsymbol{X}$, i.e., that it has an inverse $\boldsymbol{F}^{-1}$. However, the existence of $\boldsymbol{F}^{-1}$ at a particular $\boldsymbol{X}$ does not imply that χ_κ is invertible at all $\boldsymbol{X}$. Rather, it implies only that χ_κ is invertible in a neighborhood of the point considered. A necessary and sufficient condition for the invertibility of $\boldsymbol{F}$ is that its determinant, $\det \boldsymbol{F}$, be non-zero, where (see Appendix A)

$$\det \boldsymbol{F} = \frac{[\boldsymbol{F}\boldsymbol{a}, \boldsymbol{F}\boldsymbol{b}, \boldsymbol{F}\boldsymbol{c}]}{[\boldsymbol{a}, \boldsymbol{b}, \boldsymbol{c}]} \tag{2.47}$$

for an arbitrary linearly independent set $\{\boldsymbol{a}, \boldsymbol{b}, \boldsymbol{c}\}$ in $\hat{E}^3$.

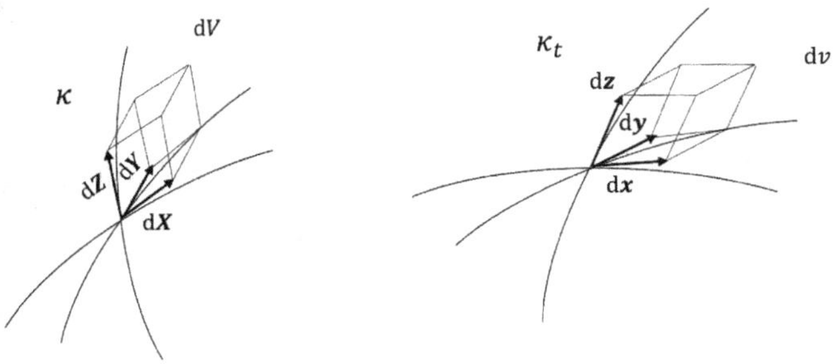

Figure 2.8. Three material curves intersecting at a material point.

We have previously discussed determinants of tensors that map a vector space to itself. However, this formula is also applicable to two-point tensors because the numerator and denominator each involve vectors belonging to a single vector space. Accordingly, the scalar triple products are well defined.

Consider three material curves passing through $\boldsymbol{X} \in \kappa$ (Figure 2.8). The three vectors $d\boldsymbol{X}$, $d\boldsymbol{Y}$ and $d\boldsymbol{Z}$ in Figure 2.8 are oriented in such a way as to form a parallelepiped with positive volume

$$dV = [d\boldsymbol{X}, d\boldsymbol{Y}, d\boldsymbol{Z}]. \tag{2.48}$$

If this configuration is an *occupiable* one, i.e., one that could, in principle, be occupied by the body B whether or not it is actually occupied in the course of its motion, then we would expect the material in this volume to inhabit a positive volume

$$dv = [d\boldsymbol{x}, d\boldsymbol{y}, d\boldsymbol{z}] \tag{2.49}$$

in the configuration κ_t, where $d\boldsymbol{x} = \boldsymbol{F}(\boldsymbol{X}, t)d\boldsymbol{X}$, $d\boldsymbol{y} = \boldsymbol{F}(\boldsymbol{X}, t)d\boldsymbol{Y}$ and $d\boldsymbol{z} = \boldsymbol{F}(\boldsymbol{X}, t)d\boldsymbol{Z}$. Taking $\{\boldsymbol{a}, \boldsymbol{b}, \boldsymbol{c}\} = \{d\boldsymbol{X}, d\boldsymbol{Y}, d\boldsymbol{Z}\}$ in (2.47) then yields

$$dv = JdV, \quad \text{where} \quad J = \det \boldsymbol{F} > 0. \tag{2.50}$$

The occupiability condition is automatically satisfied by taking $\kappa = \kappa_{t_0}$, the actual configuration at the instant t_0. However, occupiability of the reference configuration κ is by no means a general

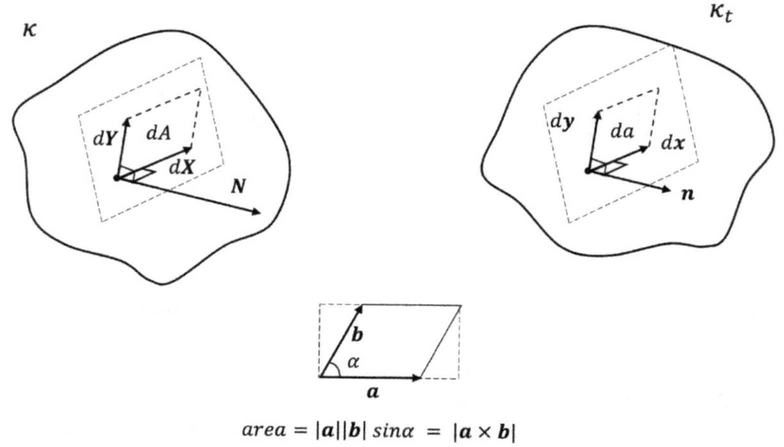

Figure 2.9. Deformation of oriented surface measure.

requirement. All that is required is that it be in one-to-one correspondence with B in the sense that $\boldsymbol{X} = \boldsymbol{\kappa}(p)$ be invertible. If κ is not occupiable, then $\det \boldsymbol{F} < 0$ and $(2.50)_2$ is replaced by $J = |\det \boldsymbol{F}|$.

We can think of (2.50) as the formula for the deformation of local volume measure. We will also have the need for the local area deformation of a material surface, i.e., a surface consisting of a fixed set of material points. This surface is fixed, independent of time, in the reference configuration κ. Let $d\boldsymbol{X}$ and $d\boldsymbol{Y}$ span the tangent plane to this surface at the point with location $\boldsymbol{X}$ (Figure 2.9).

Let dA be the area of the parallelogram formed by these vectors. Then, $dA = |d\boldsymbol{X} \times d\boldsymbol{Y}|$, and the vector $\boldsymbol{N}$, defined by

$$\boldsymbol{N} dA = d\boldsymbol{X} \times d\boldsymbol{Y}, \tag{2.51}$$

is a unit normal to the tangent plane. In the current configuration κ_t, the same material points are convected to a surface whose tangent plane at $\boldsymbol{x} = \boldsymbol{\chi}_\kappa(\boldsymbol{X}, t)$ is spanned by $d\boldsymbol{x} = \boldsymbol{F}(\boldsymbol{X}, t) d\boldsymbol{X}$ and $d\boldsymbol{y} = \boldsymbol{F}(\boldsymbol{X}, t) d\boldsymbol{Y}$, forming a parallelogram of area $da = |d\boldsymbol{x} \times d\boldsymbol{y}|$ and unit normal $\boldsymbol{n}$, defined by $\boldsymbol{n} da = d\boldsymbol{x} \times d\boldsymbol{y}$. Thus,

$$\boldsymbol{n} da = \boldsymbol{F} d\boldsymbol{X} \times \boldsymbol{F} d\boldsymbol{Y} = \boldsymbol{F}^*(d\boldsymbol{X} \times d\boldsymbol{Y}), \tag{2.52}$$

where $\boldsymbol{F}^*$ is the *cofactor* of $\boldsymbol{F}$, defined by (see (1.86) and Appendix A):

$$\boldsymbol{F}^*(\boldsymbol{a} \times \boldsymbol{b}) = \boldsymbol{F}\boldsymbol{a} \times \boldsymbol{F}\boldsymbol{b} \quad \text{for all vectors} \quad \boldsymbol{a}, \boldsymbol{b} \in \hat{E}^3. \tag{2.53}$$

We thus arrive at the Piola–Nanson formula

$$\boldsymbol{n}da = \boldsymbol{F}^*\boldsymbol{N}dA \tag{2.54}$$

for the deformation of the vector-valued area measure. The component form of this formula is $n_i da = F^*_{iB} N_B dA$.

It is straightforward to show that the result of Problem 22(c) of Chapter 1 is also valid for two-point tensors. Thus,

$$\boldsymbol{F}^* = J\boldsymbol{F}^{-t}. \tag{2.55}$$

2.5 Mass and density

Consider a sub-body $S \subset B$, occupying a part $P \subset \kappa$ of the reference configuration and a part $P_t \subset \kappa_t$ of the current configuration at time t. We write $P = \boldsymbol{\kappa}(S)$ and $P_t = \boldsymbol{\chi}(S, t)$ to indicate that these parts are the images of S under the maps $\boldsymbol{\kappa}(\cdot)$ and $\boldsymbol{\chi}(\cdot, t)$ taking the body into the reference and current configurations, respectively. We may similarly write $P_t = \boldsymbol{\chi}_\kappa(P, t)$.

Conventionally, the mass of S is assumed to be an absolutely continuous function of volume. Roughly, this means the more the volume, the greater the mass. The mass of a material region of positive volume is assumed to be strictly positive. These assumptions imply that there exist positively valued *mass density* functions $\rho(\boldsymbol{x}, t)$ and $\rho_\kappa(\boldsymbol{X}, t)$ such that the mass $m(S, t)$ of S is expressible in the forms

$$m(S,t) = \int_{P_t} \rho(\boldsymbol{x}, t) dv = \int_P \rho_\kappa(\boldsymbol{X}, t) dV. \tag{2.56}$$

Later, we discuss the principle of conservation of mass, which stipulates that $m(S, t)$ be independent of t, i.e., the mass of a fixed set of material points is invariant in time.

Using (2.5) and (2.50) to change variables, we have

$$\int_{P_t} \rho(\boldsymbol{x}, t) dv = \int_P \rho(\boldsymbol{\chi}_\kappa(\boldsymbol{X}, t), t) J(\boldsymbol{X}, t) dV, \tag{2.57}$$

allowing us to reduce the second part of the previous equation to

$$\int_P [\rho_\kappa(\boldsymbol{X}, t) - \rho(\boldsymbol{\chi}_\kappa(\boldsymbol{X}, t), t) J(\boldsymbol{X}, t)] dV = 0, \quad \text{for all} \quad P \subset \kappa. \tag{2.58}$$

Clearly, for this to be satisfied, it is sufficient that

$$\rho_\kappa = \rho J \quad \text{at every} \quad \boldsymbol{X} \in \kappa. \tag{2.59}$$

In fact, this is also necessary provided that the integrand is a continuous function of $\boldsymbol{X}$. To show this, we first establish the following.

Localization Theorem [3]: Let $\phi(\boldsymbol{X})$ be a continuous function, and suppose that

$$\int_P \phi(\boldsymbol{X})dV = 0, \quad \text{for all} \quad P \subset \kappa. \tag{2.60}$$

Then,

$$\phi(\boldsymbol{X}) = 0 \quad \text{at every} \quad \boldsymbol{X} \in \kappa. \tag{2.61}$$

The converse statement is obvious.

To prove the theorem, we follow [3] and consider $P = R_\varepsilon$, a sphere of radius ε and volume V_ε, centered at $\boldsymbol{X}_0 \in \kappa$ and contained in κ. Let

$$I_\varepsilon = \left| \frac{1}{V_\varepsilon} \int_{R_\varepsilon} [\phi(\boldsymbol{X}_0) - \phi(\boldsymbol{X})]dV \right|. \tag{2.62}$$

Then,

$$\begin{aligned} I_\varepsilon &\leq \frac{1}{V_\varepsilon} \int_{R_\varepsilon} |\phi(\boldsymbol{X}_0) - \phi(\boldsymbol{X})|\, dV \\ &\leq \frac{1}{V_\varepsilon} \int_{R_\varepsilon} \max_{\boldsymbol{X} \in R_\varepsilon} |\phi(\boldsymbol{X}_0) - \phi(\boldsymbol{X})|\, dV \\ &= \max_{\boldsymbol{X} \in R_\varepsilon} |\phi(\boldsymbol{X}_0) - \phi(\boldsymbol{X})|, \end{aligned} \tag{2.63}$$

where we have invoked the compactness of R_ε, i.e., its closedness and boundedness, and the fact that a continuous function defined on a compact set attains its maximum (and minimum) on that set [2]. Accordingly, by the continuity of ϕ,

$$0 \leq I_\varepsilon \leq \max_{\boldsymbol{X} \in R_\varepsilon} |\phi(\boldsymbol{X}_0) - \phi(\boldsymbol{X})| \to 0 \quad \text{as} \quad \varepsilon \to 0, \tag{2.64}$$

and thus, $I_\varepsilon \to 0$ as $\varepsilon \to 0$. We then have

$$\phi(\boldsymbol{X}_0) = \lim_{\varepsilon \to 0} \frac{1}{V_\varepsilon} \int_{R_\varepsilon} \phi(\boldsymbol{X})dV = 0 \tag{2.65}$$

because the integral vanishes by hypothesis. The result (2.61) then follows from the arbitrariness of $\boldsymbol{X}_0$.

Therefore, (2.59) follows from (2.58) provided that ρ, ρ_κ and J are continuous functions of $\boldsymbol{X}$. In this regard, we note that ρ is a continuous function of $\boldsymbol{X}$ provided that it is a continuous function of $\boldsymbol{x}$ and χ_κ is a continuous function of $\boldsymbol{X}$, whereas the latter is ensured by the differentiability of χ_κ with respect to $\boldsymbol{X}$, which we have already assumed. This in turn implies that $\boldsymbol{F}$ is continuous and hence, with $(2.50)_2$, that J is continuous.

2.6 Material curves and the shifter

Consider again the material curve $C \subset \kappa$ and its convected image $C_{t_*} = \chi_\kappa(C, t_*) \subset \kappa_{t_*}$ at the instant t_* (Figure 2.10). The material point located at arclength station S on C occupies the position $\boldsymbol{x}_*(S) = \chi_\kappa(\boldsymbol{X}(S), t_*)$ on C_{t_*}.

The derivative

$$\boldsymbol{x}_*'(S) = \boldsymbol{F}(\boldsymbol{X}(S), t_*)\boldsymbol{M}(S), \quad \text{where} \quad \boldsymbol{M}(S) = \boldsymbol{X}'(S) \tag{2.66}$$

is a unit tangent to C at the point $\boldsymbol{X}(S)$, is tangential to the curve C_{t_*} at $\boldsymbol{x}_*(S)$. Accordingly, $\boldsymbol{x}_*' = \mu\boldsymbol{m}$, where $\boldsymbol{m}$ is a *unit* tangent to C_{t_*} ($|\boldsymbol{m}| = 1$) and $\mu = |\boldsymbol{x}_*'|$ is the local *stretch* of the curve. Thus,

$$\mu\boldsymbol{m} = \boldsymbol{F}\boldsymbol{M}. \tag{2.67}$$

We seek an interpretation of this formula in the case when the body is undeformed. If κ is identified with κ_{t_0}, for example, then it would be natural to identify this undeformed configuration with κ

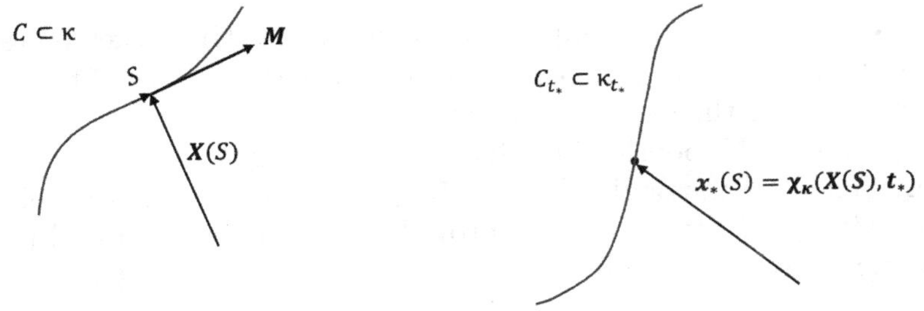

Figure 2.10. Orientation of material curves at point p with arclength S in κ.

and regard $\boldsymbol{X} \in \kappa$ as the position of $p \in B$ in this configuration. That is, it would be natural to assume that $\boldsymbol{X} \in \kappa_{t_0}$. However, this raises a subtle conceptual difficulty. We have already stipulated that $\boldsymbol{X}$, regarded as a position, is an element of $\hat{\mathcal{E}}^3$, or an element of $\hat{E}^3$ if interpreted as a position vector, whereas $\kappa_t \subset \mathcal{E}^3$ for all t, including t_0. To be consistent, we should then regard the 'reference' position *vector* as belonging to E^3.

To see how this can be accomplished, we introduce fixed orthonormal vectors $\boldsymbol{E}_A \in E^3$ and a *uniform* two-point tensor

$$\mathbf{1} = \boldsymbol{E}_A \otimes \hat{\boldsymbol{E}}_A \tag{2.68}$$

called the *shifter*. If the bases $\{\hat{\boldsymbol{E}}_A\}$ and $\{\boldsymbol{E}_A\}$ have the same (right- or left-) handedness, then the shifter rotates the former to the latter and may thus be regarded as a two-point rotation tensor. It is otherwise orthogonal. The rotation is trivial if $\boldsymbol{E}_A$ are aligned with their counterparts $\hat{\boldsymbol{E}}_A \in \hat{E}^3$. In this case, its effect is to parallel transport a vector from one vector space to another without otherwise altering it. It is then appropriate to think of the shifter as a two-point identity tensor.

Since $\boldsymbol{E}_A \in E^3$, we may resolve them in the basis $\{\boldsymbol{e}_i\}$ in accordance with (1.14). Thus,

$$\boldsymbol{E}_A = \delta_{iA}\boldsymbol{e}_i, \quad \text{where} \quad \delta_{iA} = \boldsymbol{e}_i \cdot \boldsymbol{E}_A. \tag{2.69}$$

The dot products are well defined because the factors belong to the same vector space. We then have the standard two-point decomposition

$$\mathbf{1} = \delta_{iA}\boldsymbol{e}_i \otimes \hat{\boldsymbol{E}}_A. \tag{2.70}$$

Evidently, δ_{iA} are the components of the usual Kronecker delta if, and only if, $\{\boldsymbol{E}_A\} = \{\boldsymbol{e}_i\}$. Otherwise, they are the cosines of the fixed angles between the elements of the two sets.

Let $\boldsymbol{I}$ and $\hat{\boldsymbol{I}}$ be the identity tensors for E^3 and $\hat{E}^3$, respectively. Then, in particular, $\boldsymbol{E}_A = \boldsymbol{I}\boldsymbol{E}_A$ and $\hat{\boldsymbol{E}}_A = \hat{\boldsymbol{I}}\hat{\boldsymbol{E}}_A$. Combining these with (2.68) and the rules proved in Problems 17(b) and (c) of Chapter 1, we find that

$$\mathbf{1} = \mathbf{1}\hat{\boldsymbol{I}} = \boldsymbol{I}\mathbf{1}, \tag{2.71}$$

and, from (2.68),

$$\begin{aligned}\mathbf{1}^t\mathbf{1} &= (\hat{\boldsymbol{E}}_A \otimes \boldsymbol{E}_A)(\boldsymbol{E}_B \otimes \hat{\boldsymbol{E}}_B)\\ &= (\boldsymbol{E}_A \cdot \boldsymbol{E}_B)\hat{\boldsymbol{E}}_A \otimes \hat{\boldsymbol{E}}_B\\ &= \delta_{AB}\hat{\boldsymbol{E}}_A \otimes \hat{\boldsymbol{E}}_B\\ &= \hat{\boldsymbol{I}},\end{aligned} \tag{2.72}$$

and, similarly, that

$$\mathbf{1}\mathbf{1}^t = \boldsymbol{I}. \tag{2.73}$$

In terms of components,

$$\delta_{iA}\delta_{iB} = \delta_{AB} \quad \text{and} \quad \delta_{iA}\delta_{jA} = \delta_{ij}, \tag{2.74}$$

respectively.

A suitable referential position vector is obtained by operating on (2.23) with the shifter:

$$\mathbf{1}\boldsymbol{X} = X_A\mathbf{1}\hat{\boldsymbol{E}}_A = X_A\boldsymbol{E}_A \in E^3 \tag{2.75}$$

and interpreting the statement $\kappa = \kappa_{t_0}$ (cf. (2.4)) to mean

$$\chi_\kappa(\boldsymbol{X}, t_0) = \mathbf{1}\boldsymbol{X}, \tag{2.76}$$

assuming, for the sake of convenience and without loss of generality, that the origins coincide in Figure 2.6. It is then appropriate to introduce the *displacement* $\boldsymbol{u} \in E^3$ of the material point p relative to its position in κ_{t_0} defined by

$$\boldsymbol{u}(\boldsymbol{X}, t) = \chi_\kappa(\boldsymbol{X}, t) - \mathbf{1}\boldsymbol{X}. \tag{2.77}$$

This makes sense because the right-hand side is the difference between two vectors belonging to the same space. The component form is given simply by

$$\hat{u}_i(X_B, t) = \chi_i(X_B, t) - \delta_{iA}X_A. \tag{2.78}$$

In contrast, the standard definition found most often in the literature omits the shifter and is therefore not well defined because, with $\mathbf{1}$ omitted, the right-hand side of (2.77) entails the impermissible subtraction of vectors belonging to different spaces. This omission has given rise to confusion surrounding the use of the displacement to characterize deformation.

We shall have occasion to utilize the shifter and the displacement further in Chapters 7 and 12.

Returning to our task of interpreting (2.67) when the body is undeformed in accordance with (2.76), we observe that the latter yields

$$\boldsymbol{F}(\boldsymbol{X}, t_0) = \mathbf{1} \tag{2.79}$$

by virtue of the uniformity of the shifter, the component form of which is simply $F_{iA} = \delta_{iA}$. Then, (2.67) reduces to

$$\mu \boldsymbol{m} = \mathbf{1}\boldsymbol{M}. \tag{2.80}$$

Since the shifter is an orthogonal tensor, we conclude that

$$\mu = \sqrt{\mathbf{1}\boldsymbol{M} \cdot \mathbf{1}\boldsymbol{M}} = \sqrt{\boldsymbol{M} \cdot \boldsymbol{M}} = 1 \tag{2.81}$$

when the body is undeformed.

In the general case, (2.67) implies that $\mu > 0$; otherwise, $\boldsymbol{F}\boldsymbol{M} = \mathbf{0}$ with $\boldsymbol{M} \neq \mathbf{0}$ (since $|\boldsymbol{M}| = 1$), and this is impossible because $\boldsymbol{F}$ is invertible.

2.7 The Cauchy–Green deformation tensors and the Lagrange–Euler strain tensors

The stretch is the positive square root of

$$\begin{aligned} \mu^2 = |\mu \boldsymbol{m}|^2 &= \boldsymbol{F}\boldsymbol{M} \cdot \boldsymbol{F}\boldsymbol{M} \\ &= \boldsymbol{M} \cdot \boldsymbol{F}^t(\boldsymbol{F}\boldsymbol{M}) = \boldsymbol{M} \cdot (\boldsymbol{F}^t\boldsymbol{F})\boldsymbol{M}, \end{aligned} \tag{2.82}$$

where we have used (2.67), and in the second line, we have used the definitions of the transpose of a tensor and the product of two tensors. Thus,

$$0 < \mu^2 = \boldsymbol{M} \cdot \boldsymbol{C}\boldsymbol{M}, \quad \text{where} \quad \boldsymbol{C} = \boldsymbol{F}^t\boldsymbol{F} \tag{2.83}$$

is the *right Cauchy–Green deformation tensor*. Since the material curve C (and hence its local unit tangent $\boldsymbol{M}$) is arbitrary, it follows that the tensor $\boldsymbol{C}$ is *positive definite.* That is, for any nonzero $\boldsymbol{a} \in \hat{E}^3$,

$$\boldsymbol{a} \cdot \boldsymbol{C}\boldsymbol{a} = \boldsymbol{a} \cdot \boldsymbol{F}^t\boldsymbol{F}\boldsymbol{a} = \boldsymbol{F}\boldsymbol{a} \cdot \boldsymbol{F}\boldsymbol{a} = |\boldsymbol{F}\boldsymbol{a}|^2 > 0 \tag{2.84}$$

on account of the fact that $\boldsymbol{F}$ is invertible, with determinant $\det \boldsymbol{F} \neq 0$.

Since $\boldsymbol{F}$ maps $\hat{E}^3$ to E^3, it follows that $\boldsymbol{F}^t$ maps E^3 to $\hat{E}^3$ and hence that $\boldsymbol{C}$ maps $\hat{E}^3$ to itself. We say that $\boldsymbol{C}$ is a *referential* tensor. It is clearly symmetric:

$$\boldsymbol{C}^t = (\boldsymbol{F}^t\boldsymbol{F})^t = \boldsymbol{F}^t(\boldsymbol{F}^t)^t = \boldsymbol{F}^t\boldsymbol{F} = \boldsymbol{C}, \tag{2.85}$$

where use has been made of the result of Problem 20 in Chapter 1. In terms of components, on making use of (2.39) and Problem 17(a) therein,

$$\begin{aligned}
\boldsymbol{C} &= (F_{iA}\hat{\boldsymbol{E}}_A \otimes \boldsymbol{e}_i)(F_{jB}\boldsymbol{e}_j \otimes \hat{\boldsymbol{E}}_B) \\
&= F_{iA}F_{jB}(\boldsymbol{e}_i \cdot \boldsymbol{e}_j)\hat{\boldsymbol{E}}_A \otimes \hat{\boldsymbol{E}}_B \\
&= F_{iA}F_{jB}\delta_{ij}\hat{\boldsymbol{E}}_A \otimes \hat{\boldsymbol{E}}_B \\
&= C_{AB}\hat{\boldsymbol{E}}_A \otimes \hat{\boldsymbol{E}}_B,
\end{aligned} \tag{2.86}$$

where

$$C_{AB} = F_{iA}F_{iB} = F_{iB}F_{iA} = C_{BA}. \tag{2.87}$$

Thus,

$$\begin{aligned}
\mu^2 &= \boldsymbol{M} \cdot (C_{AB}\hat{\boldsymbol{E}}_A \otimes \hat{\boldsymbol{E}}_B)\boldsymbol{M} \\
&= \boldsymbol{M} \cdot (C_{AB}M_B\hat{\boldsymbol{E}}_A) = C_{AB}M_AM_B,
\end{aligned} \tag{2.88}$$

and so, (C_{AB}) is a positive-definite matrix.

From (2.67), we have

$$\mu\boldsymbol{F}^{-1}\boldsymbol{m} = \boldsymbol{F}^{-1}(\boldsymbol{F}\boldsymbol{M}) = (\boldsymbol{F}^{-1}\boldsymbol{F})\boldsymbol{M} = \hat{\boldsymbol{I}}\boldsymbol{M} = \boldsymbol{M}, \tag{2.89}$$

or, more simply,

$$\mu^{-1}\boldsymbol{M} = \boldsymbol{F}^{-1}\boldsymbol{m}. \tag{2.90}$$

This affords an alternative calculation of μ:

$$\begin{aligned}\mu^{-2} = \left|\mu^{-1}\boldsymbol{M}\right|^2 &= \boldsymbol{F}^{-1}\boldsymbol{m}\cdot\boldsymbol{F}^{-1}\boldsymbol{m}\\ &= \boldsymbol{m}\cdot(\boldsymbol{F}^{-1})^t\boldsymbol{F}^{-1}\boldsymbol{m}\\ &= \boldsymbol{m}\cdot(\boldsymbol{F}^{-t}\boldsymbol{F}^{-1})\boldsymbol{m}\\ &= \boldsymbol{m}\cdot\boldsymbol{B}^{-1}\boldsymbol{m},\end{aligned} \tag{2.91}$$

where

$$\boldsymbol{B} = \boldsymbol{F}\boldsymbol{F}^t \tag{2.92}$$

is the *left Cauchy–Green deformation tensor*. This too is symmetric.

In terms of components,

$$\begin{aligned}\boldsymbol{B} &= (F_{iA}\boldsymbol{e}_i\otimes\hat{\boldsymbol{E}}_A)(F_{jB}\hat{\boldsymbol{E}}_B\otimes\boldsymbol{e}_j)\\ &= F_{iA}F_{jB}(\hat{\boldsymbol{E}}_A\cdot\hat{\boldsymbol{E}}_B)\boldsymbol{e}_i\otimes\boldsymbol{e}_j\\ &= F_{iA}F_{jB}\delta_{AB}\boldsymbol{e}_i\otimes\boldsymbol{e}_j\\ &= B_{ij}\boldsymbol{e}_i\otimes\boldsymbol{e}_j,\end{aligned} \tag{2.93}$$

where

$$B_{ij} = F_{iA}F_{jA} = F_{jA}F_{iA} = B_{ji}. \tag{2.94}$$

Clearly, $\boldsymbol{B}$ maps E^3 to itself. Such objects are called *spatial tensors.*

Moreover, for any non-zero $\boldsymbol{a}\in E^3$,

$$\begin{aligned}\boldsymbol{a}\cdot\boldsymbol{B}^{-1}\boldsymbol{a} = \boldsymbol{a}\cdot(\boldsymbol{F}\boldsymbol{F}^t)^{-1}\boldsymbol{a} = \boldsymbol{a}\cdot\boldsymbol{F}^{-t}\boldsymbol{F}^{-1}\boldsymbol{a} &= \boldsymbol{F}^{-1}\boldsymbol{a}\cdot\boldsymbol{F}^{-1}\boldsymbol{a}\\ = \left|\boldsymbol{F}^{-1}\boldsymbol{a}\right|^2 > 0&\end{aligned} \tag{2.95}$$

because $\boldsymbol{F}^{-1}$ is invertible, with inverse $\boldsymbol{F}$. Thus, $\boldsymbol{B}^{-1}$ is also positive definite. In Chapter 3, we will show that this is equivalent to the positive definiteness of $\boldsymbol{B}$.

The difference $\mu^2 - 1$ is a measure of the deviation of the stretch from unity, its value in an undeformed state. We can relate it to the right Cauchy–Green tensor via (2.83). Thus,

$$\frac{1}{2}(\mu^2 - 1) = \frac{1}{2}(\boldsymbol{M}\cdot\boldsymbol{C}\boldsymbol{M} - 1) = \boldsymbol{M}\cdot\boldsymbol{E}\boldsymbol{M}, \tag{2.96}$$

where

$$\boldsymbol{E} = \frac{1}{2}(\boldsymbol{C} - \hat{\boldsymbol{I}}) \tag{2.97}$$

is the *Lagrange strain tensor.* The factor $\frac{1}{2}$ is a convention motivated by the observation that, for $\mu \simeq 1$, we have $\mu^2 - 1 \simeq 2(\mu - 1)$. The extension $\mu - 1$ is then approximated by $\boldsymbol{M} \cdot \boldsymbol{EM}$.

Alternatively, (2.91) yields

$$\frac{1}{2}(1 - \mu^{-2}) = \boldsymbol{m} \cdot \boldsymbol{em}, \tag{2.98}$$

where

$$\boldsymbol{e} = \frac{1}{2}(\boldsymbol{I} - \boldsymbol{B}^{-1}) \tag{2.99}$$

is the *Euler strain tensor*, yielding $\mu - 1 \simeq \boldsymbol{m} \cdot \boldsymbol{em}$ when μ is close to unity.

We note that the names we have attached to these tensors are not universal in the literature. Further, there are infinitely many equivalent ways to define strain tensors. Some of the possibilities are discussed in Ogden's book [4].

The referential Lagrange strain and spatial Euler strain are symmetric tensors. We emphasize the fact that the notions of symmetry and skew-symmetry are meaningful only for tensors that map a vector space to itself. Accordingly, two-point tensors can be neither symmetric nor skew.

Having found formulas for the local stretch of a material curve C in terms of the Cauchy–Green tensors, it is possible to compute the entire arclength of its image C_{t_*} in κ_{t_*} (Figure 2.11). To this end, we introduce the arclength coordinate $s(S, t_*)$ on C_{t_*} corresponding to the material point labeled S on C.

The associated arclength measure is

$$\begin{aligned} ds &= |d\boldsymbol{x}| = \sqrt{\boldsymbol{F}d\boldsymbol{X} \cdot \boldsymbol{F}d\boldsymbol{X}} \\ &= \sqrt{\boldsymbol{M} \cdot \boldsymbol{CM}}dS = \mu(S, t_*)dS, \end{aligned} \tag{2.100}$$

where we have used the fact that $d\boldsymbol{X}(S) = \boldsymbol{M}dS$, and hence, $dS = |d\boldsymbol{X}|$, on C. Consequently,

$$\frac{ds}{dS} = \mu(S, t_*), \tag{2.101}$$

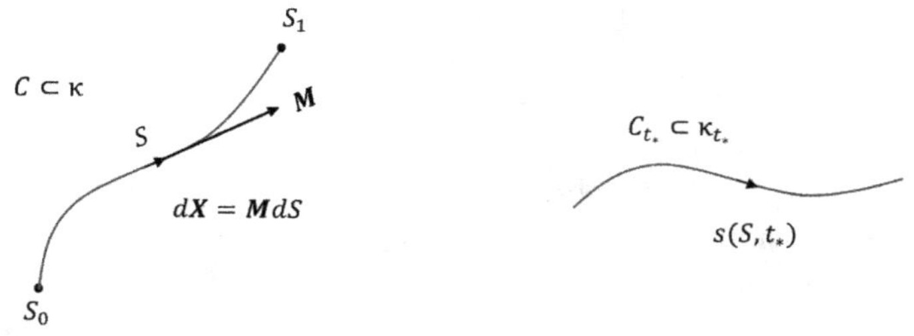

Figure 2.11. Arclength of a material point on material curves in κ and κ_{t_*}.

Figure 2.12. Intersecting material curves in κ and κ_{t_*}.

and the total length of C_{t_*} is

$$l(C_{t_*}) = \int_{S_0}^{S_1} \mu(S, t_*) dS. \tag{2.102}$$

The change in length due to deformation is given simply by

$$l(C_{t_*}) - l(C) = \int_{S_0}^{S_1} [\mu(S, t_*) - 1] dS. \tag{2.103}$$

The Cauchy–Green tensors also afford a way to compute the angle changes between intersecting material curves induced by a deformation (Figure 2.12).

For example, with reference to the figure, we have the curve stretches and orientations

$$\mu_1 \boldsymbol{m}_1 = \boldsymbol{F}\boldsymbol{M}_1 \quad \text{and} \quad \mu_2 \boldsymbol{m}_2 = \boldsymbol{F}\boldsymbol{M}_2 \tag{2.104}$$

at the point $\boldsymbol{X}$ where the curves intersect, at angle Θ, in the reference configuration κ. After deformation the curves intersect at angle θ, say, in κ_t, where $\cos\theta = \boldsymbol{m}_1 \cdot \boldsymbol{m}_2$. Thus,

$$\mu_1\mu_2 \cos\theta = \boldsymbol{F}\boldsymbol{M}_1 \cdot \boldsymbol{F}\boldsymbol{M}_2 = \boldsymbol{M}_1 \cdot \boldsymbol{C}\boldsymbol{M}_2, \tag{2.105}$$

yielding

$$\cos\theta = \frac{\boldsymbol{M}_1 \cdot \boldsymbol{C}\boldsymbol{M}_2}{\sqrt{(\boldsymbol{M}_1 \cdot \boldsymbol{C}\boldsymbol{M}_1)(\boldsymbol{M}_2 \cdot \boldsymbol{C}\boldsymbol{M}_2)}}, \tag{2.106}$$

in which $\boldsymbol{M}_1 \cdot \boldsymbol{C}\boldsymbol{M}_2 = \boldsymbol{M}_2 \cdot \boldsymbol{C}\boldsymbol{M}_1$ by virtue of the symmetry of $\boldsymbol{C}$, i.e., $\boldsymbol{M}_1 \cdot \boldsymbol{C}\boldsymbol{M}_2 = \boldsymbol{C}^t\boldsymbol{M}_1 \cdot \boldsymbol{M}_2 = \boldsymbol{M}_2 \cdot \boldsymbol{C}\boldsymbol{M}_1$.

In the same way, using

$$\mu_1^{-1}\boldsymbol{M}_1 = \boldsymbol{F}^{-1}\boldsymbol{m}_1 \quad \text{and} \quad \mu_2^{-1}\boldsymbol{M}_2 = \boldsymbol{F}^{-1}\boldsymbol{m}_2 \tag{2.107}$$

gives

$$\cos\Theta = \frac{\boldsymbol{m}_1 \cdot \boldsymbol{B}^{-1}\boldsymbol{m}_2}{\sqrt{(\boldsymbol{m}_1 \cdot \boldsymbol{B}^{-1}\boldsymbol{m}_1)(\boldsymbol{m}_2 \cdot \boldsymbol{B}^{-1}\boldsymbol{m}_2)}}, \tag{2.108}$$

with $\boldsymbol{m}_1 \cdot \boldsymbol{B}^{-1}\boldsymbol{m}_2 = \boldsymbol{m}_2 \cdot \boldsymbol{B}^{-1}\boldsymbol{m}_1$, where $\cos\Theta = \boldsymbol{M}_1 \cdot \boldsymbol{M}_2$.

Since the Cauchy–Green tensors fully specify the stretches of material curves and the angles between them, we say that they characterize the state of *distortion* at the material point p. Clearly, the strain tensors furnish equivalent measures of distortion. The reason for introducing strain tensors is that they vanish when the material is undeformed, i.e., when $\boldsymbol{F} = \boldsymbol{1}$. In this case, we have $\boldsymbol{C} = \boldsymbol{1}^t\boldsymbol{1} = \hat{\boldsymbol{I}}$ and $\boldsymbol{B} = \boldsymbol{1}\boldsymbol{1}^t = \boldsymbol{I}$, yielding $\boldsymbol{E} = \hat{\boldsymbol{O}}$ and $\boldsymbol{e} = \boldsymbol{O}$, the zero tensors for $\hat{E}^3$ and E^3, respectively. Henceforth, we will suppress the superposed caret on the referential zero tensor and rely on the context to convey the intended meaning.

2.8 Problems

1. Show that the components of the cofactor of the deformation gradient are

$$F^*_{iA} = \frac{1}{2} e_{ijk} e_{ABC} F_{jB} F_{kC}.$$

2. Show the following:
   (a) $F^*_{iA} = G_{iAB,B}$, where $G_{iAB} = \frac{1}{2}e_{ijk}e_{ABC}\chi_j\chi_{k,C}$ and $x_i = \chi_i(X_A, t)$ are the Cartesian coordinates of the deformation function.
   (b) $G_{iAB} = -G_{iBA}$. Use this to prove that $F^*_{iA,A} = 0$, i.e., the divergence of $\boldsymbol{F}^*$ (with respect to reference coordinates) is zero. This is called the *Piola identity.*
   (c) $\det \boldsymbol{F} = \frac{1}{3}F_{iA}F^*_{iA}$, and hence, there is a vector field $\boldsymbol{H}(\boldsymbol{X}, t) = H_A\hat{\boldsymbol{E}}_A$ with (referential) divergence equal to $\det \boldsymbol{F}$, i.e., $\det \boldsymbol{F} = H_{A,A}$.
   (d) $\det \boldsymbol{F}^t = \det \boldsymbol{F}$.
3. Let $\{\hat{\boldsymbol{E}}_A\}$ and $\{\hat{\boldsymbol{E}}'_A\}$ be orthonormal bases in $\hat{E}^3$, and let $\{\boldsymbol{e}_i\}$ and $\{\boldsymbol{e}'_i\}$ be orthonormal bases in E^3. Then, $\boldsymbol{F} = F_{iA}\boldsymbol{e}_i \otimes \hat{\boldsymbol{E}}_A = F'_{iA}\boldsymbol{e}'_i \otimes \hat{\boldsymbol{E}}'_A$. Determine the components F'_{iA} in terms of the components F_{iA}.

Chapter 3

Further Preliminaries, with Applications to the Analysis of Deformation

Some additional mathematical preliminaries are needed before we continue with the analysis of deformation.

3.1 Principal values and vectors: Spectral representation of a symmetric tensor

A unit vector $\boldsymbol{u}$ is a *principal vector* of a tensor $\boldsymbol{A}$ if there is a scalar λ, a *principal value*, such that

$$\boldsymbol{A}\boldsymbol{u} = \lambda \boldsymbol{u}, \quad \text{or } (\boldsymbol{A} - \lambda \boldsymbol{I})\boldsymbol{u} = \boldsymbol{0}. \tag{3.1}$$

Here, for definiteness, vectors are regarded as elements of E^3 and tensors as linear maps from E^3 to itself. Our results remain valid if E^3 is replaced by $\hat{E}^3$, for example, with obvious notational adjustments.

Because $\boldsymbol{u}$, a unit vector, is non-zero, λ must be such that (see Problem 23 of Chapter 1)

$$0 = \det(\boldsymbol{A} - \lambda \boldsymbol{I}) = -\lambda^3 + I_1(\boldsymbol{A})\lambda^2 - I_2(\boldsymbol{A})\lambda + I_3(\boldsymbol{A}), \tag{3.2}$$

where (Problem 22, Chapter 1)

$$I_1(\boldsymbol{A}) = tr\boldsymbol{A}, \quad I_2(\boldsymbol{A}) = \frac{1}{2}[(tr\boldsymbol{A})^2 - tr(\boldsymbol{A}^2)] \quad \text{and} \quad I_3(\boldsymbol{A}) = \det \boldsymbol{A} \tag{3.3}$$

are the principal invariants of $\boldsymbol{A}$. These are independent of any basis in E^3 and hence so too are the roots λ of (3.2). This equation has either three real roots or only one, i.e., every tensor has at least one real principal value.

If $\boldsymbol{A}$ is symmetric, then its principal values are all real. To see this we suppose that λ and $\boldsymbol{u}$ are complex, i.e.,

$$\lambda = a + ib \quad \text{and} \quad \boldsymbol{u} = u_j \boldsymbol{e}_j \quad \text{with } u_j = a_j + ib_j, \tag{3.4}$$

where i is the complex unit ($i^2 = -1$) and a, b, a_j, b_j are all real-valued. Then, (3.1) implies that

$$\boldsymbol{A}\bar{\boldsymbol{u}} = \bar{\lambda}\bar{\boldsymbol{u}}, \tag{3.5}$$

in which the overbar is used to denote the complex conjugate:

$$\bar{\lambda} = a - ib \quad \text{and} \quad \bar{\boldsymbol{u}} = \bar{u}_j \boldsymbol{e}_j \quad \text{with } u_j = a_j - ib_j, \tag{3.6}$$

and we have used the fact that the conjugate of a product is the product of the conjugates. Thus,

$$\bar{\boldsymbol{u}} \cdot \boldsymbol{A}\boldsymbol{u} = \lambda \bar{\boldsymbol{u}} \cdot \boldsymbol{u}, \quad \text{whereas } \boldsymbol{u} \cdot \boldsymbol{A}\bar{\boldsymbol{u}} = \bar{\lambda}\bar{\boldsymbol{u}} \cdot \boldsymbol{u}. \tag{3.7}$$

The symmetry of $\boldsymbol{A}$ implies that $\boldsymbol{u} \cdot \boldsymbol{A}\bar{\boldsymbol{u}} = \bar{\boldsymbol{u}} \cdot \boldsymbol{A}\boldsymbol{u}$ and hence that

$$(\lambda - \bar{\lambda})\bar{\boldsymbol{u}} \cdot \boldsymbol{u} = 0, \tag{3.8}$$

where $\bar{\boldsymbol{u}} \cdot \boldsymbol{u} = a_j a_j + b_j b_j$. This is strictly positive unless $\boldsymbol{u}$ is the zero vector. This, however, is disallowed by the hypothesis; therefore $\lambda = \bar{\lambda}$ and λ is real.

We can also show that there exists an orthonormal basis for E^3 consisting of the principal vectors of any symmetric tensor. Let $\boldsymbol{u}_1$ and $\boldsymbol{u}_2$ be the principal vectors of a symmetric tensor $\boldsymbol{A}$ with associated principal values λ_1 and λ_2, respectively:

$$\boldsymbol{A}\boldsymbol{u}_1 = \lambda_1 \boldsymbol{u}_1 \quad \text{and} \quad \boldsymbol{A}\boldsymbol{u}_2 = \lambda_2 \boldsymbol{u}_2. \tag{3.9}$$

Then,

$$\lambda_1 \boldsymbol{u}_1 \cdot \boldsymbol{u}_2 = \boldsymbol{A}\boldsymbol{u}_1 \cdot \boldsymbol{u}_2 = \boldsymbol{u}_1 \cdot \boldsymbol{A}^t \boldsymbol{u}_2 = \boldsymbol{u}_1 \cdot \boldsymbol{A}\boldsymbol{u}_2 = \lambda_2 \boldsymbol{u}_1 \cdot \boldsymbol{u}_2, \tag{3.10}$$

and so $\boldsymbol{u}_1 \cdot \boldsymbol{u}_2 = 0$ if $\lambda_1 \neq \lambda_2$. It follows that $\boldsymbol{u}_i$, $i \in \{1, 2, 3\}$, are mutually orthogonal if all three roots λ_i of (3.2) are distinct.

In general, every symmetric tensor possesses three orthonormal principal vectors, but these are uniquely determined, apart from a multiplicative factor ± 1, only when its principal values are all distinct. This is demonstrated in Appendix D. Accordingly any such set $\{\boldsymbol{u}_i\}$ forms an orthonormal basis for E^3. The identity tensor may thus be expressed in the form

$$\boldsymbol{I} = \boldsymbol{u}_i \otimes \boldsymbol{u}_i. \tag{3.11}$$

Using the result of Problem 17(b) of Chapter 1 we then have the *spectral representation* of a symmetric tensor $\boldsymbol{A}$:

$$\boldsymbol{A} = \boldsymbol{AI} = \boldsymbol{A}(\boldsymbol{u}_i \otimes \boldsymbol{u}_i) = (\boldsymbol{Au}_i) \otimes \boldsymbol{u}_i = \sum_{i=1}^{3} \lambda_i \boldsymbol{u}_i \otimes \boldsymbol{u}_i, \tag{3.12}$$

in which the explicit summation sign is inserted to avoid a conflict with the standard summation convention.

The matrix of $\boldsymbol{A}$ in the basis $\{\boldsymbol{u}_i \otimes \boldsymbol{u}_j\}$ is simply $(A_{ij}) = diag(\lambda_1, \lambda_2, \lambda_3)$. Thus, the matrix of a symmetric tensor can be *diagonalized* by decomposing it on a basis consisting of principal vectors.

3.2 Positive definiteness and the extremal property of principal values

A tensor $\boldsymbol{A}$ is *positive definite* if

$$\boldsymbol{a} \cdot \boldsymbol{Aa} > 0 \quad \text{for all} \quad \boldsymbol{a} \neq \boldsymbol{0}. \tag{3.13}$$

With reference to Problems 6(a) and 14 of Chapter 1, we observe that this inequality involves only the symmetric part of $\boldsymbol{A}$, and, further, that a skew tensor cannot be positive definite. Accordingly we assume, with no loss of generality, that $\boldsymbol{A}$ is symmetric. In particular, if $\boldsymbol{u}$, say, is a principal vector with associated principal value λ, then

$$0 < \boldsymbol{u} \cdot \boldsymbol{Au} = \boldsymbol{u} \cdot (\lambda \boldsymbol{u}) = \lambda. \tag{3.14}$$

Thus,

$$\lambda_i > 0; \quad i = 1, 2, 3. \tag{3.15}$$

Conversely, if all the principal values are positive, then, by the spectral representation,

$$\boldsymbol{a} \cdot \boldsymbol{Aa} = \boldsymbol{a} \cdot \left(\sum_{i=1}^{3} \lambda_i \boldsymbol{u}_i \otimes \boldsymbol{u}_i \right) \boldsymbol{a} = \sum_{i=1}^{3} \lambda_i (\boldsymbol{a} \cdot \boldsymbol{u}_i)^2 > 0 \quad \text{for all } \boldsymbol{a} \neq \boldsymbol{0}, \tag{3.16}$$

in which the inequality follows from (3.15) and that fact that $\{\boldsymbol{u}_i\}$ is a basis; hence at least one of the components $\boldsymbol{a} \cdot \boldsymbol{u}_i$ is unequal to zero. Further,

$$\det \boldsymbol{A} = [\boldsymbol{A}\boldsymbol{u}_1, \boldsymbol{A}\boldsymbol{u}_2, \boldsymbol{A}\boldsymbol{u}_3]/[\boldsymbol{u}_1, \boldsymbol{u}_2, \boldsymbol{u}_3] = \lambda_1\lambda_2\lambda_3 > 0. \qquad (3.17)$$

We can also show that the smallest and largest principal values bound the diagonal components A_{11}, A_{22} and A_{33} of a symmetric, positive definite tensor $\boldsymbol{A}$ relative to any orthonormal basis $\{\boldsymbol{e}_i \otimes \boldsymbol{e}_j\}$. To prove this claim we fix $\boldsymbol{A}$ and define a function

$$f(\boldsymbol{u}) = \boldsymbol{u} \cdot \boldsymbol{A}\boldsymbol{u} \quad \text{with } |\boldsymbol{u}| = 1. \qquad (3.18)$$

Clearly, $f(\boldsymbol{u})$ is continuous and $f(\boldsymbol{u}) > 0$ for all $\boldsymbol{u} \in S = \{\boldsymbol{v} : |\boldsymbol{v}| = 1\}$, the surface of the unit sphere. Because this is a compact (closed and bounded) set, it follows [2] that there exist $\boldsymbol{u}_1, \boldsymbol{u}_2 \in S$ such that

$$0 < \lambda_1 = \min_{\boldsymbol{u}\in S} f(\boldsymbol{u}) \quad \text{and} \quad 0 < \lambda_2 = \max_{\boldsymbol{u}\in S} f(\boldsymbol{u}),$$
$$\text{where } \lambda_1 = f(\boldsymbol{u}_1) \quad \text{and} \quad \lambda_2 = f(\boldsymbol{u}_2). \qquad (3.19)$$

To compute λ_1 and λ_2 we minimize and maximize f. In both cases, we seek those $\boldsymbol{u}$ that render f stationary, i.e.,

$$0 = df(\boldsymbol{u}) = d\boldsymbol{u} \cdot \boldsymbol{A}\boldsymbol{u} + \boldsymbol{u} \cdot \boldsymbol{A}(d\boldsymbol{u}) = 2\boldsymbol{A}\boldsymbol{u} \cdot d\boldsymbol{u}, \qquad (3.20)$$

where we have invoked $\boldsymbol{u} \cdot \boldsymbol{A}(d\boldsymbol{u}) = \boldsymbol{A}^t\boldsymbol{u} \cdot d\boldsymbol{u} = d\boldsymbol{u} \cdot \boldsymbol{A}\boldsymbol{u}$. Thus we seek $\boldsymbol{u}$ such that

$$\boldsymbol{A}\boldsymbol{u} \cdot d\boldsymbol{u} = 0. \qquad (3.21)$$

The constraint $\boldsymbol{u} \cdot \boldsymbol{u} = 1$ admits all $d\boldsymbol{u}$ such that $\boldsymbol{u} \cdot d\boldsymbol{u} = 0$. Accordingly,

$$\boldsymbol{A}\boldsymbol{u} = \mu\boldsymbol{u} \qquad (3.22)$$

where μ is a scalar (Figure 3.1).

Applying this result to $\boldsymbol{u} = \boldsymbol{u}_1$ gives $\mu = \boldsymbol{u}_1 \cdot \boldsymbol{A}\boldsymbol{u}_1 = \lambda_1$. Therefore, to obtain λ_1 we solve the problem

$$\boldsymbol{A}\boldsymbol{u}_1 = \lambda_1\boldsymbol{u}_1. \qquad (3.23)$$

In the same way, applying our result to $\boldsymbol{u}_2$ yields

$$\boldsymbol{A}\boldsymbol{u}_2 = \lambda_2\boldsymbol{u}_2. \qquad (3.24)$$

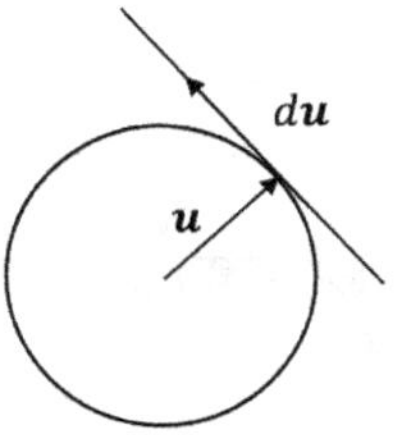

Figure 3.1. The unit sphere S.

Thus, λ_1 and λ_2 are principal values of $\boldsymbol{A}$. Altogether, then, we have the chain of inequalities

$$\lambda_1 = f(\boldsymbol{u}_1) \leq f(\boldsymbol{u}) \leq f(\boldsymbol{u}_2) = \lambda_2. \tag{3.25}$$

Finally, with $f(\boldsymbol{e}_1) = \boldsymbol{e}_1 \cdot \boldsymbol{A}\boldsymbol{e}_1 = A_{11}$, etc., on applying this result to $\boldsymbol{u} = \boldsymbol{e}_1$, etc., we establish the extremal property of the principal values:

$$\lambda_1 \leq \min\{A_{11}, A_{22}, A_{33}\} \leq \max\{A_{11}, A_{22}, A_{33}\} \leq \lambda_2. \tag{3.26}$$

3.3 Square root theorem

Suppose $\boldsymbol{A}$ is symmetric and positive definite. Then there is a unique symmetric, positive definite $\boldsymbol{U}$ ($\boldsymbol{U}^t = \boldsymbol{U}$ and $\boldsymbol{a} \cdot \boldsymbol{U}\boldsymbol{a} > 0$ for all $\boldsymbol{a} \neq \boldsymbol{0}$) such that $\boldsymbol{A} = \boldsymbol{U}^2$, where $\boldsymbol{U}^2 = \boldsymbol{U}\boldsymbol{U}$. We express this result symbolically as $\boldsymbol{U} = \sqrt{\boldsymbol{A}}$.

Our proof of this claim follows that given in [3, 5]. We write $\boldsymbol{A}$ in spectral form:

$$\boldsymbol{A} = \sum \lambda_i \boldsymbol{u}_i \otimes \boldsymbol{u}_i \quad \text{with } \lambda_i > 0, \tag{3.27}$$

and define

$$\boldsymbol{U} = \sum \sqrt{\lambda_i} \boldsymbol{u}_i \otimes \boldsymbol{u}_i. \tag{3.28}$$

This is clearly symmetric and positive definite, and

$$\begin{aligned} \boldsymbol{U}^2 &= \boldsymbol{U}\left(\sum \sqrt{\lambda_i} \boldsymbol{u}_i \otimes \boldsymbol{u}_i\right) = \sum \sqrt{\lambda_i} (\boldsymbol{U}\boldsymbol{u}_i \otimes \boldsymbol{u}_i) \\ &= \sum \lambda_i \boldsymbol{u}_i \otimes \boldsymbol{u}_i = \boldsymbol{A}, \end{aligned} \tag{3.29}$$

so that this $\boldsymbol{U}$ has the stated properties.

Evidently $\boldsymbol{U}^2\boldsymbol{u} = \lambda\boldsymbol{u}$, or

$$\mathbf{0} = (\boldsymbol{U}^2 - \lambda\boldsymbol{I})\boldsymbol{u} = (\boldsymbol{U} + \sqrt{\lambda}\boldsymbol{I})\boldsymbol{v}, \quad \text{where } \boldsymbol{v} = (\boldsymbol{U} - \sqrt{\lambda}\boldsymbol{I})\boldsymbol{u}, \tag{3.30}$$

wherein $\boldsymbol{u} \in \{\boldsymbol{u}_i\}$ and $\lambda \in \{\lambda_i\}$. Thus,

$$\boldsymbol{U}\boldsymbol{v} = -\sqrt{\lambda}\boldsymbol{v}, \tag{3.31}$$

and this requires that $\boldsymbol{v} = \mathbf{0}$ because the alternative yields the conclusion that $-\sqrt{\lambda}$ is a principal value of $\boldsymbol{U}$. This is impossible, of course, because $\boldsymbol{U}$ is positive definite, and so $\boldsymbol{U}\boldsymbol{u} = \sqrt{\lambda}\boldsymbol{u}$. This also follows directly from (3.28).

To establish uniqueness, suppose $\tilde{\boldsymbol{U}}^2 = \boldsymbol{A}$ with $\tilde{\boldsymbol{U}}$ symmetric and positive definite. Then, $\tilde{\boldsymbol{U}}^2\boldsymbol{u} = \lambda\boldsymbol{u}$, and the same argument leads to $\tilde{\boldsymbol{U}}\boldsymbol{u} = \sqrt{\lambda}\boldsymbol{u}$. Thus $\tilde{\boldsymbol{U}}\boldsymbol{u}_i = \boldsymbol{U}\boldsymbol{u}_i$ and hence $w_i\tilde{\boldsymbol{U}}\boldsymbol{u}_i = w_i\boldsymbol{U}\boldsymbol{u}_i$, implying, as $\{\boldsymbol{u}_i\}$ is a basis, that $\tilde{\boldsymbol{U}}\boldsymbol{w} = \boldsymbol{U}\boldsymbol{w}$ for arbitrary $\boldsymbol{w}(= w_i\boldsymbol{u}_i)$. Therefore, $\tilde{\boldsymbol{U}} = \boldsymbol{U}$ and the theorem is proved.

3.4 Polar decomposition theorem

Consider a tensor $\boldsymbol{F}$ with $\det \boldsymbol{F} > 0$. Here we regard $\boldsymbol{F}$ as a two-point tensor, in anticipation of applications involving the deformation gradient. The polar decomposition theorem is the assertion that there exist unique symmetric, positive definite tensors $\boldsymbol{U}$ and $\boldsymbol{V}$, and a unique rotation tensor $\boldsymbol{R}$, such that

$$\boldsymbol{F} = \boldsymbol{R}\boldsymbol{U} = \boldsymbol{V}\boldsymbol{R}. \tag{3.32}$$

Recall that $\boldsymbol{C} = \boldsymbol{F}^t\boldsymbol{F}$ is symmetric and positive definite. The square root theorem furnishes a unique symmetric, positive definite — and hence invertible — tensor $\boldsymbol{U}$ such that $\boldsymbol{U}^2 = \boldsymbol{C}$, where $\boldsymbol{C}$ is a referential tensor. It then follows from the analysis of the previous section that $\boldsymbol{U}$ is likewise referential. Thus, $\boldsymbol{R}$ is a two-point tensor and $\boldsymbol{V}$ is a spatial tensor. In terms of components,

$$F_{iA} = R_{iB}U_{BA} = V_{ij}R_{jA}. \tag{3.33}$$

If $\boldsymbol{F}$ is a deformation gradient, then $\boldsymbol{U}$ and $\boldsymbol{V}$ respectively are the *Right-* and *Left-Stretch Tensors.* They occur on the right and left of $\boldsymbol{R}$, respectively, in the polar decomposition of $\boldsymbol{F}$. This terminology, in turn, furnishes post facto justification for designating $\boldsymbol{C}(=\boldsymbol{U}^2)$ and $\boldsymbol{B}(=\boldsymbol{V}^2)$ respectively as the right- and left- Cauchy–Green tensors.

Certain results pertaining to purely spatial or purely referential rotation tensors are discussed in Appendix C. These do not apply to two-point rotation tensors. However, every two-point rotation can be obtained by pre-multiplying a referential rotation, or post-multiplying a spatial rotation, by the shifter.

Let $\{\boldsymbol{a}, \boldsymbol{b}, \boldsymbol{c}\}$ be a linearly independent set in $\hat{E}^3$. Then,

$$\begin{aligned}(\det \boldsymbol{F})[\boldsymbol{a}, \boldsymbol{b}, \boldsymbol{c}] &= [\boldsymbol{R}(\boldsymbol{U}\boldsymbol{a}), \boldsymbol{R}(\boldsymbol{U}\boldsymbol{b}), \boldsymbol{R}(\boldsymbol{U}\boldsymbol{c})] \\ &= (\det \boldsymbol{R})[\boldsymbol{U}\boldsymbol{a}, \boldsymbol{U}\boldsymbol{b}, \boldsymbol{U}\boldsymbol{c}] \\ &= (\det \boldsymbol{R})(\det \boldsymbol{U})[\boldsymbol{a}, \boldsymbol{b}, \boldsymbol{c}], \end{aligned} \tag{3.34}$$

where we have used the fact that $\boldsymbol{U}\boldsymbol{a}, \boldsymbol{U}\boldsymbol{b}, \boldsymbol{U}\boldsymbol{c} \in \hat{E}^3$. Thus the result of Problem 21(b) in Chapter 1 extends to the case in which the first factor is a two-point tensor and the second factor is referential, i.e.,

$$\det \boldsymbol{F} = (\det \boldsymbol{R})(\det \boldsymbol{U}). \tag{3.35}$$

Because $\det \boldsymbol{U} > 0$ it follows that the sign of $\det \boldsymbol{R}$ is the same as the sign of $\det \boldsymbol{F}$. Accordingly, our restriction on $\boldsymbol{F}$ yields

$$\det \boldsymbol{R} > 0. \tag{3.36}$$

In the same way, using the fact that $\boldsymbol{R}\boldsymbol{a}, \boldsymbol{R}\boldsymbol{b}, \boldsymbol{R}\boldsymbol{c} \in E^3$, we easily establish that $\det \boldsymbol{F} = (\det \boldsymbol{V})(\det \boldsymbol{R})$ and hence that $\det \boldsymbol{V} > 0$.

To establish (3.32), let $\boldsymbol{R} = \boldsymbol{F}\boldsymbol{U}^{-1}$. Then,

$$\boldsymbol{R}^t\boldsymbol{R} = \boldsymbol{U}^{-t}\boldsymbol{F}^t\boldsymbol{F}\boldsymbol{U}^{-1} = \boldsymbol{U}^{-1}\boldsymbol{C}\boldsymbol{U}^{-1} = \boldsymbol{U}^{-1}\boldsymbol{U}\boldsymbol{U}\boldsymbol{U}^{-1} = \hat{\boldsymbol{I}}, \tag{3.37}$$

and $\boldsymbol{R}$ is orthogonal, i.e., $\boldsymbol{R}^{-1} = \boldsymbol{R}^t\colon E^3 \to \hat{E}^3$. Here, we have used the fact that $\boldsymbol{U}^{-1}$ is symmetric, which follows from the spectral representation of $\boldsymbol{U}$: Its principal values are simply the reciprocals of those of $\boldsymbol{U}$. Accordingly $\boldsymbol{U}^{-1}$ is also positive definite. This, incidentally, establishes the truth of the claim made following (2.95).

The spectral representations of $\boldsymbol{U}$ and $\boldsymbol{C}$ are

$$\boldsymbol{U} = \sum \lambda_i \boldsymbol{u}_i \otimes \boldsymbol{u}_i \quad \text{and} \quad \boldsymbol{C} = \sum \lambda_i^2 \boldsymbol{u}_i \otimes \boldsymbol{u}_i, \tag{3.38}$$

respectively, where $\lambda_i(>0)$ are the *principal stretches* and $\boldsymbol{u}_i \in \hat{E}^3$ are the associated orthonormal principal vectors. Let

$$\boldsymbol{v}_i = \boldsymbol{R}\boldsymbol{u}_i \in E^3. \tag{3.39}$$

These are orthonormal on account of the orthogonality of $\boldsymbol{R}$, i.e.,

$$\boldsymbol{v}_i \cdot \boldsymbol{v}_j = \boldsymbol{R}\boldsymbol{u}_i \cdot \boldsymbol{R}\boldsymbol{u}_j = \boldsymbol{u}_i \cdot \boldsymbol{R}^t\boldsymbol{R}\boldsymbol{u}_j = \boldsymbol{u}_i \cdot \hat{\boldsymbol{I}}\boldsymbol{u}_j = \boldsymbol{u}_i \cdot \boldsymbol{u}_j = \delta_{ij}. \tag{3.40}$$

Thus,

$$\boldsymbol{I} = \boldsymbol{v}_i \otimes \boldsymbol{v}_i = \boldsymbol{R}\boldsymbol{u}_i \otimes \boldsymbol{R}\boldsymbol{u}_i = \boldsymbol{R}(\boldsymbol{u}_i \otimes \boldsymbol{u}_i)\boldsymbol{R}^t = \boldsymbol{R}\hat{\boldsymbol{I}}\boldsymbol{R}^t = \boldsymbol{R}\boldsymbol{R}^t. \tag{3.41}$$

Regarding $(3.32)_2$ as the definition of $\boldsymbol{V}$, with $\boldsymbol{V} = \boldsymbol{V}\boldsymbol{I} = \boldsymbol{V}\boldsymbol{R}\boldsymbol{R}^t$ we then have

$$\begin{aligned}\boldsymbol{V} &= \boldsymbol{R}\left(\sum \lambda_i \boldsymbol{u}_i \otimes \boldsymbol{u}_i\right)\boldsymbol{R}^t \\ &= \sum \lambda_i \boldsymbol{R}\boldsymbol{u}_i \otimes \boldsymbol{R}\boldsymbol{u}_i \\ &= \sum \lambda_i \boldsymbol{v}_i \otimes \boldsymbol{v}_i.\end{aligned} \tag{3.42}$$

This is clearly symmetric and positive definite, and furnishes the spectral decomposition $\boldsymbol{B} = \sum \lambda_i^2 \boldsymbol{v}_i \otimes \boldsymbol{v}_i$ of the left Cauchy–Green tensor. Moreover,

$$\begin{aligned}\boldsymbol{F} &= \boldsymbol{R}\left(\sum \lambda_i \boldsymbol{u}_i \otimes \boldsymbol{u}_i\right) \\ &= \sum \lambda_i \boldsymbol{R}\boldsymbol{u}_i \otimes \boldsymbol{u}_i \\ &= \sum \lambda_i \boldsymbol{v}_i \otimes \boldsymbol{u}_i.\end{aligned} \tag{3.43}$$

It remains only to prove that $\det \boldsymbol{R} = 1$. That this is so follows from (3.36), (3.39) and the orthonormality of the principal vectors. Thus, $(\det \boldsymbol{R})[\boldsymbol{u}_1, \boldsymbol{u}_2, \boldsymbol{u}_3] = [\boldsymbol{R}\boldsymbol{u}_1, \boldsymbol{R}\boldsymbol{u}_2, \boldsymbol{R}\boldsymbol{u}_3] = [\boldsymbol{v}_1, \boldsymbol{v}_2, \boldsymbol{v}_3]$, and (3.36) implies that $[\boldsymbol{u}_1, \boldsymbol{u}_2, \boldsymbol{u}_3]$ and $[\boldsymbol{v}_1, \boldsymbol{v}_2, \boldsymbol{v}_3]$ have the same sign, whereas $[\boldsymbol{u}_1, \boldsymbol{u}_2, \boldsymbol{u}_3] = \pm 1$ and $[\boldsymbol{v}_1, \boldsymbol{v}_2, \boldsymbol{v}_3] = \pm 1$ by virtue of orthonormality. Thus, $\det \boldsymbol{R} = 1$, $\boldsymbol{R}$ is a two-point rotation tensor and the polar decomposition theorem is proved.

These results imply that $\boldsymbol{R}$ rotates principal vectors of $\boldsymbol{U}$ to principal vectors of $\boldsymbol{V}$. These are not material vectors. That is, the $\{\boldsymbol{u}_i\}$ are properties of the tensor $\boldsymbol{U}$, and thus generally depend on time at any particular material point. In contrast, a material vector $\boldsymbol{M}$, say, is independent of t when expressed as a function of $\boldsymbol{X}$ and t. Nevertheless, at any *fixed* instant t_*, say, we can pick out a material curve — in fact, infinitely many — with unit tangent $\boldsymbol{u} \in \{\boldsymbol{u}_i\}$ at the point with referential position $\boldsymbol{X}$; that is, we can select a material curve with the property that $\boldsymbol{M} = \boldsymbol{u}$ at $\boldsymbol{X}$, at the time t_*. From (2.67), $(3.32)_1$, $(3.38)_1$ and (3.39) we then have

$$\begin{aligned}\mu\boldsymbol{m} &= \boldsymbol{F}\boldsymbol{M} = \boldsymbol{F}\boldsymbol{u}\\ &= \boldsymbol{R}(\boldsymbol{U}\boldsymbol{u})\\ &= \boldsymbol{R}(\lambda\boldsymbol{u})\\ &= \lambda\boldsymbol{R}\boldsymbol{u}\\ &= \lambda\boldsymbol{v},\end{aligned} \tag{3.44}$$

where $\boldsymbol{v} \in \{\boldsymbol{v}_i\}$. Thus, $\boldsymbol{v} = \boldsymbol{m}$, the local unit tangent to the material curve in κ_{t_*}, and the local stretch μ of the material curve is simply λ. This furnishes the justification for referring to the λ_i as principal *stretches*.

The passage from the second line of (3.44) to the third may be interpreted as a stretch of $\boldsymbol{u}$ to $\lambda\boldsymbol{u}$; and the passage from the fourth line to the fifth as a rotation of $\lambda\boldsymbol{u}$ to $\lambda\boldsymbol{v}$. That is, the deformation is a stretch followed by a rotation. Alternatively, $(3.32)_2$ furnishes

$$\begin{aligned}\mu\boldsymbol{m} &= \boldsymbol{F}\boldsymbol{M} = \boldsymbol{F}\boldsymbol{u}\\ &= \boldsymbol{V}(\boldsymbol{R}\boldsymbol{u})\\ &= \boldsymbol{V}\boldsymbol{v}\\ &= \lambda\boldsymbol{v},\end{aligned} \tag{3.45}$$

so that it is also permissible to view the deformation as a rotation followed by a stretch.

3.5 Examples

3.5.1 *Simple shear*

Consider the deformation

$$\boldsymbol{x} = \mathbf{1}\boldsymbol{X} + \boldsymbol{u}, \tag{3.46}$$

where $\mathbf{1}$ is the shifter from $\hat{E}^3$ to E^3, and $\boldsymbol{u}$, the displacement, is given, in the notation of Section 2.6, by

$$\begin{aligned}\boldsymbol{u} &= \gamma(\boldsymbol{E}_2 \cdot \mathbf{1}\boldsymbol{X})\boldsymbol{e}_1 \\ &= \gamma(\mathbf{1}^t\boldsymbol{E}_2 \cdot \boldsymbol{X})\boldsymbol{e}_1 \\ &= \gamma(\hat{\boldsymbol{E}}_2 \cdot \boldsymbol{X})\boldsymbol{e}_1,\end{aligned} \tag{3.47}$$

in which γ, the amount of shear, is a dimensionless constant.

The Cartesian coordinates of a material point after deformation are $x_i = \boldsymbol{e}_i \cdot \boldsymbol{x}$. Thus,

$$x_1 = \boldsymbol{e}_1 \cdot \mathbf{1}\boldsymbol{X} + \gamma\boldsymbol{E}_2 \cdot \mathbf{1}\boldsymbol{X}, \quad x_2 = \boldsymbol{e}_2 \cdot \mathbf{1}\boldsymbol{X} \quad \text{and} \quad x_3 = \boldsymbol{e}_3 \cdot \mathbf{1}\boldsymbol{X}. \tag{3.48}$$

Using $\mathbf{1}\boldsymbol{X} = X_A\boldsymbol{E}_A$, where X_A are the referential Cartesian coordinates of the same point, and choosing $\{\boldsymbol{E}_A\} = \{\boldsymbol{e}_i\}$, we obtain

$$x_1 = X_1 + \gamma X_2, \quad x_2 = X_2 \quad \text{and} \quad x_3 = X_3. \tag{3.49}$$

The situation is depicted in Figure 3.2.

The deformation gradient follows from

$$\begin{aligned}\boldsymbol{F}d\boldsymbol{X} &= d\boldsymbol{x} \\ &= \mathbf{1}d\boldsymbol{X} + \gamma(\hat{\boldsymbol{E}}_2 \cdot d\boldsymbol{X})\boldsymbol{e}_1 \\ &= (\mathbf{1} + \gamma\boldsymbol{e}_1 \otimes \hat{\boldsymbol{E}}_2)d\boldsymbol{X},\end{aligned} \tag{3.50}$$

yielding

$$\boldsymbol{F} = \mathbf{1} + \gamma\boldsymbol{e}_1 \otimes \hat{\boldsymbol{E}}_2, \tag{3.51}$$

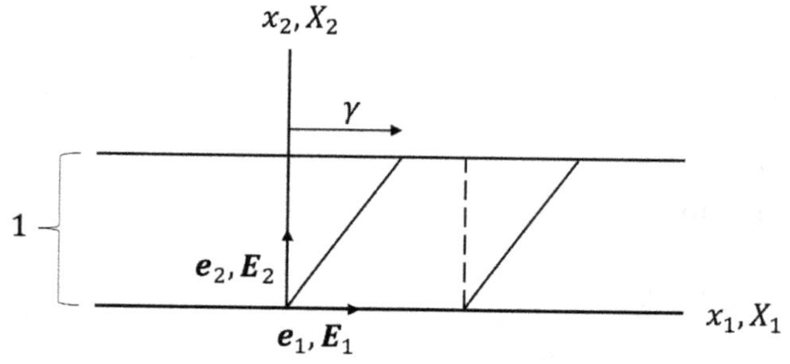

Figure 3.2. Simple shear.

with

$$\begin{aligned} J = \det \boldsymbol{F} &= [\boldsymbol{F}\hat{\boldsymbol{E}}_1, \boldsymbol{F}\hat{\boldsymbol{E}}_2, \boldsymbol{F}\hat{\boldsymbol{E}}_3]/[\hat{\boldsymbol{E}}_1, \hat{\boldsymbol{E}}_2, \hat{\boldsymbol{E}}_3] \\ &= [\boldsymbol{E}_1, \boldsymbol{E}_2 + \gamma\boldsymbol{e}_1, \boldsymbol{E}_3] \\ &= [\boldsymbol{e}_1, \boldsymbol{e}_2 + \gamma\boldsymbol{e}_1, \boldsymbol{e}_3] \\ &= \boldsymbol{e}_1 \cdot (\boldsymbol{e}_2 \times \boldsymbol{e}_3 + \gamma\boldsymbol{e}_1 \times \boldsymbol{e}_3) \\ &= 1. \end{aligned} \tag{3.52}$$

In view of (2.50) there is no change of material volume induced by this deformation, whereas (2.59) implies that the referential mass density ρ_κ coincides with the mass density ρ in the current configuration κ_t. Deformations with this property are said to be *isochoric*.

The right Cauchy–Green tensor for this deformation is

$$\begin{aligned} \boldsymbol{C} &= (\boldsymbol{1} + \gamma\boldsymbol{e}_1 \otimes \hat{\boldsymbol{E}}_2)^t(\boldsymbol{1} + \gamma\boldsymbol{e}_1 \otimes \hat{\boldsymbol{E}}_2) \\ &= (\boldsymbol{1}^t + \gamma\hat{\boldsymbol{E}}_2 \otimes \boldsymbol{e}_1)(\boldsymbol{1} + \gamma\boldsymbol{e}_1 \otimes \hat{\boldsymbol{E}}_2) \\ &= \boldsymbol{1}^t\boldsymbol{1} + \gamma(\hat{\boldsymbol{E}}_2 \otimes \boldsymbol{1}^t\boldsymbol{E}_1 + \boldsymbol{1}^t\boldsymbol{E}_1 \otimes \hat{\boldsymbol{E}}_2) + \gamma^2\hat{\boldsymbol{E}}_2 \otimes \hat{\boldsymbol{E}}_2 \\ &= \hat{\boldsymbol{I}} + \gamma(\hat{\boldsymbol{E}}_1 \otimes \hat{\boldsymbol{E}}_2 + \hat{\boldsymbol{E}}_2 \otimes \hat{\boldsymbol{E}}_1) + \gamma^2\hat{\boldsymbol{E}}_2 \otimes \hat{\boldsymbol{E}}_2 \\ &= \hat{\boldsymbol{E}}_1 \otimes \hat{\boldsymbol{E}}_1 + \gamma(\hat{\boldsymbol{E}}_1 \otimes \hat{\boldsymbol{E}}_2 + \hat{\boldsymbol{E}}_2 \otimes \hat{\boldsymbol{E}}_1) \\ &\quad + (1 + \gamma^2)\hat{\boldsymbol{E}}_2 \otimes \hat{\boldsymbol{E}}_2 + \hat{\boldsymbol{E}}_3 \otimes \hat{\boldsymbol{E}}_3. \end{aligned} \tag{3.53}$$

Similarly, the left Cauchy–Green tensor is

$$\boldsymbol{B} = (1+\gamma^2)\boldsymbol{e}_1 \otimes \boldsymbol{e}_1 + \boldsymbol{e}_2 \otimes \boldsymbol{e}_2 + \boldsymbol{e}_3 \otimes \boldsymbol{e}_3 + \gamma(\boldsymbol{e}_1 \otimes \boldsymbol{e}_2 + \boldsymbol{e}_2 \otimes \boldsymbol{e}_1). \tag{3.54}$$

The matrix of $\boldsymbol{C}$ relative to the basis $\{\hat{\boldsymbol{E}}_A \otimes \hat{\boldsymbol{E}}_B\}$ is

$$(C_{AB}) = \begin{pmatrix} 1 & \gamma & 0 \\ \gamma & 1+\gamma^2 & 0 \\ 0 & 0 & 1 \end{pmatrix}, \tag{3.55}$$

and that of $\boldsymbol{B}$ relative to $\{\boldsymbol{e}_i \otimes \boldsymbol{e}_j\}$ is

$$(B_{ij}) = \begin{pmatrix} 1+\gamma^2 & \gamma & 0 \\ \gamma & 1 & 0 \\ 0 & 0 & 1 \end{pmatrix}. \tag{3.56}$$

Evidently $\boldsymbol{C}\hat{\boldsymbol{E}}_3 = \hat{\boldsymbol{E}}_3$, so that $\hat{\boldsymbol{E}}_3$ is a principal vector of $\boldsymbol{C}$ corresponding to principal value $\lambda_3^2 = 1$. We wish to find the remaining principal values and associated principal vectors. Before proceeding, we observe that, because the deformation is isochoric,

$$\begin{aligned} 1 &= \det \boldsymbol{F} = \det(\boldsymbol{R}\boldsymbol{U}) = (\det \boldsymbol{R})(\det \boldsymbol{U}) \\ &= \det \boldsymbol{U} = [\boldsymbol{U}\boldsymbol{u}_1, \boldsymbol{U}\boldsymbol{u}_2, \boldsymbol{U}\boldsymbol{u}_3]/[\boldsymbol{u}_1, \boldsymbol{u}_2, \boldsymbol{u}_3] \\ &= [\lambda_1 \boldsymbol{u}_1, \lambda_2 \boldsymbol{u}_2, \lambda_3 \boldsymbol{u}_3] = \lambda_1 \lambda_2 \lambda_3. \end{aligned} \tag{3.57}$$

Accordingly, in the present example we have $\lambda_2 = \lambda_1^{-1}$.

To determine the $\lambda's$ we solve $\det(\boldsymbol{C} - \lambda^2 \hat{\boldsymbol{I}}) = 0$, i.e.,

$$\begin{aligned} 0 &= \begin{vmatrix} 1-\lambda^2 & \gamma & 0 \\ \gamma & 1+\gamma^2-\lambda^2 & 0 \\ 0 & 0 & 1-\lambda^2 \end{vmatrix} \\ &= (1-\lambda^2)[(1-\lambda^2)(1+\gamma^2-\lambda^2) - \gamma^2] \\ &= (1-\lambda^2)[(1-\lambda^2)^2 - \gamma^2\lambda^2]. \end{aligned} \tag{3.58}$$

One root is $\lambda_3^2 = 1$, as we have already noted, and the others satisfy

$$4\lambda_{1,2}^2 = 4 + 2\gamma^2 \pm 2\gamma\sqrt{4+\gamma^2}. \tag{3.59}$$

On identifying λ_1 with the "+" root, we conclude that the principal stretches are

$$\lambda_1 = \frac{1}{2}(\gamma + \sqrt{4+\gamma^2}), \quad \lambda_2 = \lambda_1^{-1} \quad \text{and} \quad \lambda_3 = 1, \tag{3.60}$$

and the spectral representation of $\boldsymbol{C}$ is

$$\boldsymbol{C} = \lambda_1^2 \boldsymbol{u}_1 \otimes \boldsymbol{u}_1 + \lambda_1^{-2} \boldsymbol{u}_2 \otimes \boldsymbol{u}_2 + \hat{\boldsymbol{E}}_3 \otimes \hat{\boldsymbol{E}}_3, \tag{3.61}$$

where $\boldsymbol{u}_1$ and $\boldsymbol{u}_2$ are orthonormal vectors lying in the plane perpendicular to $\hat{\boldsymbol{E}}_3$. To determine them we write

$$\boldsymbol{u}_1 = \cos A \hat{\boldsymbol{E}}_1 + \sin A \hat{\boldsymbol{E}}_2, \quad \boldsymbol{u}_2 = -\sin A \hat{\boldsymbol{E}}_1 + \cos A \hat{\boldsymbol{E}}_2, \tag{3.62}$$

and take the scalar product of the equation $\boldsymbol{C}\boldsymbol{u}_1 = \lambda_1^2 \boldsymbol{u}_1$ with $\hat{\boldsymbol{E}}_1$, reaching

$$\boldsymbol{u}_1 \cdot \boldsymbol{C}\hat{\boldsymbol{E}}_1 = \hat{\boldsymbol{E}}_1 \cdot \boldsymbol{C}\boldsymbol{u}_1 = \lambda_1^2 \hat{\boldsymbol{E}}_1 \cdot \boldsymbol{u}_1. \tag{3.63}$$

On combining this with (3.53), we arrive at $\cos A + \gamma \sin A = \lambda_1^2 \cos A$, i.e.,

$$\tan A = \frac{1}{\gamma}(\lambda_1^2 - 1) = \frac{1}{2}(\gamma + \sqrt{4+\gamma^2}). \tag{3.64}$$

This determines $\boldsymbol{u}_1$ and $\boldsymbol{u}_2$ in terms of the amount of shear. Observe that $\tan A \to 1$, and hence that $A \to \pi/4$, as $\gamma \to 0$; for small shears the principal axes are oriented at approximately $\pm 45°$ relative to the direction of shear.

We remark that the equation obtained by taking the scalar product with $\hat{\boldsymbol{E}}_2$ is redundant. This is due to the fact that $\det(\boldsymbol{C} - \lambda_1^2 \hat{\boldsymbol{I}}) = 0$. This is an example of a more general result, which we pause to discuss here before proceeding further.

Thus, suppose that $\boldsymbol{A}$ is such that $\det \boldsymbol{A} = 0$; equivalently, $\boldsymbol{Au} = \boldsymbol{0}$ for some non-zero $\boldsymbol{u}$. This is equivalent to the statement: $\boldsymbol{a} \cdot \boldsymbol{Au} = 0$ for arbitrary $\boldsymbol{a}$. Expanding this statement in an orthonormal basis $\{\boldsymbol{e}_i\}$, we have that

$$a_1\boldsymbol{e}_1 \cdot \boldsymbol{Au} + a_2\boldsymbol{e}_2 \cdot \boldsymbol{Au} + a_3\boldsymbol{e}_3 \cdot \boldsymbol{Au} = 0 \tag{3.65}$$

for arbitrary a_1, a_2, a_3. Therefore if, for example, we have solved the equations $\boldsymbol{e}_1 \cdot \boldsymbol{Au} = 0$ and $\boldsymbol{e}_2 \cdot \boldsymbol{Au} = 0$, then the equation $\boldsymbol{e}_3 \cdot \boldsymbol{Au} = 0$ is automatically satisfied and hence redundant.

Returning to the simple-shear problem, on observing that $\boldsymbol{Be}_3 = \boldsymbol{e}_3$ we have the spectral decomposition

$$\boldsymbol{B} = \lambda_1^2 \boldsymbol{v}_1 \otimes \boldsymbol{v}_1 + \lambda_1^{-2} \boldsymbol{v}_2 \otimes \boldsymbol{v}_2 + \boldsymbol{e}_3 \otimes \boldsymbol{e}_3 \tag{3.66}$$

of the left Cauchy–Green tensor, where

$$\boldsymbol{v}_1 = \cos\alpha \boldsymbol{e}_1 + \sin\alpha \boldsymbol{e}_2 \quad \text{and} \quad \boldsymbol{v}_2 = -\sin\alpha \boldsymbol{e}_1 + \cos\alpha \boldsymbol{e}_2 \tag{3.67}$$

for some angle α. To evaluate it we scalar multiply the equation $\boldsymbol{Bv}_1 = \lambda_1^2 \boldsymbol{v}_1$ by $\boldsymbol{e}_2$ and obtain

$$1 + \gamma \cot\alpha = \lambda_1^2. \tag{3.68}$$

Thus,

$$\tan\alpha = \frac{2}{\gamma + \sqrt{4+\gamma^2}} = \cot A; \quad \text{hence} \quad A = \frac{\pi}{2} - \alpha. \tag{3.69}$$

Accordingly, $\sin A = \cos\alpha$, $\cos A = \sin\alpha$ and $\boldsymbol{v}_1, \boldsymbol{v}_2$ are determined.

Finally, the rotation factor in the polar decomposition is obtained from (3.39):

$$R = e_3 \otimes \hat{E}_3 + \sin 2\alpha(e_1 \otimes \hat{E}_1 + e_2 \otimes \hat{E}_2) + \cos 2\alpha(e_1 \otimes \hat{E}_2 - e_2 \otimes \hat{E}_1). \tag{3.70}$$

As a check on this result, note that $\boldsymbol{R}$ rotates the right-handed orthonormal set $\{\hat{\boldsymbol{E}}_A\}$ to the right-handed orthonormal set $\{\boldsymbol{w}_A\}$, where

$$w_1 = \sin 2\alpha e_1 - \cos 2\alpha e_2, \quad w_2 = \cos 2\alpha e_1 + \sin 2\alpha e_2, \quad w_3 = e_3, \tag{3.71}$$

i.e., $\boldsymbol{w}_A = \boldsymbol{R}\hat{\boldsymbol{E}}_A$.

3.5.2 *Radial deformation of a right circular cylinder*

Suppose κ and κ_t are hollow right circular cylinders. We describe position in κ_t using the polar-coordinate parametrization (1.117). Thus,

$$x = re_r(\theta) + ze_3, \tag{3.72}$$

where $\boldsymbol{e}_r(\theta) = \cos\theta\boldsymbol{e}_1 + \sin\theta\boldsymbol{e}_2$. Position in κ is given similarly by

$$X = R\hat{E}_R(\Theta) + Z\hat{E}_3, \tag{3.73}$$

where $\hat{\boldsymbol{E}}_R(\Theta) = \cos\Theta\hat{\boldsymbol{E}}_1 + \sin\Theta\hat{\boldsymbol{E}}_2$ (Figure 3.3).

To specify the deformation we suppose that a circle $R = const.$ is mapped to a circle $r = const.$, and that the azimuthal and axial

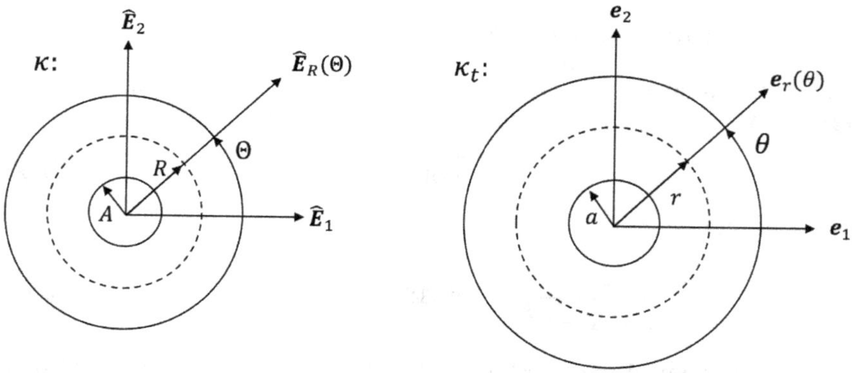

Figure 3.3. Annular cross-section of a right circular cylinder in κ and κ_t.

coordinates of a material point are unchanged; i.e., that $\theta = \Theta$ and $z = Z$. Thus, $r = f(R)$ for some function R. This function determines the deformation completely. An expression for the deformation gradient follows from

$$\begin{aligned} \boldsymbol{F}d\boldsymbol{X} &= d\boldsymbol{x} = dr\boldsymbol{e}_r(\theta) + rd\theta\boldsymbol{e}_\theta(\theta) + dz\boldsymbol{e}_3 \\ &= f'(R)dR\boldsymbol{e}_r(\theta) + f(R)d\Theta\boldsymbol{e}_\theta(\theta) + dZ\boldsymbol{e}_3 \\ &= f'(R)\boldsymbol{e}_r(\theta)[\hat{\boldsymbol{E}}_R(\Theta)\cdot d\boldsymbol{X}] + (f/R)\boldsymbol{e}_\theta(\theta)[\hat{\boldsymbol{E}}_\Theta(\Theta)\cdot d\boldsymbol{X}] \\ &\quad + \boldsymbol{e}_3[\hat{\boldsymbol{E}}_3\cdot d\boldsymbol{X}], \end{aligned} \tag{3.74}$$

where $\boldsymbol{e}_\theta(\theta) = \boldsymbol{e}_r'(\theta)$ and $\hat{\boldsymbol{E}}_\Theta(\Theta) = \hat{\boldsymbol{E}}_R'(\Theta)$ respectively are the azimuthal unit vectors in κ_t and κ. Recalling that $\Theta = \theta$ in this deformation, we arrive at

$$\boldsymbol{F} = f'(R)\boldsymbol{e}_r(\theta)\otimes\hat{\boldsymbol{E}}_R(\theta) + (f/R)\boldsymbol{e}_\theta(\theta)\otimes\hat{\boldsymbol{E}}_\Theta(\theta) + \boldsymbol{e}_3\otimes\hat{\boldsymbol{E}}_3. \tag{3.75}$$

Assuming κ to be occupiable, we have

$$\begin{aligned} 0 < J &= \det\boldsymbol{F} \\ &= [\boldsymbol{F}\hat{\boldsymbol{E}}_R, \boldsymbol{F}\hat{\boldsymbol{E}}_\Theta, \boldsymbol{F}\hat{\boldsymbol{E}}_3]/[\hat{\boldsymbol{E}}_R, \hat{\boldsymbol{E}}_\Theta, \hat{\boldsymbol{E}}_3] \\ &= [f'(R)\boldsymbol{e}_r, (f/R)\boldsymbol{e}_\theta, \boldsymbol{e}_3] \\ &= ff'/R. \end{aligned} \tag{3.76}$$

This implies that $f'(R) > 0$. Thus, if $R_2 > R_1$, then $f(R_2) > f(R_1)$. Concentric circles are mapped to concentric circles in such a way that the outer (resp., inner) circle in κ is mapped to the outer (resp., inner) circle in κ_t.

If the deformation is isochoric, then $J = 1$ and $(\frac{1}{2}f^2)' = R$. Recalling that $f(R) = r$, this integrates to

$$r^2 - a^2 = R^2 - A^2, \quad \text{where } a = f(A) \tag{3.77}$$

and A is a positive constant. Thus, the area of the annulus $[A, R]$ is preserved by the deformation. This, of course, is an immediate consequence of the fact that volume is conserved in the absence of axial deformation.

The transpose of $\boldsymbol{F}$ is

$$\boldsymbol{F}^t = f'(R)\hat{\boldsymbol{E}}_R(\theta) \otimes \boldsymbol{e}_r(\theta) + (f/R)\hat{\boldsymbol{E}}_\Theta(\theta) \otimes \boldsymbol{e}_\theta(\theta) + \hat{\boldsymbol{E}}_3 \otimes \boldsymbol{e}_3, \quad (3.78)$$

and the right- and left-Cauchy–Green tensors, $\boldsymbol{C} = \boldsymbol{F}^t\boldsymbol{F}$ and $\boldsymbol{B} = \boldsymbol{F}\boldsymbol{F}^t$, are

$$\boldsymbol{C} = (f')^2\hat{\boldsymbol{E}}_R(\theta) \otimes \hat{\boldsymbol{E}}_R(\theta) + (f/R)^2\hat{\boldsymbol{E}}_\Theta(\theta) \otimes \hat{\boldsymbol{E}}_\Theta(\theta) + \hat{\boldsymbol{E}}_3 \otimes \hat{\boldsymbol{E}}_3, \quad \text{and}$$
$$\boldsymbol{B} = (f')^2\boldsymbol{e}_r(\theta) \otimes \boldsymbol{e}_r(\theta) + (f/R)^2\boldsymbol{e}_\theta(\theta) \otimes \boldsymbol{e}_\theta(\theta) + \boldsymbol{e}_3 \otimes \boldsymbol{e}_3. \quad (3.79)$$

Clearly these are spectral representations, with principal stretches $f'(R)$, f/R and 1. It follows immediately that the right and left stretch tensors are

$$\boldsymbol{U} = f'\hat{\boldsymbol{E}}_R(\theta) \otimes \hat{\boldsymbol{E}}_R(\theta) + (f/R)\hat{\boldsymbol{E}}_\Theta(\theta) \otimes \hat{\boldsymbol{E}}_\Theta(\theta) + \hat{\boldsymbol{E}}_3 \otimes \hat{\boldsymbol{E}}_3, \quad \text{and}$$
$$\boldsymbol{V} = f'\boldsymbol{e}_r(\theta) \otimes \boldsymbol{e}_r(\theta) + (f/R)\boldsymbol{e}_\theta(\theta) \otimes \boldsymbol{e}_\theta(\theta) + \boldsymbol{e}_3 \otimes \boldsymbol{e}_3. \quad (3.80)$$

Finally, the rotation $\boldsymbol{R}$ in the polar factorization of $\boldsymbol{F}$ is

$$\boldsymbol{R} = \boldsymbol{F}\boldsymbol{U}^{-1} = \boldsymbol{V}^{-1}\boldsymbol{F} = \boldsymbol{e}_r(\theta) \otimes \hat{\boldsymbol{E}}_R(\theta) + \boldsymbol{e}_\theta(\theta) \otimes \hat{\boldsymbol{E}}_\Theta(\theta) + \boldsymbol{e}_3 \otimes \hat{\boldsymbol{E}}_3. \quad (3.81)$$

We introduce the shifted vectors $\boldsymbol{E}_R(\Theta) = \boldsymbol{1}\hat{\boldsymbol{E}}_R(\Theta)$, $\boldsymbol{E}_\Theta(\Theta) = \boldsymbol{1}\hat{\boldsymbol{E}}_\Theta(\Theta)$ and $\boldsymbol{E}_3 = \boldsymbol{1}\hat{\boldsymbol{E}}_3$. In terms of these the rotation is found, on making repeated use of Problem 17(c) of Chapter 1, to be

$$\boldsymbol{R} = [\boldsymbol{e}_r(\theta) \otimes \boldsymbol{E}_R(\theta) + \boldsymbol{e}_\theta(\theta) \otimes \boldsymbol{E}_\Theta(\theta) + \boldsymbol{e}_3 \otimes \boldsymbol{E}_3]\boldsymbol{1}. \quad (3.82)$$

The choices $\boldsymbol{E}_R(\theta) = \boldsymbol{e}_r(\theta)$, $\boldsymbol{E}_\Theta(\theta) = \boldsymbol{e}_\theta(\theta)$, $\boldsymbol{E}_3 = \boldsymbol{e}_3$ reduce the bracketed expression to $\boldsymbol{I}$, the spatial identity, yielding

$$\boldsymbol{R} = \boldsymbol{I}\boldsymbol{1} = \boldsymbol{1}. \quad (3.83)$$

3.5.3 *Azimuthal shear*

In this deformation of the cylinder the radial and axial coordinates remain unchanged, while the azimuthal coordinate changes by an amount that depends on R. Thus,

$$r = R, \quad \theta = \Theta + \phi(R) \quad \text{and} \quad z = Z, \tag{3.84}$$

for some function ϕ. Proceeding as in the previous example, we have

$$\boldsymbol{F}d\boldsymbol{X} = [\boldsymbol{e}_r(\theta) + R\phi'(R)\boldsymbol{e}_\theta(\theta)]dR + R\boldsymbol{e}_\theta(\theta)d\Theta + \boldsymbol{e}_3 dZ. \tag{3.85}$$

Thus,

$$\boldsymbol{F} = \boldsymbol{Q} + R\phi'(R)\boldsymbol{e}_\theta(\theta) \otimes \hat{\boldsymbol{E}}_R(\Theta), \tag{3.86}$$

where

$$\boldsymbol{Q} = \boldsymbol{e}_r(\theta) \otimes \hat{\boldsymbol{E}}_R(\Theta) + \boldsymbol{e}_\theta(\theta) \otimes \hat{\boldsymbol{E}}_\Theta(\Theta) + \boldsymbol{e}_3 \otimes \hat{\boldsymbol{E}}_3 \tag{3.87}$$

is a two-point rotation tensor. Then, with $\boldsymbol{e}_\theta(\theta) = \boldsymbol{Q}\hat{\boldsymbol{E}}_\Theta(\Theta)$,

$$\boldsymbol{F} = \boldsymbol{Q}\hat{\boldsymbol{G}}, \quad \text{where } \hat{\boldsymbol{G}} = \hat{\boldsymbol{I}} + R\phi'(R)\hat{\boldsymbol{E}}_\Theta(\Theta) \otimes \hat{\boldsymbol{E}}_R(\Theta). \tag{3.88}$$

Alternatively, with $\hat{\boldsymbol{E}}_R(\Theta) = \boldsymbol{Q}^t\boldsymbol{e}_r(\theta)$ we have

$$\boldsymbol{F} = \boldsymbol{G}\boldsymbol{Q}, \quad \text{where } \boldsymbol{G} = \boldsymbol{I} + R\phi'(R)\boldsymbol{e}_\theta(\theta) \otimes \boldsymbol{e}_r(\theta). \tag{3.89}$$

It is straightforward to show, using either representation, that the deformation is isochoric: $\det \boldsymbol{F} = 1$.

To aid in the interpretation of this deformation, we select a radial material curve in κ, with unit tangent $\boldsymbol{M} = \hat{\boldsymbol{E}}_R(\Theta)$. This is mapped by the deformation to the spiral curve with unit tangent $\boldsymbol{m}$ given by

$$\mu\boldsymbol{m} = \boldsymbol{F}\hat{\boldsymbol{E}}_R(\Theta) = \boldsymbol{e}_r(\theta) + R\phi'(R)\boldsymbol{e}_\theta(\theta); \quad \mu = \sqrt{1 + (R\phi')^2}. \tag{3.90}$$

The Cauchy–Green deformation tensors are

$$\boldsymbol{C} = \hat{\boldsymbol{G}}^t\boldsymbol{Q}^t\boldsymbol{Q}\hat{\boldsymbol{G}} = \hat{\boldsymbol{G}}^t\hat{\boldsymbol{I}}\hat{\boldsymbol{G}} = \hat{\boldsymbol{G}}^t\hat{\boldsymbol{G}} \quad \text{and}$$

$$\boldsymbol{B} = \boldsymbol{G}\boldsymbol{Q}\boldsymbol{Q}^t\boldsymbol{G}^t = \boldsymbol{G}\boldsymbol{I}\boldsymbol{G}^t = \boldsymbol{G}\boldsymbol{G}^t. \tag{3.91}$$

Thus,

$$\boldsymbol{C} = \hat{\boldsymbol{I}} + \gamma(R)[\hat{\boldsymbol{E}}_R(\Theta) \otimes \hat{\boldsymbol{E}}_\Theta(\Theta) + \hat{\boldsymbol{E}}_\Theta(\Theta) \otimes \hat{\boldsymbol{E}}_R(\Theta)] + \gamma^2(R)\hat{\boldsymbol{E}}_R(\Theta) \otimes \hat{\boldsymbol{E}}_R(\Theta) \tag{3.92}$$

and

$$\boldsymbol{B} = \boldsymbol{I} + \gamma(R)[\boldsymbol{e}_r(\theta) \otimes \boldsymbol{e}_\theta(\theta) + \boldsymbol{e}_\theta(\theta) \otimes \boldsymbol{e}_r(\theta))] + \gamma^2(R)\boldsymbol{e}_\theta(\theta) \otimes \boldsymbol{e}_\theta(\theta), \tag{3.93}$$

where $\gamma(R) = R\phi'(R)$. On replacing $\hat{\boldsymbol{E}}_1$ by $\hat{\boldsymbol{E}}_\Theta(\Theta)$, $\hat{\boldsymbol{E}}_2$ by $\hat{\boldsymbol{E}}_R(\Theta)$ and $\hat{\boldsymbol{E}}_3$ by $-\hat{\boldsymbol{E}}_3$, $\boldsymbol{C}$ is seen to have exactly the same form as its counterpart (3.53) in the simple shear problem. Likewise, $\boldsymbol{B}$ is seen to coincide with (3.54) if $\boldsymbol{e}_1$ is replaced by $\boldsymbol{e}_\theta(\theta)$ and $\boldsymbol{e}_2$ by $\boldsymbol{e}_r(\theta)$. Accordingly $\gamma(R)$ represents a non-uniform shear on the relevant axes. The right and left stretch tensors $\boldsymbol{U}$ and $\boldsymbol{V}$ are obtained as they were in Section 3.5.1 with the indicated replacements.

Further, it follows from (3.91) that the right polar factorization of $\hat{\boldsymbol{G}}$ and the left polar factorization of $\boldsymbol{G}$ are

$$\hat{\boldsymbol{G}} = \boldsymbol{R}_{\hat{G}}\boldsymbol{U} \quad \text{and} \quad \boldsymbol{G} = \boldsymbol{V}\boldsymbol{R}_G, \tag{3.94}$$

where $\boldsymbol{R}_{\hat{G}}$ and $\boldsymbol{R}_G$ respectively are referential and spatial rotation tensors. These are computed in the same way, in terms of the appropriate bases, as the rotation $\boldsymbol{R}$ of (3.70) was computed. On combining these with (3.88), (3.89) and the uniqueness of the polar decomposition, and invoking the easily proved rule that the product of rotations is a rotation, the rotation $\boldsymbol{R}$ in the polar factorization of $\boldsymbol{F}$ is found to be

$$\boldsymbol{R} = \boldsymbol{Q}\boldsymbol{R}_{\hat{G}} = \boldsymbol{R}_G\boldsymbol{Q}. \tag{3.95}$$

3.5.4 *Torsion*

Consider the deformation of a cylinder described by

$$r = R, \quad \theta = \Theta + \tau Z, \quad z = Z, \tag{3.96}$$

where τ is the constant twist, i.e., the rate of change of azimuth with respect to the axial coordinate. The effect of this deformation is to rotate a cross section $Z = const.$ through an angle that varies linearly with Z. Different sections are rotated by different amounts to produce a torsional deformation of the cylinder.

The usual procedure leads again to (3.88) and (3.89), but with $\hat{\boldsymbol{G}}$ and $\boldsymbol{G}$ now given respectively by

$$\hat{\boldsymbol{G}} = \hat{\boldsymbol{I}} + R\tau\hat{\boldsymbol{E}}_\Theta(\Theta) \otimes \hat{\boldsymbol{E}}_3 \quad \text{and} \quad \boldsymbol{G} = \boldsymbol{I} + R\tau\boldsymbol{e}_\theta(\theta) \otimes \boldsymbol{e}_3. \tag{3.97}$$

This is again a shearing deformation, with variable amount of shear $\gamma(R) = R\tau$.

This deformation maps a generator of the cylinder, with unit tangent $\boldsymbol{M} = \hat{\boldsymbol{E}}_3$, to a helix with unit tangent $\boldsymbol{m}$ given by

$$\mu\boldsymbol{m} = \boldsymbol{e}_3 + R\tau\boldsymbol{e}_\theta(\theta); \quad \mu = \sqrt{1 + (R\tau)^2}. \tag{3.98}$$

3.6 Homogeneous deformations

A deformation is *homogeneous* if its gradient is the same at every material point, i.e., if $\boldsymbol{F}$ is independent of $\boldsymbol{X}$. In this case we have

$$d\boldsymbol{x} = \boldsymbol{F}d\boldsymbol{X} = d(\boldsymbol{F}\boldsymbol{X}), \tag{3.99}$$

which integrates at once to furnish the general form

$$\boldsymbol{x} = \chi_\kappa(\boldsymbol{X}, t) = \boldsymbol{F}(t)\boldsymbol{X} + \boldsymbol{c}(t) \tag{3.100}$$

of a homogeneous deformation, where $\boldsymbol{c}(t) \in E^3$ is an arbitrary function of time. The simple shear problem of Section 3.5.1 affords an example of a homogeneous deformation, whereas the deformations considered in Sections 3.5.2–3.5.4 are not homogeneous.

For homogeneous deformations, it follows from (3.100) that

$$\chi_\kappa(\boldsymbol{Y}, t) = \chi_\kappa(\boldsymbol{X}, t) + \boldsymbol{F}(t)(\boldsymbol{Y} - \boldsymbol{X}), \tag{3.101}$$

whereas, for a general differential deformation, (1.127) and (1.128) imply that

$$\chi_\kappa(\boldsymbol{Y}, t) = \chi_\kappa(\boldsymbol{X}, t) + \boldsymbol{F}(\boldsymbol{X}, t)(\boldsymbol{Y} - \boldsymbol{X}) + o(\boldsymbol{Y} - \boldsymbol{X}). \tag{3.102}$$

Thus, every smooth deformation is approximately homogeneous in any sufficiently small neighborhood of a material point. Accordingly, homogeneous deformations offer insight into the local features of any such deformation. This is our main reason for studying them here.

3.6.1 *Some properties of homogeneous deformations*

(a) Material planes deform to planes; parallel planes map to parallel planes

Consider a fixed plane in κ with (uniform) unit normal $\boldsymbol{N}$. The differential equation of the plane is (Figure 3.4)

$$0 = d\boldsymbol{X} \cdot \boldsymbol{N} = d(\boldsymbol{X} \cdot \boldsymbol{N}), \tag{3.103}$$

which integrates immediately to furnish the equation of the plane:

$$\boldsymbol{X} \cdot \boldsymbol{N} = D. \tag{3.104}$$

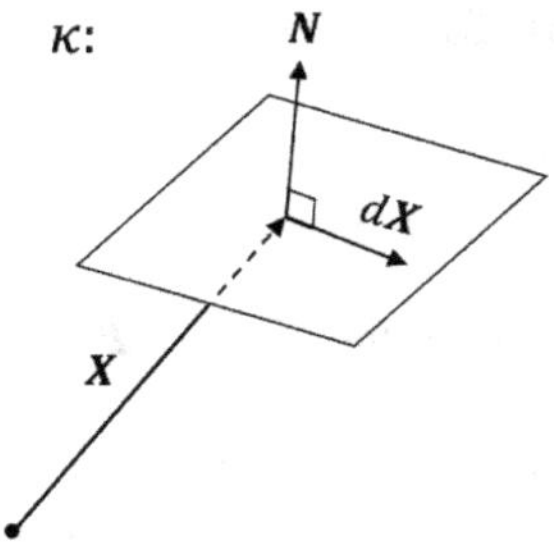

Figure 3.4. A material plane in κ.

Here D, a constant, is the perpendicular distance from the origin to the plane. Planes parallel to this plane are distinguished by the values of this constant; all have the same unit normal.

Substituting $\boldsymbol{X} = \boldsymbol{F}^{-1}(\boldsymbol{x} - \boldsymbol{c})$ from (3.100) gives

$$\boldsymbol{x} \cdot \boldsymbol{F}^{-t}\boldsymbol{N} = d, \quad \text{where } d = D + \boldsymbol{c} \cdot \boldsymbol{F}^{-t}\boldsymbol{N}. \tag{3.105}$$

The Piola–Nanson formulas (2.54) and (2.55) imply that this is the equation of a plane in κ_t with unit normal parallel to $\boldsymbol{F}^{-t}\boldsymbol{N}$. Further, parallel planes in κ map to parallel planes having a common normal in κ_t.

(b) Straight lines deform to straight lines; parallel lines map to parallel lines

Consider a fixed straight line in κ with (uniform) unit tangent $\boldsymbol{M}$. The differential equation of this line is (Figure 3.5)

$$d\boldsymbol{X} = \boldsymbol{M}dS = d(\boldsymbol{M}S), \tag{3.106}$$

where S is arclength along the line. This integrates to

$$\boldsymbol{X}(S) = \boldsymbol{M}S + \boldsymbol{X}_0, \quad \text{where } \boldsymbol{X}_0 = \boldsymbol{X}(0), \tag{3.107}$$

and substitution into (3.100) furnishes

$$\boldsymbol{x}(s) = s\boldsymbol{m} + \boldsymbol{x}_0, \tag{3.108}$$

where $s = \mu S$, with $\mu = |\boldsymbol{F}\boldsymbol{M}|$, measures arclength on the deformed line with uniform tangent $\boldsymbol{m} = \mu^{-1}\boldsymbol{F}\boldsymbol{M}$, and $\boldsymbol{x}_0 = \boldsymbol{F}\boldsymbol{X}_0 + \boldsymbol{c}$.

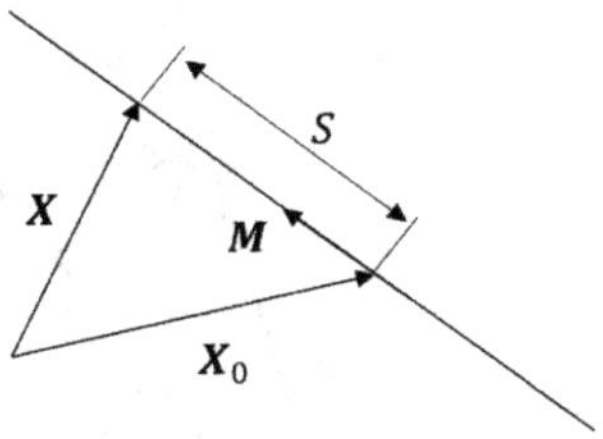

Figure 3.5. A material line in κ.

Clearly parallel lines, all having the same unit tangent $\boldsymbol{M}$, are mapped to parallel lines with unit tangent $\boldsymbol{m}$.

(c) A spherical material surface is mapped to the surface of an ellipsoid

Consider a unit sphere centered at $\boldsymbol{X}$ in κ. Let $\boldsymbol{Y}$ be the position of a point on the surface of the sphere. The radius vector from the center to this point is $\hat{\boldsymbol{r}} = \boldsymbol{Y} - \boldsymbol{X}$, and $|\hat{\boldsymbol{r}}| = 1$. After deformation the center and the point on the surface deform to $\boldsymbol{x} = \chi_\kappa(\boldsymbol{X}, t)$ and $\boldsymbol{y} = \chi_\kappa(\boldsymbol{Y}, t)$, respectively, and (3.100) yields

$$\boldsymbol{r} = \boldsymbol{F}\hat{\boldsymbol{r}}, \quad \text{where } \boldsymbol{r} = \boldsymbol{y} - \boldsymbol{x}. \tag{3.109}$$

We have

$$1 = \hat{\boldsymbol{r}} \cdot \hat{\boldsymbol{r}} = \boldsymbol{F}^{-1}\boldsymbol{r} \cdot \boldsymbol{F}^{-1}\boldsymbol{r} = \boldsymbol{r} \cdot \boldsymbol{F}^{-t}\boldsymbol{F}^{-1}\boldsymbol{r} = \boldsymbol{r} \cdot \boldsymbol{B}^{-1}\boldsymbol{r}, \tag{3.110}$$

where $\boldsymbol{B}$ is the left Cauchy–Green tensor.

Substituting the spectral representation

$$\boldsymbol{B}^{-1} = \sum \lambda_i^{-2} \boldsymbol{v}_i \otimes \boldsymbol{v}_i \tag{3.111}$$

results in

$$1 = (r_1/\lambda_1)^2 + (r_2/\lambda_2)^2 + (r_3/\lambda_3)^2, \tag{3.112}$$

where $r_i = \boldsymbol{v}_i \cdot \boldsymbol{r}$ are the coordinates of $\boldsymbol{y}$ relative to $\boldsymbol{x}$ in the basis $\{\boldsymbol{v}_i\}$. This is the equation of the surface of an ellipsoid with axes $\boldsymbol{v}_i$ and semi-axis lengths λ_i, centered at $\boldsymbol{x}$ (Figure 3.6).

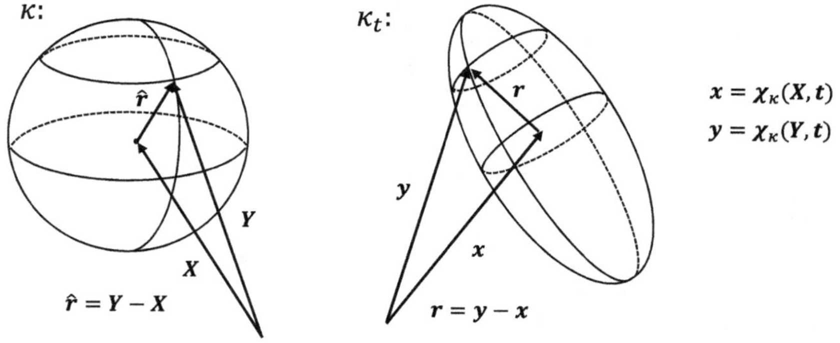

Figure 3.6. A sphere in κ is mapped to an ellipsoid in κ_t.

3.7 Rigid-body motions

A motion is *rigid* if it preserves the distance between every pair of material points $\boldsymbol{X}, \boldsymbol{Y} \in \kappa$, i.e., if $|\boldsymbol{\chi}_\kappa(\boldsymbol{Y}, t) - \boldsymbol{\chi}_\kappa(\boldsymbol{X}, t)| = |\boldsymbol{Y} - \boldsymbol{X}|$, or

$$[\boldsymbol{\chi}_\kappa(\boldsymbol{Y}, t) - \boldsymbol{\chi}_\kappa(\boldsymbol{X}, t)] \cdot [\boldsymbol{\chi}_\kappa(\boldsymbol{Y}, t) - \boldsymbol{\chi}_\kappa(\boldsymbol{X}, t)] = (\boldsymbol{Y} - \boldsymbol{X}) \cdot (\boldsymbol{Y} - \boldsymbol{X}) \tag{3.113}$$

for all $\boldsymbol{X}, \boldsymbol{Y}$ (Figure 3.7). This implies that

$$[\boldsymbol{\chi}_\kappa(\boldsymbol{Y}, t) - \boldsymbol{\chi}_\kappa(\boldsymbol{X}, t)] \cdot d[\boldsymbol{\chi}_\kappa(\boldsymbol{Y}, t) - \boldsymbol{\chi}_\kappa(\boldsymbol{X}, t)] = (\boldsymbol{Y} - \boldsymbol{X}) \cdot d(\boldsymbol{Y} - \boldsymbol{X}), \tag{3.114}$$

in which the derivative is with respect to $\boldsymbol{X}$ or $\boldsymbol{Y}$.

Fixing $\boldsymbol{X}$, we have

$$[\boldsymbol{\chi}_\kappa(\boldsymbol{Y}, t) - \boldsymbol{\chi}_\kappa(\boldsymbol{X}, t)] \cdot \boldsymbol{F}(\boldsymbol{Y}, t) d\boldsymbol{Y} = (\boldsymbol{Y} - \boldsymbol{X}) \cdot d\boldsymbol{Y} \tag{3.115}$$

for all $d\boldsymbol{Y}$, yielding

$$\boldsymbol{F}^t(\boldsymbol{Y}, t)[\boldsymbol{\chi}_\kappa(\boldsymbol{Y}, t) - \boldsymbol{\chi}_\kappa(\boldsymbol{X}, t)] = \boldsymbol{Y} - \boldsymbol{X}. \tag{3.116}$$

Fixing $\boldsymbol{Y}$ and taking the derivative with respect to $\boldsymbol{X}$, we arrive at

$$\boldsymbol{F}^t(\boldsymbol{Y}, t)\boldsymbol{F}(\boldsymbol{X}, t) = \hat{\boldsymbol{I}} \quad \text{for all } \boldsymbol{X}, \boldsymbol{Y} \in \kappa. \tag{3.117}$$

Thus,

$$\boldsymbol{F}^t(\boldsymbol{X}, t)\boldsymbol{F}(\boldsymbol{X}, t) = \hat{\boldsymbol{I}} \quad \text{for all } \boldsymbol{X} \in \kappa, \tag{3.118}$$

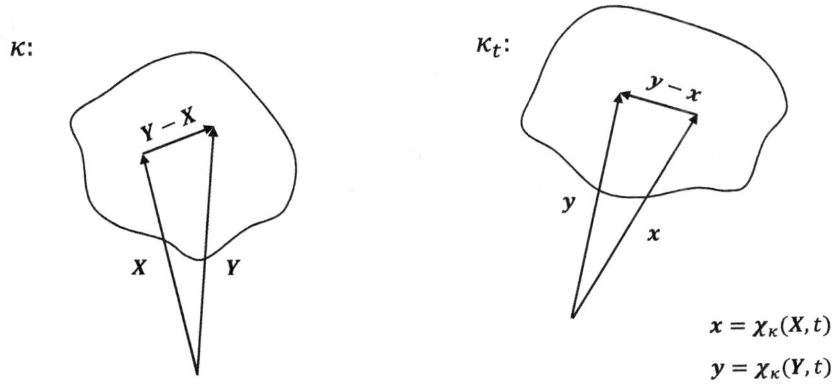

Figure 3.7. Rigid-body motion.

and $\boldsymbol{F}(\boldsymbol{X},t)$ is orthogonal at every $\boldsymbol{X}$. Therefore $\boldsymbol{F}^t(\boldsymbol{Y},t) = \boldsymbol{F}^{-1}(\boldsymbol{Y},t)$ and (3.117) gives

$$\boldsymbol{F}(\boldsymbol{Y},t) = \boldsymbol{F}(\boldsymbol{X},t) \quad \text{for all } \boldsymbol{X},\boldsymbol{Y} \in \kappa. \tag{3.119}$$

We conclude that $\boldsymbol{F}(\boldsymbol{X},t)$ is independent of $\boldsymbol{X}$, i.e.,

$$\boldsymbol{F}(\boldsymbol{X},t) = \boldsymbol{Q}(t), \tag{3.120}$$

where $\boldsymbol{Q}$ is a uniform orthogonal tensor. If κ is occupiable, then $\boldsymbol{Q}$ is a rotation.

It follows that rigid-body motions are homogeneous, i.e.,

$$\boldsymbol{x} = \chi_\kappa(\boldsymbol{X},t) = \boldsymbol{Q}(t)\boldsymbol{X} + \boldsymbol{c}(t). \tag{3.121}$$

Conversely, this implies that (3.113) is satisfied, and thus furnishes the necessary and sufficient condition for rigidity of the motion.

The polar decomposition of the deformation gradient is trivial in this case:

$$\boldsymbol{R} = \boldsymbol{Q}, \quad \boldsymbol{U} = \hat{\boldsymbol{I}} \quad \text{and} \quad \boldsymbol{V} = \boldsymbol{I}, \tag{3.122}$$

and, of course, $\boldsymbol{C} = \hat{\boldsymbol{I}}$ and $\boldsymbol{B} = \boldsymbol{I}$.

3.8 Problems

1. Consider the deformation defined, in terms of Cartesian coordinates, by $x_1 = \lambda_1 X_1$, $x_2 = \lambda_2 X_2$, $x_3 = \lambda_3 X_3$, where the λ_i are constants and $\{\boldsymbol{e}_i\} = \{\boldsymbol{E}_A\}$.

(a) Obtain the deformation gradient and state a restriction on the λ_i for this deformation to be physically possible.
(b) Consider the case in which $\lambda_3 = \lambda_2$ and suppose the deformation is isochoric. Consider a plane surface in the reference configuration with unit normal inclined at angle ϕ to the 1-axis. Find an expression, in terms of λ_1 and ϕ, for the ratio of the area of this material surface after deformation to that before deformation.
(c) What is the *unit* normal to this material surface after deformation?

2. Prove that the product of two orthogonal tensors is orthogonal, and that the product of two rotations is a rotation.
3. Evaluate the left and right Cauchy–Green deformation tensors for the torsion problem. Find the principal stretches in terms of the constant twist τ per unit length along the axis of the cylinder. Verify that azimuthal shear and torsion are isochoric.
4. Consider the static deformation $\boldsymbol{x} = \boldsymbol{\chi}_{\kappa}(\boldsymbol{X})$, where

$$\boldsymbol{\chi}_{\kappa}(\boldsymbol{X}) = \lambda(R)\mathbf{1}\boldsymbol{X}, \quad \text{where } R = |\boldsymbol{X}|$$

and $\lambda(R)$ is a given function.

(a) Describe this deformation in terms of what it does to a spherical annulus defined by $A \leq R \leq B$, centered at the origin, where A and B are positive constants.
(b) Derive an expression for the deformation gradient $\boldsymbol{F}$. What restriction(s) must be imposed on the function $\lambda(R)$ for this deformation to be physically possible?
(c) Find the function $\lambda(R)$ if the deformation is *isochoric* with $\lambda(A) = a/A$, where a is a positive constant.

5. Consider a deformation $\boldsymbol{X} \to \boldsymbol{x}$ of the form

$$\boldsymbol{x} = \boldsymbol{\chi}_{\kappa}(\boldsymbol{X}, t) = \mathbf{1}\boldsymbol{X} + w(R, t)\boldsymbol{e}_3, \quad \text{with} \quad \boldsymbol{X} = R\hat{\boldsymbol{E}}_R(\Theta) + Z\hat{\boldsymbol{E}}_3,$$

in a system of polar coordinates, where R, Θ and Z are the radial, azimuthal and axial coordinates of a material point in the reference configuration.

(a) Show that the deformation gradient $\boldsymbol{F}$ is

$$\boldsymbol{F} = \mathbf{1} + \gamma(R,t)\boldsymbol{e}_3 \otimes \hat{\boldsymbol{E}}_R(\Theta), \quad \text{where } \gamma(R,t) = \frac{\partial}{\partial R}w(R,t),$$

and derive expressions for the right and left Cauchy–Green deformation tensors $\boldsymbol{C}$ and $\boldsymbol{B}$. This deformation is called *telescopic shear*, or *anti-plane shear.* Justify this terminology.

(b) What is the stretch μ of the helical material curve that has unit tangent $\boldsymbol{M} = \frac{\sqrt{2}}{2}[\hat{\boldsymbol{E}}_\Theta(\Theta) + \hat{\boldsymbol{E}}_3]$ in the reference configuration? What is the unit tangent $\boldsymbol{m}$ to this material curve in the current configuration? Same question for the curve with $\boldsymbol{M} = \frac{\sqrt{2}}{2}[\hat{\boldsymbol{E}}_R(\Theta) + \hat{\boldsymbol{E}}_\Theta(\Theta)]$.

6. Consider the static deformation $\boldsymbol{X} \to \boldsymbol{x}$, with $\boldsymbol{X} = X_A\hat{\boldsymbol{E}}_A$ in terms of Cartesian coordinates, and

$$\boldsymbol{x} = \chi_\kappa(\boldsymbol{X}) = r\boldsymbol{e}_r(\theta) + z\boldsymbol{e}_3,$$

in terms of polar coordinates. Suppose that $r = f(X_1) > 0$, $\theta = g(X_2)$ and $z = Z = X_3$, where f and g are given functions.

(a) Describe the geometry, in the deformed configuration, of the vertical and horizontal material lines $X_1 = const.$ and $X_2 = const.$ in the reference configuration. Justify the interpretation of this deformation as a flexure of a block with edges parallel to the Cartesian coordinate lines in the reference configuration.

(b) Derive the expression

$$\boldsymbol{F} = f'(X_1)\boldsymbol{e}_r \otimes \hat{\boldsymbol{E}}_1 + f(X_1)g'(X_2)\boldsymbol{e}_\theta \otimes \hat{\boldsymbol{E}}_2 + \boldsymbol{e}_3 \otimes \hat{\boldsymbol{E}}_3$$

for the deformation gradient, where the primes are derivatives with respect to the indicated arguments.

(c) Show that the requirement $\det \boldsymbol{F} > 0$ is equivalent to

$$f(X_1)f'(X_1)g'(X_2) > 0.$$

(d) Recalling that $f(X_1) > 0$ and assuming that $g'(X_2) > 0$, obtain the factors $\boldsymbol{R}$, $\boldsymbol{U}$ and $\boldsymbol{V}$ in the polar decomposition *by inspection.* Find the general forms of the functions f and g if the deformation is isochoric.

7. Let $\boldsymbol{A}$ be a symmetric tensor.
   (a) Show that

$$I_1(\boldsymbol{A}) = \lambda_1 + \lambda_2 + \lambda_3, \quad I_2(\boldsymbol{A}) = \lambda_1\lambda_2 + \lambda_1\lambda_3 + \lambda_2\lambda_3 \quad \text{and}$$
$$I_3(\boldsymbol{A}) = \lambda_1\lambda_2\lambda_3,$$

   where $I_{1,2,3}(\boldsymbol{A})$ are the principal invariants of $\boldsymbol{A}$ and λ_i are the (real) principal values of $\boldsymbol{A}$.
   (b) Show that $\boldsymbol{A}$ satisfies the cubic characteristic equation for principal values, i.e.,

$$\boldsymbol{A}^3 - I_1(\boldsymbol{A})\boldsymbol{A}^2 + I_2(\boldsymbol{A})\boldsymbol{A} - I_3(\boldsymbol{A})\boldsymbol{I} = \boldsymbol{O},$$

   where $\boldsymbol{A}^2 = \boldsymbol{A}\boldsymbol{A}$ and $\boldsymbol{A}^3 = \boldsymbol{A}\boldsymbol{A}^2 = \boldsymbol{A}^2\boldsymbol{A}$. This is called the Cayley–Hamilton formula, and is also valid for non-symmetric tensors. This formula is meaningful only for tensors that map a vector space to itself. It is not valid for two-point tensors.
8. Let $\boldsymbol{A}$ be a symmetric tensor in two-dimensional space, i.e., a linear map from E^2 to itself. This has the spectral decomposition $\boldsymbol{A} = \sum_{i=1}^{2} \lambda_i \boldsymbol{u}_i \otimes \boldsymbol{u}_i$, where λ_i are the (real) principal values and $\boldsymbol{u}_i \in E^2$ are the corresponding orthonormal principal vectors. Prove the two-dimensional Cayley–Hamilton formula

$$\boldsymbol{A}^2 - (tr\boldsymbol{A})\boldsymbol{A} + (\det \boldsymbol{A})\boldsymbol{I} = \boldsymbol{O},$$

   where $\boldsymbol{I}$ is the *two-dimensional* identity for E^2.
9. Show that the two-dimensional right stretch tensor $\boldsymbol{U}$ and the two-dimensional right Cauchy–Green deformation tensor $\boldsymbol{C}$, both mapping $\hat{E}^2$ to itself, are related by

$$I\boldsymbol{U} = \boldsymbol{C} + J\hat{\boldsymbol{I}},$$

   where $I = tr\boldsymbol{U}$ and $J = \det \boldsymbol{U}$, and, of course, $\hat{\boldsymbol{I}}$ is the identity for $\hat{E}^2$. Show that $J^2 = \det \boldsymbol{C}$ and $I^2 = tr\boldsymbol{C} + 2J$. Thus, it is not necessary to compute the principal stretches and principal vectors to determine $\boldsymbol{U}$.

   Consider a simple-shear deformation in which the *two-dimensional* part of the deformation gradient is

$$\boldsymbol{F} = \mathbf{1} + \gamma \boldsymbol{e}_1 \otimes \hat{\boldsymbol{E}}_2,$$

where $\mathbf{1}$ is the shifter from $\hat{E}^2$ to E^2 and γ is the amount of shear. Show that

$$\boldsymbol{U} = \frac{1}{\sqrt{4+\gamma^2}}[2\hat{\boldsymbol{I}} + \gamma(\hat{\boldsymbol{E}}_1 \otimes \hat{\boldsymbol{E}}_2 + \hat{\boldsymbol{E}}_2 \otimes \hat{\boldsymbol{E}}_1) + \gamma^2 \hat{\boldsymbol{E}}_1 \otimes \hat{\boldsymbol{E}}_1],$$

and use this result to construct the three-dimensional right stretch tensor and rotation factor in the polar decomposition. Recover our earlier results for the principal stretches, and use an appropriate variant of the foregoing procedure to derive an expression for the left stretch $\boldsymbol{V}$ directly.

Use similar procedures to derive explicit formulas for the stretch and rotation tensors in azimuthal shear and torsion.

Chapter 4

Rate of Deformation

4.1 Velocity gradient, stretching and spin

Let $\boldsymbol{v}$ be the material velocity field. On recalling (2.19)–(2.21) and (2.37), its referential gradient is seen to be

$$Grad\boldsymbol{v} = \boldsymbol{L}\boldsymbol{F}, \quad \text{where } \boldsymbol{L} = grad\boldsymbol{v} \tag{4.1}$$

is its spatial gradient. In terms of Cartesian components,

$$\begin{aligned} Grad\boldsymbol{v} &= \hat{v}_{i,A}\boldsymbol{e}_i \otimes \hat{\boldsymbol{E}}_A \\ &= \frac{\partial}{\partial X_A}\left(\frac{\partial \chi_i}{\partial t}\right)\boldsymbol{e}_i \otimes \hat{\boldsymbol{E}}_A \\ &= \frac{\partial}{\partial t}\left(\frac{\partial \chi_i}{\partial X_A}\right)\boldsymbol{e}_i \otimes \hat{\boldsymbol{E}}_A \\ &= \dot{F}_{iA}\boldsymbol{e}_i \otimes \hat{\boldsymbol{E}}_A, \end{aligned} \tag{4.2}$$

where we have interchanged the order of differentiation. This is permissible provided that the functions $\chi_i(X_B, t)$ are twice continuously differentiable with respect to the X_B and t jointly. Thus, $Grad\boldsymbol{v} = \dot{\boldsymbol{F}}$ and (4.1) yields

$$\dot{\boldsymbol{F}} = \boldsymbol{L}\boldsymbol{F}. \tag{4.3}$$

It is of interest to relate the spatial velocity gradient to the rotation tensor and the stretch tensors, together with their material derivatives, in the polar decompositions of $\boldsymbol{F}$. To this end, we use (3.32) to obtain

$$\dot{\boldsymbol{F}} = \dot{\boldsymbol{R}}\boldsymbol{U} + \boldsymbol{R}\dot{\boldsymbol{U}} = \dot{\boldsymbol{V}}\boldsymbol{R} + \boldsymbol{V}\dot{\boldsymbol{R}}, \tag{4.4}$$

where we have invoked the product rule for tensors, which is easily proved by combining Cartesian tensor representations, for example, with the conventional product rule for ordinary functions. Accordingly,

$$\boldsymbol{L} = \dot{\boldsymbol{R}}\boldsymbol{R}^t + \boldsymbol{R}\dot{\boldsymbol{U}}\boldsymbol{U}^{-1}\boldsymbol{R}^t = \dot{\boldsymbol{V}}\boldsymbol{V}^{-1} + \boldsymbol{V}\dot{\boldsymbol{R}}\boldsymbol{R}^t\boldsymbol{V}^{-1}. \tag{4.5}$$

Let the reference configuration be the configuration κ_{t_0} occupied by the body at the instant t_0. Then, (2.79) implies, by the uniqueness of the polar factors, that $\boldsymbol{R} = \boldsymbol{1}$, $\boldsymbol{U} = \hat{\boldsymbol{I}}$ and $\boldsymbol{V} = \boldsymbol{I}$ at this instant, and therefore that

$$\boldsymbol{L} = \dot{\boldsymbol{R}}\boldsymbol{1}^t + \boldsymbol{1}\dot{\boldsymbol{U}}\boldsymbol{1}^t = \dot{\boldsymbol{V}} + \dot{\boldsymbol{R}}\boldsymbol{1}^t, \tag{4.6}$$

in which the derivatives are evaluated *at* time t_0. It is a simple matter to demonstrate, by differentiating $\boldsymbol{R}\boldsymbol{R}^t = \boldsymbol{I}$ and invoking the easily proved rule that the derivative commutes with transposition, i.e., $(\boldsymbol{R}^t)^{\cdot} = \dot{\boldsymbol{R}}^t$, that $\dot{\boldsymbol{R}}\boldsymbol{R}^t$ is a skew spatial tensor, and hence that $\dot{\boldsymbol{R}}\boldsymbol{1}^t$ is such a tensor *at* time t_0. Then, if we decompose $\boldsymbol{L}$ as a sum of its symmetric part $\boldsymbol{D}$ and its skew part $\boldsymbol{W}$, i.e.,

$$\boldsymbol{L} = \boldsymbol{D} + \boldsymbol{W}; \quad \boldsymbol{D} = Sym\boldsymbol{L} = \frac{1}{2}(\boldsymbol{L} + \boldsymbol{L}^t),$$
$$\boldsymbol{W} = Skw\boldsymbol{L} = \frac{1}{2}(\boldsymbol{L} - \boldsymbol{L}^t), \tag{4.7}$$

we conclude, by the uniqueness of the decomposition, that

$$\boldsymbol{D} = \boldsymbol{1}\dot{\boldsymbol{U}}\boldsymbol{1}^t = \dot{\boldsymbol{V}} \quad \text{and} \quad \boldsymbol{W} = \dot{\boldsymbol{R}}\boldsymbol{1}^t \tag{4.8}$$

at the instant t_0. Thus, $\boldsymbol{D}$ measures the rate of the left stretch tensor, and the shifted rate of the right stretch tensor, relative to a configuration occupied momentarily by the body, whereas $\boldsymbol{W}$ measures the shifted rate of rotation relative to the same configuration. These observations motivate calling $\boldsymbol{D}$ the *stretching tensor* — the suffix *-ing* connoting an ongoing process — and $\boldsymbol{W}$, the *spin tensor.*

In terms of Cartesian components,

$$\boldsymbol{D} = D_{ij}\boldsymbol{e}_i \otimes \boldsymbol{e}_j; \quad D_{ij} = \frac{1}{2}(v_{i,j} + v_{j,i}) \quad \text{and}$$

$$\boldsymbol{W} = W_{ij}\boldsymbol{e}_i \otimes \boldsymbol{e}_j; \quad W_{ij} = \frac{1}{2}(v_{i,j} - v_{j,i}). \tag{4.9}$$

The relationships, holding at arbitrary times, connecting the stretching and spin tensors to the factors in the polar decomposition follow from (4.5) and (4.7).

We can also connect the spatial velocity gradient to the material derivatives of the Cauchy–Green deformation tensors. For example,

$$\begin{aligned}\dot{\boldsymbol{C}} &= (\boldsymbol{F}^t\boldsymbol{F})^{\cdot} = (\boldsymbol{F}^t)^{\cdot}\boldsymbol{F} + \boldsymbol{F}^t\dot{\boldsymbol{F}} \\ &= \dot{\boldsymbol{F}}^t\boldsymbol{F} + \boldsymbol{F}^t\dot{\boldsymbol{F}} = (\boldsymbol{L}\boldsymbol{F})^t\boldsymbol{F} + \boldsymbol{F}^t\boldsymbol{L}\boldsymbol{F} \\ &= \boldsymbol{F}^t(\boldsymbol{L} + \boldsymbol{L}^t)\boldsymbol{F}. \end{aligned} \tag{4.10}$$

Thus,

$$\boldsymbol{F}^t\boldsymbol{D}\boldsymbol{F} = \frac{1}{2}\dot{\boldsymbol{C}} = \dot{\boldsymbol{E}}, \tag{4.11}$$

where $\boldsymbol{E}$ is the Lagrange strain. Similarly, $\dot{\boldsymbol{B}} = (\boldsymbol{F}\boldsymbol{F}^t)^{\cdot}$ yields

$$\dot{\boldsymbol{B}} = \boldsymbol{L}\boldsymbol{B} + \boldsymbol{B}\boldsymbol{L}^t. \tag{4.12}$$

Evaluating (4.11) at the instant t_0, we have that

$$\boldsymbol{D} = \boldsymbol{1}\dot{\boldsymbol{E}}\boldsymbol{1}^t, \tag{4.13}$$

where, of course, the derivative is evaluated at the same instant. We might therefore prefer to refer to $\boldsymbol{D}$ as the *straining tensor*.

In a rigid-body motion we have $\boldsymbol{F} = \boldsymbol{Q}(t)$, an orthogonal tensor. Then, $\boldsymbol{C} = \hat{\boldsymbol{I}}$ and $\boldsymbol{B} = \boldsymbol{I}$, and (4.11) or (4.12) imply that $\boldsymbol{D}$ vanishes in a rigid-body motion. Conversely, if $\boldsymbol{D}$ vanishes identically, then, using the fact that $\boldsymbol{L}$ is a gradient, it is possible to show (see [6]) that the motion is rigid. Thus, $\boldsymbol{D} = \boldsymbol{O}$ is the necessary and sufficient condition for rigidity of the motion.

4.2 Material curves

On taking the material derivative of (2.67), we have

$$\begin{aligned}\dot{\mu}\boldsymbol{m} + \mu\dot{\boldsymbol{m}} &= \dot{\boldsymbol{F}}\boldsymbol{M} \\ &= \boldsymbol{L}\boldsymbol{F}\boldsymbol{M} \\ &= \mu\boldsymbol{L}\boldsymbol{m}. \end{aligned} \tag{4.14}$$

Forming the inner product with $\boldsymbol{m}$, noting (because $\boldsymbol{m}\cdot\boldsymbol{m} = 1$) that $\boldsymbol{m}\cdot\dot{\boldsymbol{m}} = 0$, and (because $\boldsymbol{W}$ is skew) that $\boldsymbol{m}\cdot\boldsymbol{W}\boldsymbol{m} = 0$, we obtain

$$(\ln\mu)^{\cdot} = \boldsymbol{m}\cdot\boldsymbol{D}\boldsymbol{m}, \tag{4.15}$$

and substituting back into (4.14) gives

$$\dot{\boldsymbol{m}} = \boldsymbol{D}\boldsymbol{m} - (\boldsymbol{m}\cdot\boldsymbol{D}\boldsymbol{m})\boldsymbol{m} + \boldsymbol{W}\boldsymbol{m}. \tag{4.16}$$

Because $\boldsymbol{m}$ is a unit vector, any change in this vector must be due entirely to its reorientation; that is, to its spin. Evidently, then, the spin tensor $\boldsymbol{W}$ does not account fully for the spin of material vectors.

Suppose, however, that $\boldsymbol{m}$ coincides with a principal vector $\boldsymbol{\nu}$ of $\boldsymbol{D}$ at a particular instant t_*, say. Given $\boldsymbol{D}(\boldsymbol{x},t)$, there are infinitely many material curves having unit tangent $\boldsymbol{\nu}$ at $\boldsymbol{x}\in\kappa_{t_*}$. Let η be a corresponding principal value, i.e., $\boldsymbol{D}\boldsymbol{\nu} = \eta\boldsymbol{\nu}$. Then,

$$\eta = (\ln\mu)^{\cdot}_{|t_*}. \tag{4.17}$$

The principal values of $\boldsymbol{D}$ are thus the material derivatives of the logarithms of the stretches of material curves that are momentarily aligned with the principal vectors of $\boldsymbol{D}$. Moreover, at the instant t_*, we have $\boldsymbol{D}\boldsymbol{m} = \boldsymbol{D}\boldsymbol{\nu} = \eta\boldsymbol{\nu}$ and $(\boldsymbol{m}\cdot\boldsymbol{D}\boldsymbol{m})\boldsymbol{m} = (\boldsymbol{\nu}\cdot\boldsymbol{D}\boldsymbol{\nu})\boldsymbol{\nu} = \eta\boldsymbol{\nu}$, and (4.16) yields the conclusion that $\boldsymbol{W}$ characterizes the instantaneous spin of these same material vectors:

$$\dot{\boldsymbol{m}}_{|t_*} = \boldsymbol{W}\boldsymbol{m}_{|t_*}. \tag{4.18}$$

Naturally, at earlier and later times the convecting material curve in question will not, in general, be aligned with a principal vector of $\boldsymbol{D}$.

Consider two material vectors with unit tangents $\boldsymbol{M}_1$ and $\boldsymbol{M}_2$ at $\boldsymbol{X}\in\kappa$. According to (4.16),

$$\begin{aligned}\dot{\boldsymbol{m}}_1 &= \boldsymbol{L}\boldsymbol{m}_1 - (\boldsymbol{m}_1\cdot\boldsymbol{D}\boldsymbol{m}_1)\boldsymbol{m}_1 \quad \text{and}\\ \dot{\boldsymbol{m}}_2 &= \boldsymbol{L}\boldsymbol{m}_2 - (\boldsymbol{m}_2\cdot\boldsymbol{D}\boldsymbol{m}_2)\boldsymbol{m}_2.\end{aligned} \tag{4.19}$$

Let θ be the angle between $\boldsymbol{m}_1$ and $\boldsymbol{m}_2$ at $\boldsymbol{x}\in\kappa_t$: $\cos\theta = \boldsymbol{m}_1\cdot\boldsymbol{m}_2$. Then,

$$\begin{aligned}-(\sin\theta)\dot{\theta} &= \dot{\boldsymbol{m}}_1\cdot\boldsymbol{m}_2 + \boldsymbol{m}_1\cdot\dot{\boldsymbol{m}}_2\\ &= \boldsymbol{m}_2\cdot\boldsymbol{L}\boldsymbol{m}_1 - (\boldsymbol{m}_1\cdot\boldsymbol{D}\boldsymbol{m}_1)(\boldsymbol{m}_1\cdot\boldsymbol{m}_2)\\ &\quad +\boldsymbol{m}_1\cdot\boldsymbol{L}\boldsymbol{m}_2 - (\boldsymbol{m}_2\cdot\boldsymbol{D}\boldsymbol{m}_2)(\boldsymbol{m}_1\cdot\boldsymbol{m}_2).\end{aligned} \tag{4.20}$$

If $\boldsymbol{m}_1$ and $\boldsymbol{m}_2$ enclose a right angle at the instant t_*, say, then $\sin\theta = 1$ at this instant, and

$$\dot{\theta}_{|t_*} = -(\boldsymbol{m}_1 \cdot \boldsymbol{L}\boldsymbol{m}_2 + \boldsymbol{m}_1 \cdot \boldsymbol{L}^t\boldsymbol{m}_2) = -2\boldsymbol{m}_1 \cdot \boldsymbol{D}\boldsymbol{m}_2. \tag{4.21}$$

It follows that the off-diagonal components of the matrix of $\boldsymbol{D}$ relative to an orthonormal basis are proportional to the material derivatives of the angles between material curves that are momentarily aligned with the elements of the basis.

Recall that for any skew tensor $\boldsymbol{W}$, say, there is a vector $\boldsymbol{w}$ such that $\boldsymbol{W}\boldsymbol{u} = \boldsymbol{w} \times \boldsymbol{u}$ for arbitrary vectors $\boldsymbol{u}$ in the same vector space. This is the content of Problem 16 in Chapter 1. The vector $\boldsymbol{w}$ is called the *axial vector* of $\boldsymbol{W}$. If $\boldsymbol{W}$ is the spin tensor, its axial vector is the *vorticity* vector. In terms of Cartesian components,

$$\begin{aligned} w_i &= \frac{1}{2}e_{ijk}W_{kj} = \frac{1}{4}e_{ijk}(v_{k,j} - v_{j,k}) \\ &= \frac{1}{4}(e_{ijk}v_{k,j} + e_{ikj}v_{j,k}) \\ &= \frac{1}{4}(e_{ijk}v_{k,j} + e_{ijk}v_{k,j}) \\ &= \frac{1}{2}e_{jki}v_{k,j}, \end{aligned} \tag{4.22}$$

where, in passing from the first line to the second, we have used $e_{ijk} = -e_{ikj}$; in passing from the second line to the third, we have relabeled the dummy indices; and in passing from the third line to the fourth, we have used the cyclic symmetry $e_{ijk} = e_{kij} = e_{jki}$ of the permutation symbol. On comparing with (1.165), we conclude that

$$\boldsymbol{w} = \frac{1}{2}curl\boldsymbol{v}, \tag{4.23}$$

where *curl* is the spatial curl.

4.3 Example: A spinning disc

Consider a spinning disc with center located at the fixed position $\boldsymbol{x} \in \kappa_t$. Let $\boldsymbol{y}$ be the position of a point on the edge of the disc, and let $\boldsymbol{r} = \boldsymbol{y} - \boldsymbol{x}$. The motion is described by

$$\dot{\boldsymbol{r}} = \omega\boldsymbol{n} \times \boldsymbol{r}, \tag{4.24}$$

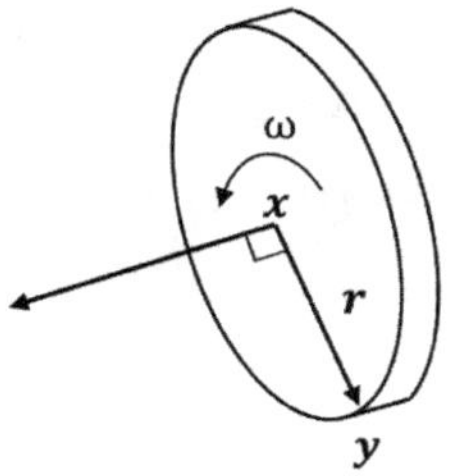

Figure 4.1. Spinning disc.

where ω is the constant rotation rate and $\boldsymbol{n}$ is a fixed unit normal to the disc (Figure 4.1).

With $\boldsymbol{v}(\boldsymbol{x}, t) = \boldsymbol{0}$ we have

$$\boldsymbol{v}(\boldsymbol{y}, t) - \boldsymbol{v}(\boldsymbol{x}, t) = \omega \boldsymbol{n} \times (\boldsymbol{y} - \boldsymbol{x}). \tag{4.25}$$

This is linear in $\boldsymbol{y} - \boldsymbol{x}$ and so there is a spatially uniform tensor $\boldsymbol{L}(t)$, the velocity gradient, such that $\boldsymbol{v}(\boldsymbol{y}, t) - \boldsymbol{v}(\boldsymbol{x}, t) = \boldsymbol{L}(t)(\boldsymbol{y} - \boldsymbol{x})$. Evidently $\boldsymbol{L}(t)$ is skew, i.e., $\boldsymbol{L}(t) = \boldsymbol{W}(t)$, and the associated vorticity vector is

$$\boldsymbol{w}(t) = \omega \boldsymbol{n}. \tag{4.26}$$

It is instructive to reconsider this problem in terms of Cartesian components. The component form of (4.25) is

$$v_i(\boldsymbol{y}, t) - v_i(\boldsymbol{x}, t) = \omega e_{ijk} n_j (y_k - x_k). \tag{4.27}$$

Thus, $L_{ij} = \omega e_{ijk} n_j$. Accordingly, $L_{ji} = -L_{ij}$, implying that $D_{ij} = 0$ and $W_{ij} = \omega e_{ijk} n_j$. Thus, the motion is rigid.

The vorticity vector is

$$\begin{aligned} \boldsymbol{w} = w_j \boldsymbol{e}_j &= \frac{1}{2} e_{jki} v_{i,k} \boldsymbol{e}_j \\ &= \frac{1}{2} \omega e_{jki} L_{ik} \boldsymbol{e}_j = \frac{1}{2} \omega e_{jki} e_{lki} n_l \boldsymbol{e}_j \\ &= \omega \delta_{jl} n_l \boldsymbol{e}_j = \omega n_j \boldsymbol{e}_j, \end{aligned} \tag{4.28}$$

where, in passing from the second line to the third we have invoked one of the $e - \delta$ identities together with the cyclic symmetry of the permutation symbol. Thus, we confirm (4.26).

4.4 Rigid-body motions

Recall that a general rigid-body motion is characterized by (3.121) in which $\boldsymbol{Q}(t)$ is orthogonal; it is a rotation if κ is occupiable. The deformation gradient is $\boldsymbol{Q}$, and the spatial velocity gradient is

$$\boldsymbol{L} = \dot{\boldsymbol{Q}}\boldsymbol{Q}^t. \tag{4.29}$$

This is skew; therefore, $\boldsymbol{D} = \boldsymbol{O}$ and $\boldsymbol{W} = \dot{\boldsymbol{Q}}\boldsymbol{Q}^t$. The spin is spatially uniform. To obtain the velocity field, we integrate

$$d\boldsymbol{v} = \boldsymbol{L}d\boldsymbol{x} = \boldsymbol{W}(t)d\boldsymbol{x} = d(\boldsymbol{W}\boldsymbol{x}), \tag{4.30}$$

obtaining

$$\boldsymbol{v}(\boldsymbol{x}, t) = \boldsymbol{W}(t)\boldsymbol{x} + \boldsymbol{d}(t) = \boldsymbol{w}(t) \times \boldsymbol{x} + \boldsymbol{d}(t), \tag{4.31}$$

where $\boldsymbol{d}(t) \in E^3$ is an arbitrary spatially uniform vector. Thus,

$$\boldsymbol{v}(\boldsymbol{y}, t) - \boldsymbol{v}(\boldsymbol{x}, t) = \boldsymbol{w}(t) \times (\boldsymbol{y} - \boldsymbol{x}). \tag{4.32}$$

Clearly the rigidly spinning disc is a special case.

Alternatively, we may differentiate (3.121) directly and obtain

$$\boldsymbol{v}(\boldsymbol{x}, t) = \dot{\boldsymbol{Q}}\boldsymbol{X} + \dot{\boldsymbol{c}} = \dot{\boldsymbol{Q}}\boldsymbol{Q}^t(\boldsymbol{x} - \boldsymbol{c}) + \dot{\boldsymbol{c}}, \tag{4.33}$$

which is just (4.31) with $\boldsymbol{d} = \dot{\boldsymbol{c}} - \boldsymbol{W}\boldsymbol{c}$.

4.5 Problems

1. (a) Prove the product rule for the time derivatives of tensors.
   (b) Show that time differentiation commutes with transposition.
   (c) Show that if $\boldsymbol{R}$ is orthogonal, then $\dot{\boldsymbol{R}}\boldsymbol{R}^t$ is skew.
   (d) Show that time differentiation of an invertible tensor does not commute with inversion. Obtain the derivative of the inverse of a tensor in terms of the tensor and its time derivative.
2. Recall the definition of $J = \det \boldsymbol{F}$, i.e., $J\boldsymbol{u} \cdot \boldsymbol{v} \times \boldsymbol{w} = \boldsymbol{F}\boldsymbol{u} \cdot \boldsymbol{F}\boldsymbol{v} \times \boldsymbol{F}\boldsymbol{w}$. Further, recall the definition of the trace: $(tr\boldsymbol{A})\, \boldsymbol{u} \cdot \boldsymbol{v} \times \boldsymbol{w} = \boldsymbol{A}\boldsymbol{u} \cdot \boldsymbol{v} \times \boldsymbol{w} + \boldsymbol{u} \cdot \boldsymbol{A}\boldsymbol{v} \times \boldsymbol{w} + \boldsymbol{u} \cdot \boldsymbol{v} \times \boldsymbol{A}\boldsymbol{w}$.

(a) Take the material derivative of the expression for $\det \boldsymbol{F}$, for *fixed* $\boldsymbol{u}, \boldsymbol{v}, \boldsymbol{w}$, and use the expression for the trace, to show that $\dot{J}/J = tr\boldsymbol{L}$, where $\boldsymbol{L} = \dot{\boldsymbol{F}}\boldsymbol{F}^{-1}$ is the spatial velocity gradient.

(b) Use the chain rule in the form $\dot{J} = (\partial J/\partial F_{iA})\dot{F}_{iA}$, and compare with the result of part (a) to conclude that $\partial J/\partial F_{iA} = F^*_{iA}$, where $\boldsymbol{F}^*$ is the cofactor of $\boldsymbol{F}$.

3. In the spatial description the temperature distribution in a body is $\theta(\boldsymbol{x}, t) = (A\cos t)/r$, where $r^2 = x_1^2 + x_2^2$, x_1 and x_2 are Cartesian spatial coordinates, and A is a constant. The velocity field is $\boldsymbol{v} = v_i\boldsymbol{e}_i$, where $v_1 = x_1^2x_2 + x_3^2$, $v_2 = -x_1^3 - x_1x_2^2$, $v_3 = 0$. Find the material derivative $\dot{\theta}$ as a function of $\boldsymbol{x}$ and t.
4. Show that $\dot{\boldsymbol{v}} = \boldsymbol{v}' + \frac{1}{2}grad(|\boldsymbol{v}|^2) + 2\boldsymbol{w} \times \boldsymbol{v}$, where $\boldsymbol{w}$ is the vorticity vector.
5. The components of the velocity field are given, relative to $\{\boldsymbol{e}_i\}$ and in the spatial description, by

$$v_1 = ax_2x_3, \quad v_2 = -ax_1x_3, \quad v_3 = bx_3,$$

where a, b are constants. Find:

(a) the spatial velocity gradient $\boldsymbol{L}$,
(b) the associated tensors $\boldsymbol{D}, \boldsymbol{W}$,
(c) the vorticity vector $\boldsymbol{w}$,
(d) restrictions on the constants a, b needed to ensure that the velocity field is

(i) isochoric, and
(ii) 'irrotational', i.e., $\boldsymbol{w} = \boldsymbol{0}$.

6. Consider a motion with deformation gradient $\boldsymbol{F} = \lambda\boldsymbol{e}_1 \otimes \hat{\boldsymbol{E}}_1 + \lambda^{-1/2}(\boldsymbol{e}_2 \otimes \hat{\boldsymbol{E}}_2 + \boldsymbol{e}_3 \otimes \hat{\boldsymbol{E}}_3)$, where λ is a function of t.

(a) Show that the deformation is isochoric.
(b) Show that $\boldsymbol{F}^{-1} = \lambda^{-1}\hat{\boldsymbol{E}}_1 \otimes \boldsymbol{e}_1 + \lambda^{1/2}(\hat{\boldsymbol{E}}_2 \otimes \boldsymbol{e}_2 + \hat{\boldsymbol{E}}_3 \otimes \boldsymbol{e}_3)$.
(c) Find $\boldsymbol{L}$ and show that $\boldsymbol{W} = \boldsymbol{O}$.
(d) Let $\boldsymbol{m}$ be the unit tangent to a material curve in the deformed configuration at a particular material point. This may be specified in terms of two angles θ and ϕ; i.e., $\boldsymbol{m} = \cos\phi\,\boldsymbol{e}_r(\theta) + \sin\phi\,\boldsymbol{e}_3$, where $\boldsymbol{e}_r(\theta) = \cos\theta\,\boldsymbol{e}_1 + \sin\theta\,\boldsymbol{e}_2$. Find expressions for $\dot{\theta}$ and $\dot{\phi}$ in terms of λ, $\dot{\lambda}$, θ and ϕ. This problem should convince you that $\boldsymbol{W}$ does not characterize

the spin of the material in general ($\boldsymbol{W} = \boldsymbol{O}$ but $\dot{\boldsymbol{m}} \neq \boldsymbol{0}$.) Exceptionally, there are certain angles θ and ϕ for which $\dot{\boldsymbol{m}} = \boldsymbol{0}$. Can you identify them?

7. Consider a material surface in κ_t with unit normal $\boldsymbol{n}$. Let α be the local ratio of current area to initial area of the material surface (see (2.54)).

   (a) Show that $\alpha > 0$.
   (b) Establish the relations $\dot{\alpha}/\alpha = tr\boldsymbol{D} - \boldsymbol{n} \cdot \boldsymbol{D}\boldsymbol{n}$ and $\dot{\boldsymbol{n}} = (\boldsymbol{n} \cdot \boldsymbol{D}\boldsymbol{n})\boldsymbol{n} - \boldsymbol{L}^t\boldsymbol{n}$.

8. Consider a motion in which the principal vectors $\boldsymbol{u}_i$ of $\boldsymbol{U}$ are fixed in the body, i.e., $\dot{\boldsymbol{u}}_i = \boldsymbol{0}$. Show that

$$\boldsymbol{D} = \sum (\ln \lambda_i)^{\cdot} \boldsymbol{v}_i \otimes \boldsymbol{v}_i$$

   in such a motion, where $\boldsymbol{v}_i$ are the principal vectors of $\boldsymbol{V}$.

9. Consider the velocity field $\boldsymbol{v} = (\theta/\theta_0)W\boldsymbol{e}_3$, where θ is the azimuthal coordinate in a spatial polar coordinate system and W, θ_0 are constants.

   (a) Derive expressions for the spatial velocity gradient $\boldsymbol{L}$, the stretching $\boldsymbol{D}$ and the spin $\boldsymbol{W}$.
   (b) Determine the principal values and vectors of $\boldsymbol{D}$, and the vorticity $\boldsymbol{w}$.

10. In the spatial description the velocity field is $\boldsymbol{v} = rf(r)\boldsymbol{e}_\theta$, in terms of cylindrical polar coordinates. Position in this coordinate system is given by

$$\boldsymbol{x} = r\boldsymbol{e}_r(\theta) + z\boldsymbol{e}_3.$$

   Assuming $f(r)$ to be a known function, show that

$$\boldsymbol{L} = \left[\frac{\partial}{\partial r}(rf)\right] \boldsymbol{e}_\theta \otimes \boldsymbol{e}_r - f\boldsymbol{e}_r \otimes \boldsymbol{e}_\theta,$$

   and derive expressions for stretching tensor $\boldsymbol{D}$ and the spin tensor $\boldsymbol{W}$.

Chapter 5

Transport Formulas

The basic balance laws of continuum physics typically take the form of integral statements concerning the various fields involved. These are combined with the localization theorem to arrive at local forms of these balance laws. In this chapter, we survey the basic structure of generic integral statements as a prelude to their application to balance laws in the next chapter.

5.1 The divergence theorem and its variants

The divergence theorem for smooth vector fields was mentioned in Section 1.2.7. Thus, for a vector field $\boldsymbol{a}(\boldsymbol{x})$ defined on a volume $R \subset \mathcal{E}^3$, bounded by a piecewise smooth surface ∂R,

$$\int_R div\boldsymbol{a}dv = \int_{\partial R} \boldsymbol{a} \cdot \boldsymbol{n}da, \tag{5.1}$$

where $\boldsymbol{n}$ is the piecewise continuous exterior unit-normal field to the bounding surface. In terms of Cartesian coordinates and components, this takes the form

$$\int_R a_{i,i}dv = \int_{\partial R} a_i n_i da. \tag{5.2}$$

Let $\phi(\boldsymbol{x})$ be a scalar field and let $\boldsymbol{a} = \phi\boldsymbol{c}$, where $\boldsymbol{c}$ is an arbitrary spatially uniform vector. Then, $a_i = \phi c_i$, and $\boldsymbol{c} \cdot \boldsymbol{u} = 0$, where

$\boldsymbol{u} = u_i \boldsymbol{e}_i$ is the spatially uniform vector with components

$$u_i = \int_R \phi_{,i} dv - \int_{\partial R} \phi n_i da. \tag{5.3}$$

As $\boldsymbol{c}$ is arbitrary, we may choose $\boldsymbol{c} = \boldsymbol{u}$ to obtain the necessary (and sufficient) condition $\boldsymbol{u} = \boldsymbol{0}$, i.e.,

$$\int_R \phi_{,i} dv = \int_{\partial R} \phi n_i da, \tag{5.4}$$

valid for any smooth scalar field ϕ.

In the same way, let $\boldsymbol{A}(\boldsymbol{x})$ be a smooth tensor field and let $\boldsymbol{a} = \boldsymbol{A}^t \boldsymbol{c}$, where $\boldsymbol{c}$ is again an arbitrary spatially uniform vector. Thus, $a_j = c_i A_{ij}$, and $\boldsymbol{c} \cdot \boldsymbol{u} = 0$, where $\boldsymbol{u}$ is now the vector with components

$$u_i = \int_R A_{ij,j} dv - \int_{\partial R} A_{ij} n_j da. \tag{5.5}$$

The arbitrariness of $\boldsymbol{c}$ again yields $\boldsymbol{u} = \boldsymbol{0}$, furnishing

$$\int_R A_{ij,j} dv = \int_{\partial R} A_{ij} n_j da \tag{5.6}$$

for any smooth tensor field.

We can express (5.4) and (5.6) respectively in the symbolic forms

$$\int_R grad\phi dv = \int_{\partial R} \phi \boldsymbol{n} da \quad \text{and} \quad \int_R div \boldsymbol{A} dv = \int_{\partial R} \boldsymbol{A}\boldsymbol{n} da, \tag{5.7}$$

with an important caveat, however. In physical applications we intend that integral statements such as these should have an invariant character. That is, we intend that the objects $\boldsymbol{u}$ that we have defined should be vectors, and hence independent of any basis. For example, if u_i' are the components of $\boldsymbol{u}$ relative to the fixed orthonormal basis $\{\boldsymbol{e}_i'\}$, we should require, with reference to Section 1.1.2 and Problem 8 of Chapter 1, that $u_i = a_{ij} u_j'$, where $a_{ij} = \boldsymbol{e}_i \cdot \boldsymbol{e}_j'$. Then, with $\boldsymbol{x} = x_i \boldsymbol{e}_i = x_j' \boldsymbol{e}_j'$ and hence $x_j' = a_{ij} x_i$, and with $\phi'_{,j} = \partial \phi / \partial x_j'$,

(5.3) may be written

$$\begin{aligned} u_i &= \int_R a_{ij}\phi'_{,j} dv - \int_{\partial R} a_{ij} n'_j da \\ &= a_{ij}\left(\int_R \phi'_{,j} dv - \int_{\partial R} n'_j da\right), \end{aligned} \tag{5.8}$$

yielding $u_i = a_{ij}u'_j$, as required. Accordingly, the statement $\boldsymbol{u} = \boldsymbol{0}$ has the intended invariance. That is to say, the statement $u'_i = 0$ implies that $u_i = 0$, and vice versa. Crucially, this conclusion relies on the uniformity of the components a_{ij} of the transformation matrix. In contrast, if the integrands of u'_j were replaced by components of $grad\phi$ and $\boldsymbol{n}$ relative to a polar coordinate system, for example (cf. (1.122)), then the transformation to a Cartesian system, say, would involve a non-uniform transformation matrix and would not yield an invariant integral statement. Similar remarks apply to (5.5). Accordingly, Eqs. $(5.7)_{1,2}$ have only a symbolic significance. Of course (5.1) is valid as it stands because the integrands therein, being scalar fields, are automatically invariant.

5.1.1 *Example: The Piola identity*

Consider an arbitrary subset $S \subset B$ of a material body. This subset occupies the volumes $P_t \subset \kappa_t$ and $P \subset \kappa$, where $P_t = \chi(S, t)$ and $P \subset \boldsymbol{\kappa}(S)$ respectively are the images of S in the current and reference configurations of the body, and, of course, $P_t = \chi_\kappa(P, t)$. Applying $(5.7)_1$ to P_t with $\phi = 1$, we obtain

$$\boldsymbol{0} = \int_{\partial P_t} \boldsymbol{n} da = \int_{\partial P} \boldsymbol{F}^* \boldsymbol{N} dA = \int_P Div \boldsymbol{F}^* dV, \tag{5.9}$$

where $\boldsymbol{F}^*$ is the cofactor of the deformation gradient, $\boldsymbol{N}$ is the exterior unit normal field on ∂P, the Piola–Nanson formula (2.54) has been invoked, and the referential form of the divergence formula $(5.7)_2$ has been used. Here, Div is the referential divergence operator, i.e., the divergence with respect to $\boldsymbol{X} \in \kappa$. In terms of Cartesian coordinates and components,

$$Div \boldsymbol{F}^* = F^*_{iA,A} \boldsymbol{e}_i. \tag{5.10}$$

The arbitrariness of S implies that P is an arbitrary subvolume of κ. Accordingly, applying the localization theorem, we arrive at the Piola identity (see Problem 2 of Chapter 2)

$$Div\boldsymbol{F}^* = \boldsymbol{0} \quad \text{at all } \boldsymbol{X} \in \kappa. \tag{5.11}$$

5.2 Reynolds' transport formula

Typical balance laws involve the time rate of change of a quantity of a given field pertaining to a subset $S \subset B$ of a material body. Consider a scalar field ϕ defined on the body, representing the volumetric density of the considered quantity. Thus, $\phi = \hat{\phi}(\boldsymbol{X}, t) = \tilde{\phi}(\boldsymbol{x}, t)$, with $\boldsymbol{X} \in \kappa$ and $\boldsymbol{x} = \chi_\kappa(\boldsymbol{X}, t) \in \kappa_t$. We seek a formula for the time derivative

$$\frac{d}{dt}\int_{P_t} \tilde{\phi}(\boldsymbol{x}, t)dv \quad \text{at fixed } S, \tag{5.12}$$

i.e., the derivative associated with a fixed set of material points.

Here, both the integrand and the domain of integration vary with time. To simplify matters we first convert the integral over P_t to an integral over the fixed domain P. Thus,

$$\begin{aligned}\frac{d}{dt}\int_{P_t} \tilde{\phi}(\boldsymbol{x}, t)dv &= \frac{d}{dt}\int_P \hat{\phi}(\boldsymbol{X}, t)JdV \\ &= \int_P (\phi J)^{\cdot} dV \\ &= \int_P (\dot{\phi}J + \phi\dot{J})dV,\end{aligned} \tag{5.13}$$

where we have interchanged differentiation and integration, this being permissible if the integrand ϕJ and its derivative $(\phi J)^{\cdot}$ are continuous functions of $\boldsymbol{X}$ and t and bounded above by integrable functions of $\boldsymbol{X}$ in P [2]. The formula $\dot{J} = Jtr\boldsymbol{L}$ (see Problem 2 of Chapter 4), where $\boldsymbol{L} = grad\boldsymbol{v}$ is the spatial velocity gradient, together with $tr(grad\boldsymbol{v}) = div\boldsymbol{v}$, then furnishes

$$\frac{d}{dt}\int_{P_t} \phi dv = \int_{P_t} (\dot{\phi} + \phi div\boldsymbol{v})dv. \tag{5.14}$$

Finally, we substitute $\dot{\phi} = \phi' + grad\phi \cdot \boldsymbol{v}$, where ϕ' is the time derivative at fixed $\boldsymbol{x}$, together with $grad\phi \cdot \boldsymbol{v} + \phi div\boldsymbol{v} = div(\phi\boldsymbol{v})$

(see Problem 24(c) in Chapter 1) and the divergence theorem, to arrive at Reynolds' formula

$$\frac{d}{dt}\int_{P_t}\phi dv = \int_{P_t}\phi' dv + \int_{\partial P_t}\phi \boldsymbol{v}\cdot\boldsymbol{n}da. \tag{5.15}$$

As an illustration of this formula, we apply it with $\phi = 1$ to conclude that the rate of change of a material volume is given by the flux of the material velocity through its boundary:

$$\frac{d}{dt}\int_{P_t} dv = \int_{\partial P_t}\boldsymbol{v}\cdot\boldsymbol{n}da. \tag{5.16}$$

5.3 Extensions and generalizations

5.3.1 *Reynolds' formula for vector fields*

Let $\phi = u_i$, a component of a vector field $\boldsymbol{u}$ with respect to the fixed orthonormal basis $\{\boldsymbol{e}_i\}$. Then,

$$\frac{d}{dt}\int_{P_t} u_i dv = \int_{P_t} u_i' dv + \int_{\partial P_t} u_i\boldsymbol{v}\cdot\boldsymbol{n}da; \quad i\in\{1,2,3\}. \tag{5.17}$$

Multiplying by the fixed basis vectors $\boldsymbol{e}_i$ and summing, we obtain a vector equation which we express, with the same caveat noted above, in the symbolic form

$$\frac{d}{dt}\int_{P_t}\boldsymbol{u}dv = \int_{P_t}\boldsymbol{u}' dv + \int_{\partial P_t}(\boldsymbol{u}\otimes\boldsymbol{v})\boldsymbol{n}da. \tag{5.18}$$

5.3.2 *The flux derivative*

In applications involving electromagnetic phenomena the formula

$$\frac{d}{dt}\int_{\partial P_t}\boldsymbol{u}\cdot\boldsymbol{n}da = \int_{\partial P_t}\boldsymbol{u}^*\cdot\boldsymbol{n}da \tag{5.19}$$

is often useful, where $\boldsymbol{u}^*$ is called the *flux derivative* of the field $\boldsymbol{u}$. Here the time derivative again pertains to a fixed set of material points occupying the evolving material surface ∂P_t. To establish this formula and derive an expression for $\boldsymbol{u}^*$, we again transform to a fixed domain prior to taking the derivative. Thus, on making use of the Piola–Nanson formula,

$$\begin{aligned}\frac{d}{dt}\int_{\partial P_t}\boldsymbol{u}\cdot\boldsymbol{n}da &= \frac{d}{dt}\int_{\partial P}\boldsymbol{u}\cdot\boldsymbol{F}^*\boldsymbol{N}dA\\ &= \frac{d}{dt}\int_{\partial P}J\boldsymbol{F}^{-1}\boldsymbol{u}\cdot\boldsymbol{N}dA\\ &= \int_{\partial P}(J\boldsymbol{F}^{-1}\boldsymbol{u})^{\cdot}\cdot\boldsymbol{N}dA\\ &= \int_{\partial P}[J\boldsymbol{F}^{-1}\dot{\boldsymbol{u}}+\dot{J}\boldsymbol{F}^{-1}\boldsymbol{u}+J(\boldsymbol{F}^{-1})^{\cdot}\boldsymbol{u}]\cdot\boldsymbol{N}dA.\end{aligned} \tag{5.20}$$

To reduce this to a manageable form we differentiate $\boldsymbol{F}\boldsymbol{F}^{-1}=\boldsymbol{I}$ to find that $\boldsymbol{F}(\boldsymbol{F}^{-1})^{\cdot}=-\dot{\boldsymbol{F}}\boldsymbol{F}^{-1}=-\boldsymbol{L}$, and hence that $(\boldsymbol{F}^{-1})^{\cdot}=-\boldsymbol{F}^{-1}\boldsymbol{L}$. Then,

$$\frac{d}{dt}\int_{\partial P_t}\boldsymbol{u}\cdot\boldsymbol{n}da=\int_{\partial P}[\dot{\boldsymbol{u}}+(tr\boldsymbol{L})\boldsymbol{u}-\boldsymbol{L}\boldsymbol{u}]\cdot\boldsymbol{F}^*\boldsymbol{N}dA. \tag{5.21}$$

Using (2.20), we finally arrive at (5.19), with

$$\boldsymbol{u}^*=\boldsymbol{u}'+div(\boldsymbol{u}\otimes\boldsymbol{v})-(grad\boldsymbol{v})\boldsymbol{u}. \tag{5.22}$$

5.3.3 *Non-material regions*

A formula like Reynolds' is available for non-material regions; that is, for regions that do not map back to a fixed volume in κ. Let $V(t)\subset\kappa_t$ be such a region, and suppose ϕ is defined on V. The formula to be established is

$$\frac{d}{dt}\int_V\phi dv=\int_V\phi' dv+\int_{\partial V}\phi u da, \tag{5.23}$$

where $u(\boldsymbol{x},t)$ is the local speed of ∂V in the direction of its exterior unit normal $\boldsymbol{n}$ [7]. This reduces to (5.15) if V is a material region, in which case $u=\boldsymbol{v}\cdot\boldsymbol{n}$ where $\boldsymbol{v}$ is the material velocity.

To prove this formula we revert to the basic definition

$$\frac{d}{dt}\int_{V(t)} \phi dv = \lim_{\epsilon\to 0}\frac{1}{\epsilon}\left[\int_{V(t+\epsilon)} \tilde{\phi}(\boldsymbol{x}, t+\epsilon)dv - \int_{V(t)} \tilde{\phi}(\boldsymbol{x}, t)dv\right] \quad (5.24)$$

of the derivative. This is equivalent to

$$\begin{aligned}\frac{d}{dt}\int_{V(t)} \phi dv &= \lim_{\epsilon\to 0}\frac{1}{\epsilon}\left[\int_{V(t+\epsilon)} \tilde{\phi}(\boldsymbol{x}, t+\epsilon)dv - \int_{V(t)} \tilde{\phi}(\boldsymbol{x}, t+\epsilon)dv\right] \\ &\quad + \lim_{\epsilon\to 0}\frac{1}{\epsilon}\left[\int_{V(t)} \tilde{\phi}(\boldsymbol{x}, t+\epsilon)dv - \int_{V(t)} \tilde{\phi}(\boldsymbol{x}, t)dv\right] \\ &= \lim_{\epsilon\to 0}\frac{1}{\epsilon}\left[\int_{V(t+\epsilon)\backslash V(t)} \tilde{\phi}(\boldsymbol{x}, t+\epsilon)dv\right] + \int_{V(t)} \phi' dv.\end{aligned} \quad (5.25)$$

With reference to Figure 5.1, we have that the volume measure on $V(t+\epsilon)\backslash V(t)$ is $dv = u\epsilon da + o(\epsilon)$, where da is the area measure on $\partial V(t)$.

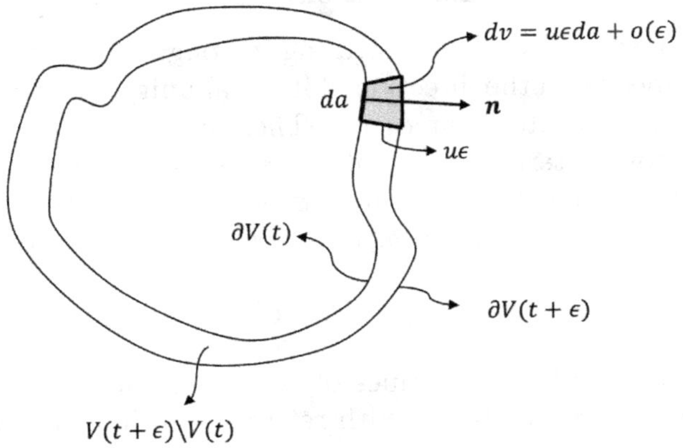

Figure 5.1. Evolution of a non-material region.

Thus,

$$\lim_{\epsilon\to 0}\frac{1}{\epsilon}\left[\int_{V(t+\epsilon)\setminus V(t)}\tilde{\phi}(\boldsymbol{x},t+\epsilon)dv\right]=\lim_{\epsilon\to 0}\left[\int_{\partial V(t)}\tilde{\phi}(\boldsymbol{x},t+\epsilon)uda+\frac{o(\epsilon)}{\epsilon}\right]$$
$$=\int_{\partial V(t)}\tilde{\phi}(\boldsymbol{x},t)uda, \tag{5.26}$$

which combines with (5.25) to give (5.23).

It is convenient to define a vector field $\boldsymbol{u}(\boldsymbol{x},t)$ on $\partial V(t)$ such that $\boldsymbol{u}\cdot\boldsymbol{n}=u$. It is only the normal component $\boldsymbol{u}\cdot\boldsymbol{n}$ that has any physical significance because the tangential part $(\boldsymbol{I}-\boldsymbol{n}\otimes\boldsymbol{n})\boldsymbol{u}$ does not contribute to the evolution of $V(t)$. We then have

$$\frac{d}{dt}\int_V\phi dv=\int_V\phi' dv+\int_{\partial V}\phi\boldsymbol{u}\cdot\boldsymbol{n}da, \tag{5.27}$$

which is directly analogous to (5.15). However, unlike the material velocity $\boldsymbol{v}$, $\boldsymbol{u}$ is defined only on ∂V.

5.4 Discontinuous fields

5.4.1 *Reynolds' formula for fields that suffer a discontinuity across a surface*

Consider a surface $s(t)$ propagating through a material volume $P_t\subset\kappa_t$ at speed u in the direction of its local unit normal $\boldsymbol{n}$. This surface partitions P_t into volumes $V^{\pm}$. These are *non*-material regions, and s is a non-material surface, if $u\neq\boldsymbol{v}\cdot\boldsymbol{n}$ (Figure 5.2).

Suppose our generic scalar field ϕ is discontinuous across s. We then refer to s as a *singular surface*. The jump of ϕ at $\boldsymbol{x}\in s$ is

$$[\![\phi]\!]=\phi^{+}-\phi^{-}, \tag{5.28}$$

where $\phi^{\pm}$ are the limiting values of ϕ as $\boldsymbol{x}$ is approached from the interiors of $V^{\pm}$, respectively. With reference to the figure we have

$$P_t=V^{+}\cup V^{-},\quad \partial P_t^{\pm}=\partial V^{\pm}\cap\partial P_t\quad\text{and}\quad \partial P_t=\partial P_t^{+}\cup\partial P_t^{-}. \tag{5.29}$$

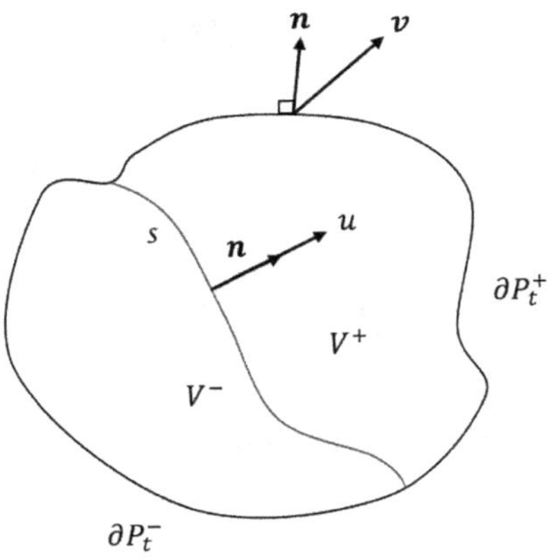

Figure 5.2. A material region $P_t \subset \kappa_t$ traversed by a propagating surface s.

Let ϕ be a smooth function of $\boldsymbol{x}$ in V^+ and V^- separately. Then, from (5.27),

$$\frac{d}{dt}\int_{V^+} \phi dv = \int_{V^+} \phi' dv + \int_{\partial P_t^+} \phi \boldsymbol{v} \cdot \boldsymbol{n} da + \int_s \phi^+(-u) da \quad \text{and}$$
$$\frac{d}{dt}\int_{V^-} \phi dv = \int_{V^-} \phi' dv + \int_{\partial P_t^-} \phi \boldsymbol{v} \cdot \boldsymbol{n} da + \int_s \phi^- u da, \tag{5.30}$$

which, when added, yield the generalization of Reynolds' formula to piecewise smooth fields:

$$\frac{d}{dt}\int_{P_t} \phi dv = \int_{P_t} \phi' dv + \int_{\partial P_t} \phi \boldsymbol{v} \cdot \boldsymbol{n} da - \int_s [\![\phi]\!] u da. \tag{5.31}$$

The extension of this formula to vector fields proceeds as in the passage from (5.15) to (5.18).

5.4.2 *The divergence theorem for discontinuous fields*

The divergence Theorem (5.1) can likewise be extended to piecewise smooth fields $\boldsymbol{a}(\boldsymbol{x}, t)$ that suffer a jump across a surface. In this formula we note that in the integrand $\boldsymbol{a} \cdot \boldsymbol{n}$ on the right-hand side, $\boldsymbol{a}$ is the limit as the boundary ∂R of the region R is approached from the interior of R. Again with reference to Figure 5.2, suppose

$\boldsymbol{a}(\boldsymbol{x}, t)$ is a smooth function of $\boldsymbol{x}$ in V^+ and V^- separately, and that it suffers a discontinuity

$$[\![\boldsymbol{a}]\!] = \boldsymbol{a}^+ - \boldsymbol{a}^- \quad \text{at } \boldsymbol{x} \in s. \tag{5.32}$$

We have

$$\int_{\partial P_t^-} \boldsymbol{a} \cdot \boldsymbol{n} da + \int_s \boldsymbol{a}^- \cdot \boldsymbol{n} da = \int_{V^-} div \boldsymbol{a} dv \quad \text{and}$$

$$\int_{\partial P_t^+} \boldsymbol{a} \cdot \boldsymbol{n} da + \int_s \boldsymbol{a}^+ \cdot (-\boldsymbol{n}) da = \int_{V^+} div \boldsymbol{a} dv, \tag{5.33}$$

and therefore

$$\int_{\partial P_t} \boldsymbol{a} \cdot \boldsymbol{n} da = \int_{P_t} div \boldsymbol{a} dv + \int_s [\![\boldsymbol{a}]\!] \cdot \boldsymbol{n} da. \tag{5.34}$$

In all of the foregoing formulas involving discontinuities the definition of a jump is defined with respect to a normal $\boldsymbol{n}$ on s directed from V^- to V^+.

5.5 Problems

1. Fill in the missing terms in the transport formula
$$\frac{d}{dt} \int_{C_t} \boldsymbol{a} \cdot d\boldsymbol{x} = \int_{C_t} (\boldsymbol{a}' + \ldots) \cdot d\boldsymbol{x},$$
where C_t is a moving material curve and $\boldsymbol{a}'$ is the time derivative of a smooth vector field $\boldsymbol{a}$ in the spatial description.
2. Show that the flux derivative of a vector field $\boldsymbol{u}$ may be written in the form
$$\boldsymbol{u}^* = \boldsymbol{u}' + (div \boldsymbol{u})\boldsymbol{v} - curl(\boldsymbol{v} \times \boldsymbol{u}).$$
3. All of the considerations in this chapter concern fields expressed in terms of the spatial description. Consider fields expressed in terms of the referential description. Derive the relevant referential statements for fields that suffer a discontinuity across a surface S propagating through a fixed region $P \subset \kappa$ at speed U in the direction of its local unit normal $\boldsymbol{N}$. [*Hint*: Use $\dot{\boldsymbol{X}} = \boldsymbol{0}$.]

Chapter 6

Balance Laws

We are now in a position to study the basic balance laws of continuum physics. These are axioms concerning the nature of mass, momentum, energy and entropy pertaining to arbitrary parts of a material body. We are concerned with scalar and vector balance laws.

6.1 Generic forms of the balance laws

We will show that scalar balance laws are all of the form

$$\frac{d}{dt}\int_{P_t} \rho\phi dv = \int_{P_t} \rho\sigma dv + \int_{\partial P_t} \boldsymbol{\pi}\cdot\boldsymbol{n}da, \tag{6.1}$$

where $P_t \subset \kappa_t$ is the region occupied by an arbitrary part S of the body B, $\boldsymbol{n}$ is the exterior unit normal to its boundary ∂P_t; $\phi = \tilde{\phi}(\boldsymbol{x}, t)$ is the mass density of the considered quantity; $\sigma = \tilde{\sigma}(\boldsymbol{x}, t)$ is the rate, per unit mass, at which the quantity is supplied; the vector field $\boldsymbol{\pi} = \tilde{\boldsymbol{\pi}}(\boldsymbol{x}, t)$ is the *flux*, or rate of supply per unit area, through the boundary; and, of course, ρ is the spatial density. The time derivative on the left-hand side pertains to a fixed set of material points. The relevant examples of this kind of balance law are the balances of mass, energy and entropy.

Vector balance laws will be shown, with the caveat noted in Chapter 5, to have the symbolic form

$$\frac{d}{dt}\int_{P_t} \rho\boldsymbol{\psi} dv = \int_{P_t} \rho\boldsymbol{r} dv + \int_{\partial P_t} \boldsymbol{A}\boldsymbol{n}da, \tag{6.2}$$

where $\boldsymbol{\psi} = \tilde{\boldsymbol{\psi}}(\boldsymbol{x}, t)$ is the mass density of the vector-valued variable in question, $\boldsymbol{r} = \tilde{\boldsymbol{r}}(\boldsymbol{x}, t)$ is the rate, per unit mass, at which it is supplied, and $\boldsymbol{A} = \tilde{\boldsymbol{A}}(\boldsymbol{x}, t)$ is its tensor-valued flux. The relevant examples are the balances of linear and rotational momentum, i.e., the equations of motion.

Before proceeding to the study of particular balance laws, we first pause to derive the local forms of the generic balances (6.1) and (6.2) holding at points $\boldsymbol{x}$ in κ_t and on any singular surfaces that may be present. Starting with (6.1), we replace the left-hand side by the expression obtained on substituting $\rho\phi$ in place of ϕ in (5.31). Thus,

$$\begin{aligned}\frac{d}{dt}\int_{P_t}\rho\phi dv &= \int_{P_t}(\rho\phi)' dv + \int_{\partial P_t}\rho\phi\boldsymbol{v}\cdot\boldsymbol{n}da - \int_s [\![\rho\phi]\!] u da\\ &= \int_{P_t}\{(\rho\phi)' + div(\rho\phi\boldsymbol{v})\}dv + \int_s [\![\rho\phi(\boldsymbol{v}-\boldsymbol{u})]\!]\cdot\boldsymbol{n}da\end{aligned} \tag{6.3}$$

where $\boldsymbol{u}$ is the vector defined by $\boldsymbol{u}\cdot\boldsymbol{n} = u$, the speed of the singular surface, and we have invoked (5.34). This combines with (6.1) to give

$$\int_{P_t}\{(\rho\phi)' + div(\rho\phi\boldsymbol{v} - \boldsymbol{\pi}) - \rho\sigma\}dv + \int_s [\![\rho\phi(\boldsymbol{v}-\boldsymbol{u}) - \boldsymbol{\pi}]\!]\cdot\boldsymbol{n}da = 0, \tag{6.4}$$

where we have again invoked (5.34) to recast the flux term. Recall that $\boldsymbol{u}\cdot\boldsymbol{n}$ is defined only on the singular surface; accordingly it can be placed outside the square brackets.

Applying this to an arbitrary part P_t that contains no singular surface, i.e., $s = \varnothing$, and assuming, as usual, that the integrand is continuous, we conclude, from the localization theorem, that

$$(\rho\phi)' + div(\rho\phi\boldsymbol{v}) = div\boldsymbol{\pi} + \rho\sigma \tag{6.5}$$

at all points $\boldsymbol{x}\in\kappa_t$ that do not belong to a singular surface. If P_t contains a part s of a singular surface, then it is divided into subvolumes $V^{\pm}$ wherein this local equation holds. It then follows from (6.4) that the integral over s vanishes. As the part P_t is otherwise arbitrary, its intersection s with the singular surface is likewise arbitrary, and we can invoke a version of the localization theorem for

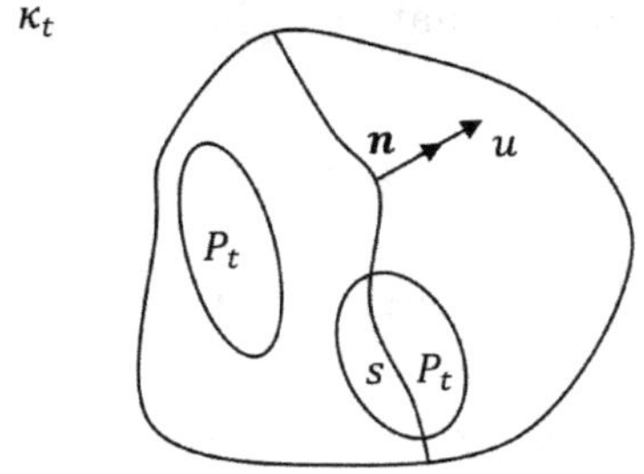

Figure 6.1. Parts of a body with and without a singular surface.

surface integrals to conclude, assuming continuity of the integrand on s, that the integrand vanishes. We thus derive the jump condition

$$[\![\rho\phi(\boldsymbol{v}-\boldsymbol{u})-\boldsymbol{\pi}]\!]\cdot\boldsymbol{n}=0 \tag{6.6}$$

holding at all points on a singular surface (Figure 6.1). This reduces to

$$[\![\boldsymbol{\pi}]\!]\cdot\boldsymbol{n}=0 \tag{6.7}$$

if the singular surface is a material surface; that is, if it moves with speed $u=\boldsymbol{v}\cdot\boldsymbol{n}$ with $[\![\boldsymbol{v}]\!]=\boldsymbol{0}$. However, as (6.6) places no restriction on the tangential discontinuity $(\boldsymbol{I}-\boldsymbol{n}\otimes\boldsymbol{n})[\![\boldsymbol{v}]\!]$, the latter condition remains valid under the weaker restrictions: $u=\boldsymbol{v}\cdot\boldsymbol{n}$ with $[\![\boldsymbol{v}]\!]\cdot\boldsymbol{n}=0$.

Turning now to (6.2), we can proceed as above, and as explained in Section 5.3.1, to cast (6.2) in the symbolic form

$$\int_{P_t}\{(\rho\boldsymbol{\psi})'+div(\rho\boldsymbol{\psi}\otimes\boldsymbol{v}-\boldsymbol{A})-\rho\boldsymbol{r}\}dv$$
$$+\int_s[\![\rho\boldsymbol{\psi}\otimes(\boldsymbol{v}-\boldsymbol{u})-\boldsymbol{A}]\!]\boldsymbol{n}da=\boldsymbol{0}, \tag{6.8}$$

use having been made of the relevant extensions of (5.31) and (5.34), namely

$$\frac{d}{dt}\int_{P_t}\rho\boldsymbol{\psi}dv=\int_{P_t}(\rho\boldsymbol{\psi})'dv+\int_{\partial P_t}(\rho\boldsymbol{\psi}\otimes\boldsymbol{v})\boldsymbol{n}da-\int_s[\![\rho\boldsymbol{\psi}]\!]uda \quad \text{and}$$

$$\int_{\partial P_t}\boldsymbol{A}\boldsymbol{n}da=\int_{P_t}div\boldsymbol{A}dv+\int_s[\![\boldsymbol{A}]\!]\boldsymbol{n}da, \tag{6.9}$$

respectively.

Accordingly, the local equation holding at points in κ_t, excluding singular surfaces, is

$$(\rho\boldsymbol{\psi})' + div(\rho\boldsymbol{\psi} \otimes \boldsymbol{v}) = div\boldsymbol{A} + \rho\boldsymbol{r}, \tag{6.10}$$

and the jump condition holding at every point on a singular surface is

$$[\![\rho\boldsymbol{\psi} \otimes (\boldsymbol{v} - \boldsymbol{u}) - \boldsymbol{A}]\!]\boldsymbol{n} = \boldsymbol{0}, \tag{6.11}$$

reducing to

$$[\![\boldsymbol{A}]\!]\boldsymbol{n} = \boldsymbol{0} \tag{6.12}$$

under the conditions stated following (6.7).

Of course the foregoing balance statements are expressed in terms of the spatial description. It is a straightforward matter, which we leave to the reader, to derive their referential counterparts.

6.2 Conservation of mass

We have already referred, in Section 2.5, to the notion that the mass of any fixed set $S \subset B$ of material points is conserved in the course of any motion that the body B may undergo, i.e.,

$$\frac{d}{dt}\int_{P_t} \rho(\boldsymbol{x}, t)dv = 0. \tag{6.13}$$

This is of the form (6.1), with $\phi = 1$, $\sigma = 0$ and $\boldsymbol{\pi} = \boldsymbol{0}$. Accordingly, (6.5) and (6.6) respectively furnish the local differential equation

$$\rho' + div(\rho\boldsymbol{v}) = 0, \tag{6.14}$$

holding in regions of smoothness of the fields involved, and the jump condition

$$[\![\rho(\boldsymbol{v} - \boldsymbol{u})]\!] \cdot \boldsymbol{n} = 0, \tag{6.15}$$

operative on singular surfaces.

Equation (6.13) must be modified in the presence of diffusion. In this case, σ is the supply of the diffusing species, per unit mass of the mixture, and $\boldsymbol{\pi}$ is the mass flux of the species through the boundary ∂P_t [8]. We will not discuss diffusion in this book, however.

Using

$$div(\rho \boldsymbol{v}) = grad\rho \cdot \boldsymbol{v} + \rho div\boldsymbol{v}, \tag{6.16}$$

together with $div\boldsymbol{v} = tr\boldsymbol{L}$, where $\boldsymbol{L}$ is the spatial velocity gradient, and recalling the result of Problem 2 in Chapter 4, we cast (6.14) in the alternative form

$$0 = \dot{\rho} + \rho div\boldsymbol{v} = \dot{\rho} + \rho(\dot{J}/J), \tag{6.17}$$

where $\dot{\rho} = \rho' + grad\rho \cdot \boldsymbol{v}$ is the material derivative of the density. Equivalently, $(\rho J)^{\cdot} = 0$. With reference to (2.59), this in turn is equivalent to the vanishing of the material derivative of the referential mass density ρ_κ:

$$\dot{\rho}_\kappa = 0. \tag{6.18}$$

Recalling that $\dot{\rho}_\kappa = \partial \rho_\kappa(\boldsymbol{X}, t)/\partial t$, it follows that the referential density, expressed as a function of $\boldsymbol{X}$ and t, is a function of $\boldsymbol{X}$ alone. We indicate this by writing $\rho_\kappa(\boldsymbol{X})$.

An alternative derivation of this result is obtained by differentiating (2.56), assuming $P \subset \kappa$ to be a region in which ρ_κ and $\dot{\rho}_\kappa$ satisfy the continuity conditions stated after (5.13). Thus,

$$0 = \frac{d}{dt}\int_P \rho_\kappa dV = \int_P \dot{\rho}_\kappa dV. \tag{6.19}$$

Choosing P to be otherwise arbitrary, we localize as usual to arrive at (6.18) directly.

The results just derived may be used to effect a significant simplification of the generic balances (6.5) and (6.10). Concerning the former, we use

$$\begin{aligned}(\rho\phi)' + div(\rho\phi\boldsymbol{v}) &= (\rho\phi)' + grad(\rho\phi)\cdot\boldsymbol{v} + \rho\phi div\boldsymbol{v}\\ &= (\rho\phi)^{\cdot} + \rho\phi div\boldsymbol{v}\\ &= (\dot{\rho} + \rho div\boldsymbol{v})\phi + \rho\dot{\phi}\\ &= \rho\dot{\phi},\end{aligned} \tag{6.20}$$

where conservation of mass in the form $(6.17)_1$ has been applied. Accordingly, (6.5) simplifies to

$$\rho\dot{\phi} = div\boldsymbol{\pi} + \rho\sigma. \tag{6.21}$$

In the same way, we use (see Problem 24(g) in Chapter 1)

$$div(\rho\boldsymbol{\psi} \otimes \boldsymbol{v}) = [grad(\rho\boldsymbol{\psi})]\boldsymbol{v} + (\rho div\boldsymbol{v})\boldsymbol{\psi}, \tag{6.22}$$

together with conservation of mass to simplify (6.10) to

$$\rho\dot{\boldsymbol{\psi}} = div\boldsymbol{A} + \rho\boldsymbol{r}. \tag{6.23}$$

Alternatively, these reduced equations may be derived directly by localizing the global statements (6.1) and (6.2) if we first replace the left-hand sides by

$$\frac{d}{dt}\int_{P_t}\rho\phi dv = \frac{d}{dt}\int_P \rho_\kappa\phi dV = \int_P \rho_\kappa\dot{\phi}dV = \int_{P_t}\rho\dot{\phi}dv \quad \text{and}$$
$$\frac{d}{dt}\int_{P_t}\rho\boldsymbol{\psi} dv = \frac{d}{dt}\int_P \rho_\kappa\boldsymbol{\psi} dV = \int_P \rho_\kappa\dot{\boldsymbol{\psi}}dV = \int_{P_t}\rho\dot{\boldsymbol{\psi}}dv, \tag{6.24}$$

respectively, where use has been made of (2.50), (2.59), and mass conservation in the form (6.18) has been invoked.

6.3 Linear and rotational momenta; forces and torques

The equations of motion for a continuum, due to Euler, are relations connecting the linear and rotational momenta of a set $S \subset B$ of material points to the forces and torques acting on it.

6.3.1 *Linear and rotational momenta*

The linear momentum of an arbitrary sub-body is the vector

$$\boldsymbol{l}(S,t) = \int_{P_t}\rho\boldsymbol{v}dv = \int_P \rho_\kappa\boldsymbol{v}dV, \tag{6.25}$$

and the rotational momentum relative to an origin located at position $\boldsymbol{o}$, also called the moment of momentum, is

$$\boldsymbol{h}_o(S,t) = \int_{P_t}\rho\boldsymbol{x}\times\boldsymbol{v}dv = \int_P \rho_\kappa\boldsymbol{x}\times\boldsymbol{v}dV, \tag{6.26}$$

where $\boldsymbol{x}$ is the position vector of a material point $p \in S$ relative to $\boldsymbol{o}$, defined by (1.95).

6.3.2 *Forces*

We assume the forces acting on S to be of two types. The first, called *body forces*, are expressible in the form

$$\boldsymbol{f}_b(S,t) = \int_{P_t} \rho \boldsymbol{b} dv = \int_P \rho_\kappa \boldsymbol{b} dV, \tag{6.27}$$

where $\boldsymbol{b} = \tilde{\boldsymbol{b}}(\boldsymbol{x},t) = \hat{\boldsymbol{b}}(\boldsymbol{X},t)$ is the body force per unit mass. For example, near a point on the surface of the Earth the body force due to gravity is $\boldsymbol{b} = -g\boldsymbol{k}$, where $g \simeq 9.81\ m/s^2$ is the gravitational constant and $\boldsymbol{k}$ is a fixed unit vector directed vertically upward from the point in question.

The second type are the *contact forces* arising from contact of S with its surroundings. These act at the surface ∂P_t and are expressed in the form

$$\boldsymbol{f}_c(S,t) = \int_{\partial P_t} \boldsymbol{t} da, \tag{6.28}$$

where $\boldsymbol{t} = \tilde{\boldsymbol{t}}(\boldsymbol{x},t)$, the force per unit area, is the Cauchy traction, or simply the traction (Figure 6.2). It is often convenient to express this force in the form of an integral over ∂P. Thus,

$$\boldsymbol{f}_c(S,t) = \int_{\partial P} \boldsymbol{p} dA, \tag{6.29}$$

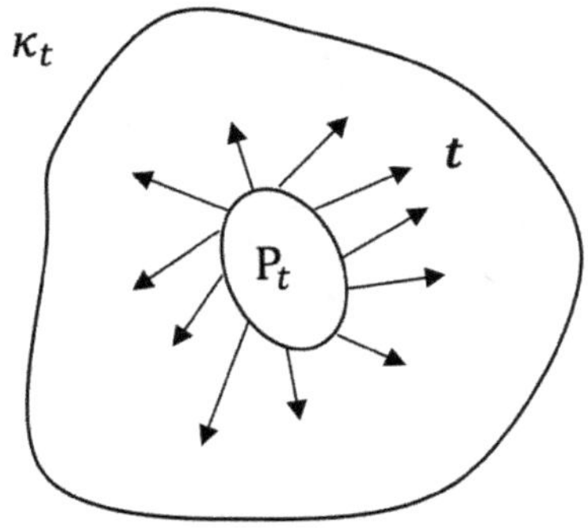

Figure 6.2. Traction acting at the surface of a part of the body.

in which $\boldsymbol{p} = \hat{\boldsymbol{p}}(\boldsymbol{X}, t)$, the Piola traction, furnishes the same force, but measured per unit area of ∂P rather than ∂P_t. That is, while $\boldsymbol{p}$ is defined on ∂P, the actual contact forces are acting on ∂P_t.

The net force acting on S is

$$\boldsymbol{f}(S,t) = \boldsymbol{f}_b(S,t) + \boldsymbol{f}_c(S,t). \tag{6.30}$$

6.3.3 *Torques*

The moment of the body forces, relative to the origin located at position $\boldsymbol{o}$, is

$$\boldsymbol{m}_b(S,t) = \int_{P_t} \rho \boldsymbol{x} \times \boldsymbol{b} dv = \int_P \rho_\kappa \boldsymbol{x} \times \boldsymbol{b} dV, \tag{6.31}$$

and that of the contact forces is

$$\boldsymbol{m}_c(S,t) = \int_{\partial P_t} \boldsymbol{x} \times \boldsymbol{t} da = \int_{\partial P} \boldsymbol{x} \times \boldsymbol{p} dA. \tag{6.32}$$

In this book, we follow the development of classical continuum mechanics as conceived by Cauchy, and thus assume that intrinsic body and contact torques are absent. These classical continua are called *non-polar* materials. As an example of a polar medium, consider a material reinforced by a continuous distribution of embedded fibers (Figure 6.3). Intrinsic contact torques are produced if the fibers offer resistance to local bending and torsion. These generate a distribution of couple traction — a typical characteristic of polar

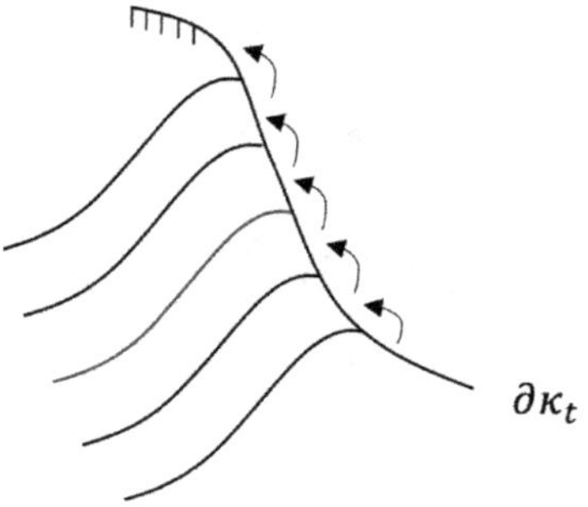

Figure 6.3. Couple traction acting at the surface of a body consisting of a fiber-reinforced polar material.

media — where they intersect the boundary of a body [9]. Assuming such effects to be absent or negligible, we have that the net torque acting on S, relative to $\boldsymbol{o}$, is

$$\boldsymbol{m}_o(S,t) = \boldsymbol{m}_b(S,t) + \boldsymbol{m}_c(S,t). \tag{6.33}$$

Further, the expression (6.26) for the rotational momentum is appropriate for non-polar media and must typically be generalized to accommodate polar media.

6.4 Euler's postulates

Euler's postulates apply to an arbitrary part S of a body. They entail the fundamental assumption that the motion of the part is governed by the same equations that govern the motion of a rigid body. Thus, the rate of change of the linear momentum of a fixed set of material points is balanced by the net force acting on it, i.e.,

$$\frac{d}{dt}\boldsymbol{l}(S,t) = \boldsymbol{f}(S,t), \tag{6.34}$$

and the rate of change of the rotational momentum is balanced by the net torque:

$$\frac{d}{dt}\boldsymbol{h}_o(S,t) = \boldsymbol{m}_o(S,t). \tag{6.35}$$

A frame of reference in which these apply, consisting of point spaces $\mathcal{E}^3$ and $\hat{\mathcal{E}}^3$, associated translation spaces E^3 and $\hat{E}^3$, respectively, and an origin located at a fixed position $\boldsymbol{o}$, is called an *inertial frame*. Thus Euler's postulates are tantamount to the assumption that an inertial frame exists. We shall prove later that there are infinitely many inertial frames, and we shall also obtain the equations holding relative to arbitrary, generally non-inertial, frames.

On invoking conservation of mass and making use of (6.24), Euler's postulates reduce to

$$\int_{P_t} \rho\dot{\boldsymbol{v}}dv = \int_{P_t} \rho\boldsymbol{b}dv + \int_{\partial P_t} \boldsymbol{t}da \tag{6.36}$$

and

$$\int_{P_t} \rho\boldsymbol{x}\times\dot{\boldsymbol{v}}dv = \int_{P_t} \rho\boldsymbol{x}\times\boldsymbol{b}dv + \int_{\partial P_t} \boldsymbol{x}\times\boldsymbol{t}da, \tag{6.37}$$

respectively, use having been made of $(\boldsymbol{x} \times \boldsymbol{v})^{\cdot} = \boldsymbol{v} \times \boldsymbol{v} + \boldsymbol{x} \times \dot{\boldsymbol{v}} = \boldsymbol{x} \times \dot{\boldsymbol{v}}$. Equivalently, the linear momentum balance may be expressed in the form

$$\int_P \rho_\kappa \dot{\boldsymbol{v}} dV = \int_P \rho_\kappa \boldsymbol{b} dV + \int_{\partial P} \boldsymbol{p} dA, \tag{6.38}$$

and the rotational momentum balance, in the form

$$\int_P \rho_\kappa \boldsymbol{x} \times \dot{\boldsymbol{v}} dV = \int_P \rho_\kappa \boldsymbol{x} \times \boldsymbol{b} dV + \int_{\partial P} \boldsymbol{x} \times \boldsymbol{p} dA, \tag{6.39}$$

with $\boldsymbol{x} = \boldsymbol{\chi}_\kappa(\boldsymbol{X}, t)$.

6.4.1 *Center of mass*

The center of mass of S, occupying position $\bar{\boldsymbol{x}}$ relative to $\boldsymbol{o}$, is located at the mass-weighted average of the position, i.e.,

$$m(S)\bar{\boldsymbol{x}}(t) = \int_{P_t} \rho \boldsymbol{x} dv = \int_P \rho_\kappa \boldsymbol{x} dV, \tag{6.40}$$

where $m(S)$ is the fixed mass of S. The position of a material point $p \in S$, relative to the center of mass, is

$$\boldsymbol{\pi} = \boldsymbol{x} - \bar{\boldsymbol{x}}. \tag{6.41}$$

Accordingly, the mass weighted average of the relative position vanishes identically:

$$\begin{aligned} \int_{P_t} \rho \boldsymbol{\pi} dv &= \int_{P_t} \rho(\boldsymbol{x} - \bar{\boldsymbol{x}}) dv \\ &= \left[m(S) - \int_{P_t} \rho dv \right] \bar{\boldsymbol{x}} = \boldsymbol{0}, \end{aligned} \tag{6.42}$$

and therefore, on invoking conservation of mass,

$$\begin{aligned} \boldsymbol{0} = \frac{d}{dt} \int_{P_t} \rho \boldsymbol{\pi} dv &= \int_{P_t} \rho \dot{\boldsymbol{\pi}} dv \\ &= \int_{P_t} \rho \boldsymbol{v} dv - m(S)(\bar{\boldsymbol{x}})^{\cdot}, \end{aligned} \tag{6.43}$$

i.e., the linear momentum of S is the identical to that of a particle of the same mass located at the center of mass. Euler's postulate (6.34) is thus equivalent, granted conservation of mass, to

$$m(S)(\bar{\boldsymbol{x}})^{\cdot\cdot} = \boldsymbol{f}(S,t), \tag{6.44}$$

where $(\bar{\boldsymbol{x}})^{\cdot\cdot}$ is the acceleration of the center of mass.

Using (6.41), we also have

$$\begin{aligned} \boldsymbol{h}_o(S,t) &= \int_{P_t} \rho\boldsymbol{x} \times \dot{\boldsymbol{x}} dv = \int_{P_t} \rho(\bar{\boldsymbol{x}} + \boldsymbol{\pi}) \times [(\bar{\boldsymbol{x}})^{\cdot} + \dot{\boldsymbol{\pi}}] dv \\ &= m(S)\, \bar{\boldsymbol{x}} \times (\bar{\boldsymbol{x}})^{\cdot} + \bar{\boldsymbol{x}} \times \int_{P_t} \rho\dot{\boldsymbol{\pi}} dv \\ &\quad + \left(\int_{P_t} \rho\boldsymbol{\pi} dv \right) \times (\bar{\boldsymbol{x}})^{\cdot} + \boldsymbol{h}(S,t), \end{aligned} \tag{6.45}$$

where

$$\boldsymbol{h}(S,t) = \int_{P_t} \rho\boldsymbol{\pi} \times \dot{\boldsymbol{\pi}} dv \tag{6.46}$$

is the rotational momentum relative to the center of mass. Noting that the integrals in (6.45) vanish, we conclude that

$$\boldsymbol{h}_o(S,t) = \bar{\boldsymbol{x}}(t) \times \boldsymbol{l}(S,t) + \boldsymbol{h}(S,t). \tag{6.47}$$

We also have

$$\begin{aligned} \boldsymbol{m}_o(S,t) &= \int_{P_t} \rho(\bar{\boldsymbol{x}} + \boldsymbol{\pi}) \times \boldsymbol{b} dv + \int_{\partial P_t} \rho(\bar{\boldsymbol{x}} + \boldsymbol{\pi}) \times \boldsymbol{t} da \\ &= \bar{\boldsymbol{x}}(t) \times \boldsymbol{f}(S,t) + \boldsymbol{m}(S,t), \end{aligned} \tag{6.48}$$

where

$$\boldsymbol{m}(S,t) = \int_{P_t} \rho\boldsymbol{\pi} \times \boldsymbol{b} dv + \int_{\partial P_t} \rho\boldsymbol{\pi} \times \boldsymbol{t} da \tag{6.49}$$

is the net torque about the center of mass. We thus recast Euler's postulate (6.35) as

$$\begin{aligned} &\bar{\boldsymbol{x}}(t) \times \boldsymbol{f}(S,t) + \boldsymbol{m}(S,t) \\ &\quad = \frac{d}{dt}\boldsymbol{h}(S,t) + [\bar{\boldsymbol{x}}(t)]^{\cdot} \times \boldsymbol{l}(S,t) + \bar{\boldsymbol{x}}(t) \times \frac{d}{dt}\boldsymbol{l}(S,t), \end{aligned} \tag{6.50}$$

which reduces, using (6.34), together with (6.43) in the form $\boldsymbol{l} = m\bar{\boldsymbol{x}}^{\cdot}$, to

$$\boldsymbol{m}(S,t) = \frac{d}{dt}\boldsymbol{h}(S,t). \tag{6.51}$$

Thus, Euler's postulate also holds relative to the center of mass, its motion in the inertial frame notwithstanding.

6.4.2 *Rigid bodies*

The foregoing development naturally subsumes the motions of rigid bodies. These are necessarily of the form (3.121), i.e.,

$$\boldsymbol{x} = \boldsymbol{Q}(t)\boldsymbol{X} + \boldsymbol{c}(t), \tag{6.52}$$

where $\boldsymbol{Q}(t)$ is a spatially uniform two-point rotation tensor and $\boldsymbol{c}(t)$ is a vector-valued function. Accordingly,

$$\boldsymbol{\pi} = \boldsymbol{Q}(t)\boldsymbol{\Pi}, \quad \text{where } \boldsymbol{\Pi} = \boldsymbol{X} - \bar{\boldsymbol{X}} \tag{6.53}$$

is the referential position of $p \in S$ relative to the center of mass of P, located at position $\bar{\boldsymbol{X}}$. Thus,

$$\dot{\boldsymbol{\pi}} = \dot{\boldsymbol{Q}}\boldsymbol{\Pi} = \boldsymbol{\Omega}(t)\boldsymbol{\pi} = \boldsymbol{\omega}(t) \times \boldsymbol{\pi}, \tag{6.54}$$

where $\boldsymbol{\omega}$ is the axial vector of the skew tensor

$$\boldsymbol{\Omega} = \dot{\boldsymbol{Q}}\boldsymbol{Q}^t, \tag{6.55}$$

this of course being simply a special case of (4.33).

Consider a material curve with unit tangent $\boldsymbol{M}$ at $\boldsymbol{X}$. From (2.67) and (3.120), we have that the unit tangent $\boldsymbol{m}$ to the image of this curve after deformation, and the associated stretch μ, satisfy

$$\mu\boldsymbol{m} = \boldsymbol{F}\boldsymbol{M} = \boldsymbol{Q}\boldsymbol{M}; \quad \text{hence,} \quad \mu = 1 \text{ and } \boldsymbol{m} = \boldsymbol{Q}\boldsymbol{M}, \tag{6.56}$$

so that

$$\dot{\boldsymbol{m}} = \boldsymbol{\Omega}\boldsymbol{m} = \boldsymbol{\omega} \times \boldsymbol{m}. \tag{6.57}$$

Let $\{\hat{\boldsymbol{E}}_i\}$ be a fixed orthonormal basis for $\hat{E}^3$ and let

$$\boldsymbol{e}_i^*(t) = \boldsymbol{Q}(t)\hat{\boldsymbol{E}}_i \in E^3. \tag{6.58}$$

Then $\{\boldsymbol{e}_i^*\}$ is fixed in the body and rotates with it. On identifying $\boldsymbol{m}$ at the point in question with an element of this basis, we have simply

$$\dot{\boldsymbol{e}}_i^* = \boldsymbol{\omega} \times \boldsymbol{e}_i^*. \tag{6.59}$$

Let $\boldsymbol{u} \in E^3$ be an arbitrary vector. We decompose it in the rotating basis: $\boldsymbol{u} = u_i \boldsymbol{e}_i^*$. Thus,

$$\begin{aligned} \dot{\boldsymbol{u}} &= \dot{u}_i \boldsymbol{e}_i^* + u_i \dot{\boldsymbol{e}}_i^* \\ &= \mathring{\boldsymbol{u}} + \boldsymbol{\omega} \times \boldsymbol{u}, \end{aligned} \tag{6.60}$$

where

$$\mathring{\boldsymbol{u}} = \dot{u}_i \boldsymbol{e}_i^*, \tag{6.61}$$

the *co-rotational derivative*, is the derivative relative to the rotating basis. Thus $\mathring{\boldsymbol{u}}$ vanishes if and only if $\boldsymbol{u}$ is fixed in the rotating body. From (6.54) it follows that $\boldsymbol{\pi}$ is such a vector.

From (6.46), we have that

$$\boldsymbol{h}(S,t) = \int_{P_t} \rho \boldsymbol{\pi} \times (\boldsymbol{\omega} \times \boldsymbol{\pi}) dv \tag{6.62}$$

for rigid bodies. We would like to exploit the spatial uniformity of the spin $\boldsymbol{\omega}$ to 'factor it out' of the integral. This may be achieved with the aid of the result of Problem 11 in Chapter 1. Thus,

$$\begin{aligned} \boldsymbol{\pi} \times (\boldsymbol{\omega} \times \boldsymbol{\pi}) &= (\boldsymbol{\pi} \cdot \boldsymbol{\pi})\boldsymbol{\omega} - (\boldsymbol{\pi} \cdot \boldsymbol{\omega})\boldsymbol{\pi} \\ &= [(\boldsymbol{\pi} \cdot \boldsymbol{\pi})\boldsymbol{I} - \boldsymbol{\pi} \otimes \boldsymbol{\pi}]\boldsymbol{\omega}, \end{aligned} \tag{6.63}$$

yielding

$$\boldsymbol{h}(S,t) = \boldsymbol{J}(S,t)\boldsymbol{\omega}(t), \tag{6.64}$$

where

$$\boldsymbol{J}(S,t) = \int_{P_t} \rho[(\boldsymbol{\pi} \cdot \boldsymbol{\pi})\boldsymbol{I} - \boldsymbol{\pi} \otimes \boldsymbol{\pi}] dv \tag{6.65}$$

is the symmetric *inertia tensor* of the rigid body.

This may be simplified on noting that $J = \det \boldsymbol{Q} = 1$, and hence that $\rho = \rho_\kappa$. Further, (6.53) implies that $\boldsymbol{\pi} \cdot \boldsymbol{\pi} = \boldsymbol{\Pi} \cdot \boldsymbol{\Pi}$, and hence that

$$\begin{aligned} \boldsymbol{J}(S,t) &= \int_P \rho_\kappa[(\boldsymbol{\Pi}\cdot\boldsymbol{\Pi})\boldsymbol{Q}\boldsymbol{Q}^t - \boldsymbol{Q}\boldsymbol{\Pi}\otimes\boldsymbol{Q}\boldsymbol{\Pi}]dV \\ &= \boldsymbol{Q}(t)\boldsymbol{J}_\kappa(S)\boldsymbol{Q}(t)^t, \end{aligned} \tag{6.66}$$

where $\boldsymbol{J}_\kappa(S)$ is the fixed referential tensor defined by

$$\boldsymbol{J}_\kappa(S) = \int_P \rho_\kappa[(\boldsymbol{\Pi}\cdot\boldsymbol{\Pi})\hat{\boldsymbol{I}} - \boldsymbol{\Pi}\otimes\boldsymbol{\Pi}]dV. \tag{6.67}$$

The inertia tensor is fixed in the body in the sense that its co-rotational derivative $\mathring{\boldsymbol{J}}$, defined by

$$\mathring{\boldsymbol{J}} = \dot{\boldsymbol{J}} - \boldsymbol{\Omega}\boldsymbol{J} + \boldsymbol{J}\boldsymbol{\Omega} \tag{6.68}$$

vanishes. This follows easily from (6.55) and (6.66).

Thus, the rotational momentum balance (6.51) combines with (6.64) to give

$$\begin{aligned} \boldsymbol{m}(S,t) &= (\boldsymbol{J}\boldsymbol{\omega})^{\cdot} = \dot{\boldsymbol{J}}\boldsymbol{\omega} + \boldsymbol{J}\dot{\boldsymbol{\omega}} \\ &= (\boldsymbol{\Omega}\boldsymbol{J} - \boldsymbol{J}\boldsymbol{\Omega})\boldsymbol{\omega} + \boldsymbol{J}\dot{\boldsymbol{\omega}}. \end{aligned} \tag{6.69}$$

This simplifies further on noting that $\boldsymbol{\Omega}\boldsymbol{\omega} = \boldsymbol{\omega}\times\boldsymbol{\omega}$ vanishes. Thus,

$$\boldsymbol{m}(S,t) = \boldsymbol{\omega}\times\boldsymbol{J}\boldsymbol{\omega} + \boldsymbol{J}\dot{\boldsymbol{\omega}}, \quad \text{with } \dot{\boldsymbol{\omega}} = \mathring{\boldsymbol{\omega}}. \tag{6.70}$$

Equations (6.70) and (6.44) are the classical equations of rigid-body dynamics for the determination of the motion of the center of mass and the rotation in terms of the net force and torque about the mass center. Conversely, these equations yield the net force and torque required to effect the motion of the center of mass and the rotation.

6.5 Forces of mutual interaction in the body and the local equations of motion

6.5.1 *Action and reaction*

Consider an arbitrary part P_t of the region κ_t occupied by the body at time t. Let σ be a surface dividing it into two sub-parts P_{t1} and P_{t2}

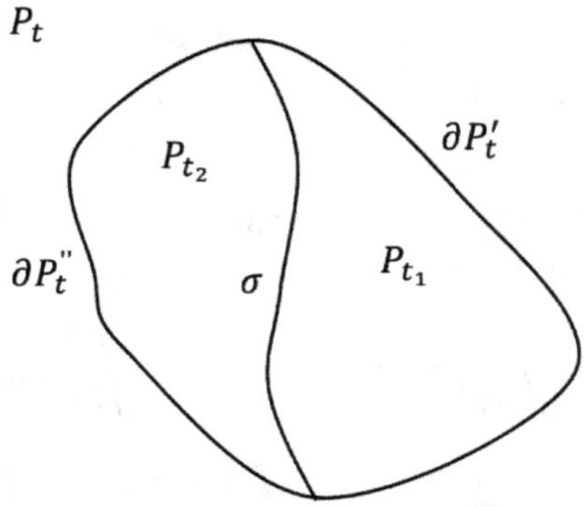

Figure 6.4. An arbitrary part of a body divided into two sub-parts by an interior surface.

(Figure 6.4). Let $\partial P_t' = \partial P_{t1} \cap \partial P_t$ and $\partial P_t'' = \partial P_{t2} \cap \partial P_t$, with $\partial P_{t1} = \partial P_t' \cup \sigma$ and $\partial P_{t2} = \partial P_t'' \cup \sigma$. Then, $P_t = P_{t1} \cup P_{t2}$ and $\partial P_t = \partial P_t' \cup \partial P_t''$.

The forces acting on P_{t1} are the body force and the contact force. We express the former as

$$\boldsymbol{f}_{b1} = \int_{P_{t1}} \rho(\bar{\boldsymbol{b}} + \boldsymbol{b}_{12})dv, \tag{6.71}$$

where $\bar{\boldsymbol{b}}$ is due to sources external to P_t and $\boldsymbol{b}_{12}$ is the body force per unit mass arising from P_{t2}. The latter may be due to the gravitational attraction of the two parts, for example. The contact force is

$$\boldsymbol{f}_{c1} = \int_{\partial P_{t1}} \boldsymbol{t} da = \int_{\partial P_t'} \boldsymbol{t} da + \int_{\sigma} \boldsymbol{t}_{12} da, \tag{6.72}$$

where $\boldsymbol{t}_{12}$ is the traction transmitted to P_{t1} by P_{t2}. The net force applied to P_{t1} is

$$\boldsymbol{f}_1 = \boldsymbol{f}_{b1} + \boldsymbol{f}_{c1}. \tag{6.73}$$

Similarly, the net force acting on P_{t2} is

$$\boldsymbol{f}_2 = \boldsymbol{f}_{b2} + \boldsymbol{f}_{c2}, \tag{6.74}$$

where

$$\boldsymbol{f}_{b2} = \int_{P_{t2}} \rho(\bar{\boldsymbol{b}} + \boldsymbol{b}_{21})dv \quad \text{and} \quad \boldsymbol{f}_{c2} = \int_{\partial P_t''} \boldsymbol{t}da + \int_{\sigma} \boldsymbol{t}_{21}da, \tag{6.75}$$

wherein $\boldsymbol{b}_{21}$ is due to P_{t1} and $\boldsymbol{t}_{21}$ is the traction exerted by P_{t1} on P_{t2}.

The linear momenta of parts P_{t1} and P_{t2} are

$$\boldsymbol{l}_1 = \int_{P_{t1}} \rho\boldsymbol{v}dv \quad \text{and} \quad \boldsymbol{l}_2 = \int_{P_{t2}} \rho\boldsymbol{v}dv, \tag{6.76}$$

respectively, and the momentum of $P_t (= P_{t1} \cup P_{t2})$ is $\boldsymbol{l} = \boldsymbol{l}_1 + \boldsymbol{l}_2$. Euler's postulate, for P_{t1} and P_{t2} separately, yields

$$\int_{P_{t1}} \rho(\bar{\boldsymbol{b}} + \boldsymbol{b}_{12})dv + \int_{\partial P_t'} \boldsymbol{t}da + \int_{\sigma} \boldsymbol{t}_{12}da = \frac{d}{dt}\boldsymbol{l}_1 \quad \text{and}$$

$$\int_{P_{t2}} \rho(\bar{\boldsymbol{b}} + \boldsymbol{b}_{21})dv + \int_{\partial P_t''} \boldsymbol{t}da + \int_{\sigma} \boldsymbol{t}_{21}da = \frac{d}{dt}\boldsymbol{l}_2. \tag{6.77}$$

On adding these and recalling that $\partial P_t = \partial P_t' \cup \partial P_t''$, we obtain

$$\int_{P_t} \rho\bar{\boldsymbol{b}}dv + \int_{P_{t1}} \rho\boldsymbol{b}_{12}dv + \int_{P_{t2}} \rho\boldsymbol{b}_{21}dv + \int_{\partial P_t} \boldsymbol{t}da + \int_{\sigma} (\boldsymbol{t}_{12} + \boldsymbol{t}_{21})da = \frac{d}{dt}\boldsymbol{l}. \tag{6.78}$$

On the other hand, for the whole of P_t, we have

$$\int_{P_t} \rho\bar{\boldsymbol{b}}dv + \int_{\partial P_t} \boldsymbol{t}da = \frac{d}{dt}\boldsymbol{l}, \tag{6.79}$$

whence it follows that the force applied to P_{t2} by P_{t1} is opposite that applied to P_{t1} by P_{t2}:

$$\int_{P_{t2}} \rho\boldsymbol{b}_{21}dv + \int_{\sigma} \boldsymbol{t}_{21}da = -\left(\int_{P_{t1}} \rho\boldsymbol{b}_{12}dv + \int_{\sigma} \boldsymbol{t}_{12}da\right). \tag{6.80}$$

In contrast to Newton's laws for a particle, for continua this "action and reaction" statement is a consequence of the balance of linear momentum and not a separate postulate.

Consider now a sequence of parts P_t, all with the same interior surface σ and collapsing onto it in the limit (Figure 6.5). Each of

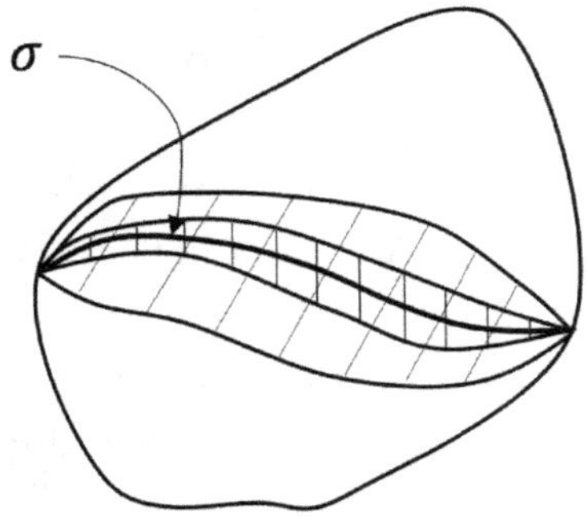

Figure 6.5. A sequence of parts, each containing the same interior surface.

these parts has associated sub-parts P_{t1} and P_{t2} with volumes that tend to zero in the limit. Assuming $|\rho \boldsymbol{b}_{21}|$ and $|\rho \boldsymbol{b}_{12}|$ to be bounded, the volume integrals in (6.80) vanish in the limit, yielding

$$\int_\sigma (\boldsymbol{t}_{21} + \boldsymbol{t}_{12}) da = \mathbf{0}. \tag{6.81}$$

Substituting back into (6.80), we conclude that

$$\int_{P_{t2}} \rho \boldsymbol{b}_{21} dv = -\int_{P_{t1}} \rho \boldsymbol{b}_{12} dv. \tag{6.82}$$

Because P_t is arbitrary, σ is an arbitrary surface in κ_t. Assuming the integrand in (6.81) to be continuous, we may invoke a version of the localization theorem to arrive at *Cauchy's lemma*, namely

$$\boldsymbol{t}_{21} = -\boldsymbol{t}_{12}. \tag{6.83}$$

6.5.2 *Cauchy's hypothesis and Noll's theorem*

Let $\boldsymbol{t}_s(\boldsymbol{x}, t)$ be the traction acting at $\boldsymbol{x} \in s$, a surface in κ_t. Here the subscript conveys the notion that the traction depends on the surface s. This idea is motivated by a simple thought experiment involving a bar subjected to an axial force F, say (Figure 6.6).

Given the force F, we expect that the force per unit area — the traction — should depend on the surface area; hence, in this instance, on the orientation of s. Thus we expect that the tractions acting on s_1 and s_2 at their common point $\boldsymbol{x}$ should be different. Because the

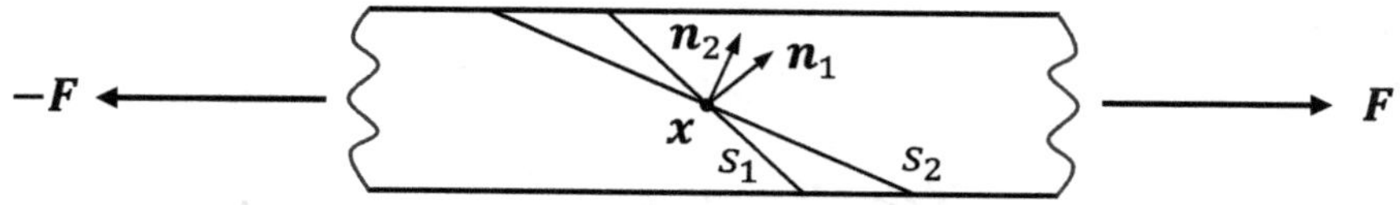

Figure 6.6. Surfaces having different orientations in a bar under axial force.

traction is defined locally at points of the surface, logically it should depend on a local measure of the surface orientation; namely, a unit normal $\boldsymbol{n}$ to the surface s at the point $\boldsymbol{x}$. This assumption is the content of the famous *Cauchy hypothesis*: There exists a function $\boldsymbol{t}(\boldsymbol{x}, t; \cdot)$ such that

$$\boldsymbol{t}_s(\boldsymbol{x}, t) = \boldsymbol{t}(\boldsymbol{x}, t; \boldsymbol{n}). \tag{6.84}$$

Cauchy's lemma (6.83) then assumes the form

$$\boldsymbol{t}(\boldsymbol{x}, t; -\boldsymbol{n}) = -\boldsymbol{t}(\boldsymbol{x}, t; \boldsymbol{n}). \tag{6.85}$$

Cauchy's hypothesis has occupied a central place in continuum mechanics since it was proposed by Cauchy in the early nineteenth century. More recently, in the mid-twentieth century, Noll strengthened it to the status of a proved theorem. He proved that it is actually a consequence of Euler's linear momentum balance [10].

Sketch of Noll's theorem

With reference to Figure 6.7, let s_1 and s_2 be two surfaces in κ_t intersecting at a point with position $\boldsymbol{x}_0$, and suppose they share a common tangent plane T, with unit normal $\boldsymbol{n}$, at this point.

Let P_{t1} be the part of the body bounded by $\partial P_{t1} = d_1 \cup f_1 \cup e$, where d_1 is a subset of s_1, f_1 is a piece of the lateral surface of a circular cylinder of radius R and axis $\boldsymbol{n}$, and e is the part of the surface of the cylinder which is common to both ∂P_{t1} and ∂P_{t2}. The part P_{t2}, which is not shown, is the region bounded on top by s_2, and $\partial P_{t2} = d_2 \cup f_2 \cup e$, with corresponding definitions of d_2 and f_2. Denoting by $A(s)$ the area of a surface s, we conclude, on noting that d_1 and d_2 approach a circular disc on T of radius R as $R \to 0$, that

$$A(d_\alpha) = \pi R^2 + o(R^2) \quad \text{and} \quad A(f_\alpha) = o(R^2); \quad \alpha = 1, 2; \tag{6.86}$$

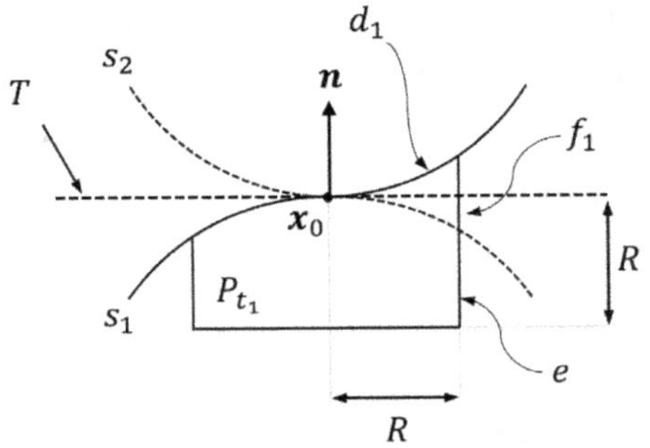

Figure 6.7. A part of a body bounded by part of a cylinder and a surface s_1.

and, if $V(P)$ is the volume of a region P, that

$$V(P_{t\alpha}) = O(R^3) = o(R^2); \quad \alpha = 1, 2. \tag{6.87}$$

The balances of linear momentum for P_{t1} and P_{t2} are given by (6.36):

$$\int_{P_{t1}} \rho(\dot{\boldsymbol{v}} - \boldsymbol{b})dv = \int_{\partial P_{t1}} \boldsymbol{t}_{\partial P_{t1}} da \ \text{ and } \int_{P_{t2}} \rho(\dot{\boldsymbol{v}} - \boldsymbol{b})dv = \int_{\partial P_{t2}} \boldsymbol{t}_{\partial P_{t2}} da, \tag{6.88}$$

which, when subtracted, give

$$\int_{d_1} \boldsymbol{t}_{d_1} da - \int_{d_2} \boldsymbol{t}_{d_2} da = \int_{P_{t1}} \rho(\dot{\boldsymbol{v}} - \boldsymbol{b})dv - \int_{P_{t2}} \rho(\dot{\boldsymbol{v}} - \boldsymbol{b})dv + \int_{f_2} \boldsymbol{t}_{f_2} da - \int_{f_1} \boldsymbol{t}_{f_1} da. \tag{6.89}$$

Assuming all the integrands to be continuous and hence bounded on their respective domains of integration, from $(6.86)_2$ and (6.87), we then have

$$\int_{d_1} \boldsymbol{t}_{d_1} da = \int_{d_2} \boldsymbol{t}_{d_2} da + \boldsymbol{o}(R^2), \tag{6.90}$$

where $\boldsymbol{o}(\epsilon)$ is a vector such that $|\boldsymbol{o}(\epsilon)| = o(\epsilon)$. Then, on dividing by πR^2 and invoking $(6.86)_1$, we obtain

$$\frac{1}{A(d_1)} \int_{d_1} \boldsymbol{t}_{d_1} da = \frac{1}{A(d_2)} \int_{d_2} \boldsymbol{t}_{d_2} da + \frac{\boldsymbol{o}(R^2)}{R^2}. \tag{6.91}$$

Under our continuity hypothesis the Mean–Value theorem [2] — essentially the same thing as the Localization Theorem — is applicable and implies that

$$\lim_{R\to 0}\frac{1}{A(d_\alpha)}\int_{d_\alpha}\boldsymbol{t}_{d_\alpha}(\boldsymbol{x},t)da=\boldsymbol{t}_{d_\alpha}(\boldsymbol{x}_0,t);\quad \alpha=1,2. \tag{6.92}$$

On passing to the limit in (6.91), we finally reach

$$\boldsymbol{t}_{d_1}(\boldsymbol{x}_0,t)=\boldsymbol{t}_{d_2}(\boldsymbol{x}_0,t), \tag{6.93}$$

and thus conclude that the traction at a point takes the same value for all surfaces that have the same unit normal at that point. Accordingly, the traction is determined by the unit normal $\boldsymbol{n}$ and (6.84) is proved. It follows that the traction acting on a surface at a particular point is identical to the traction acting on the tangent plane to the surface at the point in question. Here, for definiteness, $\boldsymbol{n}$ is taken to be the exterior unit normal to ∂P_{t1} and ∂P_{t2} at $\boldsymbol{x}_0$.

Rigorous proofs of Noll's theorem may be found in [10, 11].

6.5.3 *Cauchy's stress theorem*

This theorem establishes the existence of a tensor field $\boldsymbol{T}(\boldsymbol{x},t)$, the *Cauchy Stress Tensor*, such that

$$\boldsymbol{t}(\boldsymbol{x},t;\boldsymbol{n})=\boldsymbol{T}(\boldsymbol{x},t)\boldsymbol{n}. \tag{6.94}$$

Cauchy's theorem is a cornerstone of continuum mechanics. To prove it we apply the linear momentum balance to a part of the body in the shape of a tetrahedron $\mathcal{T}$, say, formed by three Cartesian coordinate planes intersecting at a point $\boldsymbol{x}_0\in\kappa_t$, and an oblique plane situated at a distance δ from this point (Figure 6.8).

Thus,

$$\int_{\partial\mathcal{T}}\boldsymbol{t}da=\int_{\mathcal{T}}\rho(\dot{\boldsymbol{v}}-\boldsymbol{b})dv. \tag{6.95}$$

Recall that this is symbolic notation representing the vector equation

$$\boldsymbol{e}_i\int_{\partial\mathcal{T}}t_i da=\boldsymbol{e}_i\int_{\mathcal{T}}\rho(\dot{v}_i-b_i)dv, \tag{6.96}$$

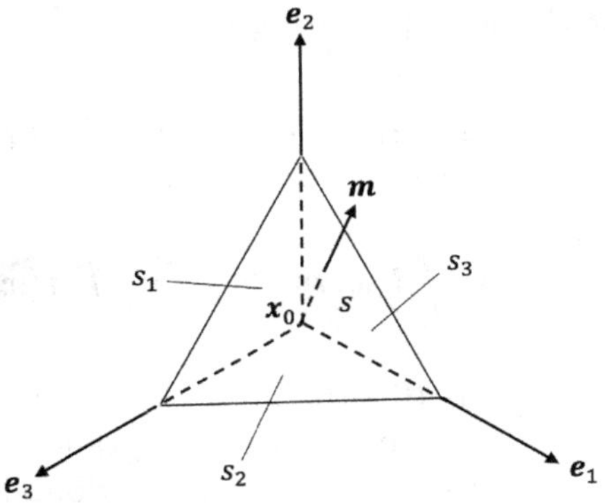

Figure 6.8. Cauchy tetrahedron with vertex at $\boldsymbol{x}_0$ and $\boldsymbol{m} \cdot \boldsymbol{e}_i > 0$.

which is equivalent to the three equations

$$\int_{\partial\tau} t_i da = \int_{\tau} \rho(\dot{v}_i - b_i) dv; \quad i = 1, 2, 3. \tag{6.97}$$

Assuming $\rho(\dot{v}_i - b_i)$ to be continuous functions of $\boldsymbol{x}$, it follows, by the compactness of τ, that

$$\int_{\partial\tau} t_i da \leq \left| \int_{\partial\tau} t_i da \right| \leq \int_{\tau} |\rho(\dot{v}_i - b_i)| \, dv \leq k_i V(\tau), \tag{6.98}$$

where $V(\tau)$ is the volume of the tetrahedron and

$$k_i = \max_{\boldsymbol{x} \in \tau} |\rho(\dot{v}_i - b_i)|. \tag{6.99}$$

Let $A(s)$ be the area of the oblique face of the tetrahedron. Then $A(s) = c_1 \delta^2$ for some positive constant c_1, whereas $V(\tau) = c_2 \delta^3$ with c_2 a positive constant. Thus,

$$\frac{1}{A(s)} \int_{\partial\tau} t_i da = O(\delta). \tag{6.100}$$

Reverting to symbolic notation, we then have that

$$\lim_{\delta \to 0} \frac{1}{A(s)} \int_{\partial\tau} \boldsymbol{t} da = \boldsymbol{0}. \tag{6.101}$$

In view of (6.84) and suppressing the passive argument t,

$$\int_{\partial\tau} \boldsymbol{t}(\boldsymbol{x};\boldsymbol{n})da = \int_s \boldsymbol{t}(\boldsymbol{x};\boldsymbol{m})da + \sum_{i=1}^{3}\int_{s_i} \boldsymbol{t}(\boldsymbol{x};-\boldsymbol{e}_i)da$$

$$= \int_s \boldsymbol{t}(\boldsymbol{x};\boldsymbol{m})da - \sum_{i=1}^{3}\int_{s_i} \boldsymbol{t}(\boldsymbol{x};\boldsymbol{e}_i)da, \qquad (6.102)$$

where Cauchy's lemma (6.85) has been invoked in the form $\boldsymbol{t}(\boldsymbol{x};-\boldsymbol{e}_i) = -\boldsymbol{t}(\boldsymbol{x};\boldsymbol{e}_i)$. Assuming the integrands to be continuous functions of $\boldsymbol{x}$ on their respective integration domains, we can invoke the mean-value theorem [2] to conclude that

$$\int_s \boldsymbol{t}(\boldsymbol{x};\boldsymbol{m})da = A(s)\boldsymbol{t}(\boldsymbol{x}^*;\boldsymbol{m}) \quad \text{for some} \quad \boldsymbol{x}^* \in s, \quad \text{and}$$

$$\int_{s_i} \boldsymbol{t}(\boldsymbol{x};\boldsymbol{e}_i)da = A_i\boldsymbol{t}(\boldsymbol{x}_i^*;\boldsymbol{e}_i); \quad i \in \{1,2,3\} \quad \text{fixed}, \quad \text{for some } \boldsymbol{x}_i^* \in s_i, \qquad (6.103)$$

where A_i is the area of the plane triangle s_i.

Here, we invoke the first equality in (5.9) to conclude, with reference to Figure 6.8 and noting that $\boldsymbol{n}$ is piecewise uniform on $\partial\tau$, that $A(s)\boldsymbol{m} - A_j\boldsymbol{e}_j = \boldsymbol{0}$. Accordingly,

$$A_i = A(s)\boldsymbol{e}_i \cdot \boldsymbol{m}, \qquad (6.104)$$

yielding

$$\frac{1}{A(s)}\int_{\partial\tau} \boldsymbol{t}(\boldsymbol{x};\boldsymbol{n})da = \boldsymbol{t}(\boldsymbol{x}^*;\boldsymbol{m}) - \sum_{i=1}^{3}(\boldsymbol{e}_i\cdot\boldsymbol{m})\boldsymbol{t}(\boldsymbol{x}_i^*;\boldsymbol{e}_i). \qquad (6.105)$$

Passing to the limit $\delta \to 0$, on noting that $\boldsymbol{x}^* \to \boldsymbol{x}_0$ and $\boldsymbol{x}_i^* \to \boldsymbol{x}_0$ we then have, from (6.101), that

$$\boldsymbol{t}(\boldsymbol{x}_0;\boldsymbol{m}) = \boldsymbol{T}(\boldsymbol{x}_0)\boldsymbol{m}, \quad \text{where } \boldsymbol{T}(\boldsymbol{x}_0) = \boldsymbol{t}(\boldsymbol{x}_0;\boldsymbol{e}_i)\otimes\boldsymbol{e}_i. \qquad (6.106)$$

This establishes (6.94) on reinstating the argument t. Thus,

$$\boldsymbol{T}(\boldsymbol{x},t) = \boldsymbol{t}_i \otimes \boldsymbol{e}_i, \quad \text{where } \boldsymbol{t}_i = \boldsymbol{t}(\boldsymbol{x},t;\boldsymbol{e}_i) \qquad (6.107)$$

is the traction acting on the ith coordinate plane at $\boldsymbol{x} \in \kappa_t$.

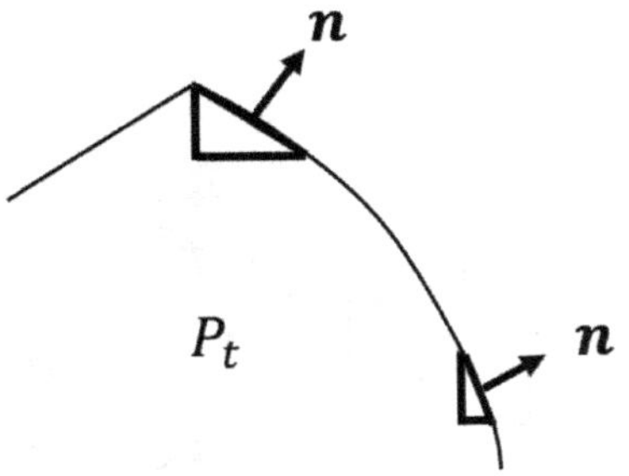

Figure 6.9. Tetrahedron-like regions adjoining the boundary of a part of a body.

We can repeat the foregoing construction by adjoining a small tetrahedron-like region with a curved oblique surface to the surface of an arbitrary part P_t of a body with piecewise smooth boundary ∂P_t (Figure 6.9).

This region approaches a tetrahedron in the limit as it collapses onto a point, and we again recover (6.94) at the boundary. We are thus able to express Euler's linear momentum balance (6.36) in the form

$$\int_{P_t} \rho \dot{\boldsymbol{v}} dv = \int_{P_t} \rho \boldsymbol{b} dv + \int_{\partial P_t} \boldsymbol{T}\boldsymbol{n} da. \tag{6.108}$$

Comparison with (6.2) and (6.23) immediately delivers the local form of the linear momentum balance:

$$\rho \dot{\boldsymbol{v}} = \rho \boldsymbol{b} + div \boldsymbol{T}, \tag{6.109}$$

holding in a region of smoothness, or

$$\rho \dot{v}_i = \rho b_i + T_{ij,j}, \tag{6.110}$$

in terms of Cartesian coordinates, where $T_{ij} = \boldsymbol{e}_i \cdot \boldsymbol{T}\boldsymbol{e}_j = \boldsymbol{e}_i \cdot \boldsymbol{t}(\boldsymbol{x}, t; \boldsymbol{e}_j) = \boldsymbol{e}_i \cdot \boldsymbol{t}_j$. Thus, the traction on a surface with normal $\boldsymbol{e}_j$ is $\boldsymbol{t}_j = (\boldsymbol{e}_i \cdot \boldsymbol{t}_j)\boldsymbol{e}_i = T_{ij}\boldsymbol{e}_i$ (Figure 6.10).

The relevant jump condition at a singular surface is given by (6.11):

$$[\![\rho \boldsymbol{v} \otimes (\boldsymbol{v} - \boldsymbol{u}) - \boldsymbol{T}]\!]\boldsymbol{n} = \boldsymbol{0}. \tag{6.111}$$

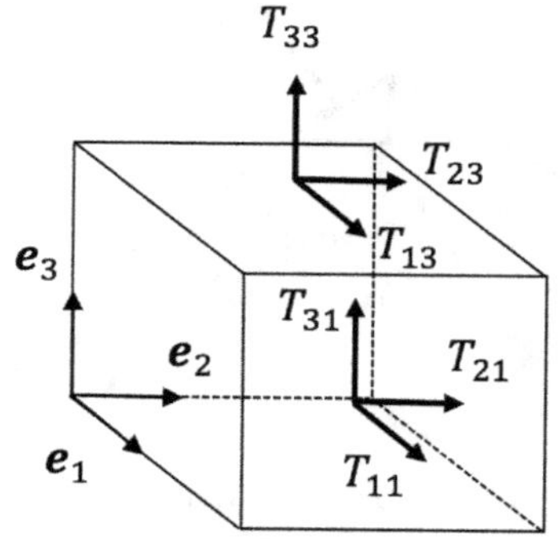

Figure 6.10. Components of traction on two faces of a block.

6.5.4 *Rotational momentum balance*

The combination of (6.94) with $\boldsymbol{x} \times \dot{\boldsymbol{v}} = (\boldsymbol{x} \times \boldsymbol{v})^{\cdot}$ reduces the rotational momentum balance (6.37) to

$$\int_{P_t} \rho(\boldsymbol{x} \times \boldsymbol{v})^{\cdot} dv = \int_{P_t} \rho \boldsymbol{x} \times \boldsymbol{b} dv + \int_{\partial P_t} \boldsymbol{x} \times \boldsymbol{T}\boldsymbol{n} da, \tag{6.112}$$

which is symbolic notation for the vector equation

$$\boldsymbol{e}_i e_{ijk} \int_{P_t} \rho(x_j v_k)^{\cdot} dv = \boldsymbol{e}_i e_{ijk} \int_{P_t} \rho x_j b_k dv + \boldsymbol{e}_i e_{ijk} \int_{\partial P_t} x_j T_{kl} n_l da, \tag{6.113}$$

where e_{ijk} is the permutation symbol. This is of the form (6.2). Taking $(6.24)_2$ into account, we have the identifications

$$\psi_i = e_{ijk} x_j v_k, \quad r_i = e_{ijk} x_j b_k \quad \text{and} \quad A_{il} = e_{ijk} x_j T_{kl}. \tag{6.114}$$

Inserting into the local balance law (6.23), i.e.,

$$\rho \dot{\psi}_i = \rho r_i + A_{il,l}, \tag{6.115}$$

expanding the derivatives, and using $e_{ijk} v_j v_k = 0$ and $x_{j,l} = \delta_{jl}$, we reduce this to

$$e_{ijk} x_j \{\rho(\dot{v}_k - b_k) + T_{kl,l}\} = e_{ijk} T_{kj}. \tag{6.116}$$

The terms in the braces vanish by (6.110), leaving the algebraic restriction $e_{ijk}T_{kj} = 0$. Recalling Problem 5 of Chapter 1, we conclude that $\boldsymbol{T}$ is a symmetric spatial tensor, i.e.,

$$\boldsymbol{T} = \boldsymbol{T}^t. \tag{6.117}$$

The jump condition (6.11) at a singular surface is

$$\begin{aligned} \mathbf{0} &= \boldsymbol{e}_i [\![\rho\psi_i(v_l - u_l) - A_{il}]\!] n_l \\ &= \boldsymbol{e}_i e_{ijk} [\![\rho x_j v_k(v_l - u_l) - x_j T_{kl}]\!] n_l. \end{aligned} \tag{6.118}$$

If the position $\boldsymbol{x}$ of a material point is continuous across the singular surface, i.e., if the motion of the body introduces no gaps or fissures in the material, then the $[\![x_j]\!]$ vanish and this reduces further to

$$\boldsymbol{x} \times \{[\![\rho\boldsymbol{v} \otimes (\boldsymbol{v} - \boldsymbol{u}) - \boldsymbol{T}]\!]\boldsymbol{n}\} = \mathbf{0}, \tag{6.119}$$

in which the term in braces vanishes by (6.111). In this case the jump condition is identically satisfied. In the event that position is discontinuous, however, (6.118) yields a non-trivial jump condition.

6.5.5 *Decomposition of the stress into normal and shear components*

Consider the traction acting on the tangent plane to a surface $s \subset \kappa_t$ at the point $\boldsymbol{x}$ (Figure 6.11). The traction on this plane at the point $\boldsymbol{x}$ is $\boldsymbol{t} = \boldsymbol{T}\boldsymbol{n}$, where $\boldsymbol{n}$ is a unit normal to the tangent plane.

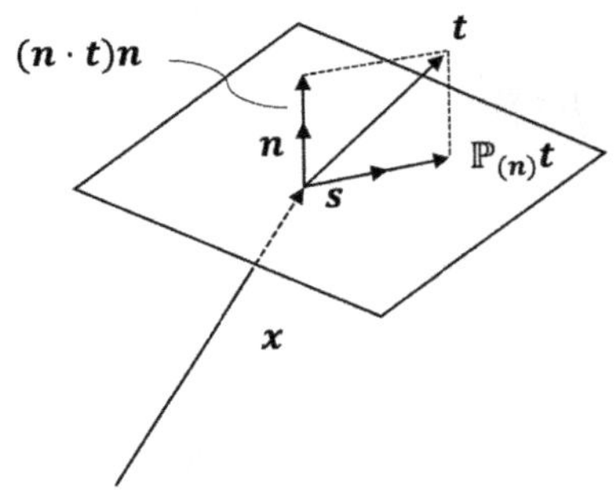

Figure 6.11. Resolution of traction into normal and shear components.

Evidently,

$$t = It = (n \otimes n)t + \mathbb{P}_{(n)}t, \tag{6.120}$$

where I is the spatial identity and

$$\mathbb{P}_{(n)} = I - n \otimes n \tag{6.121}$$

is the projection onto the tangent plane (see Problem 10 in Chapter 1). We write this in the form

$$t = \sigma n + \tau s, \tag{6.122}$$

where

$$\sigma = n \cdot t \quad \text{and} \quad \tau s = \mathbb{P}_{(n)}t, \tag{6.123}$$

with $|s| = 1$ and

$$\tau = \left|\mathbb{P}_{(n)}t\right|. \tag{6.124}$$

We refer to σ and τ respectively as the *normal* and *shear* components of the traction. These are given in terms of the stress tensor by

$$\sigma = n \cdot Tn \quad \text{and} \quad \tau = s \cdot Tn. \tag{6.125}$$

The normal part can be positive, negative or zero, whereas the shear part is non-negative. We observe that $s \cdot Tn = n \cdot T^t s = n \cdot Ts$. This is called the *complementary property of shear.* It is also clear that the shear traction is determined entirely by T and n. Thus,

$$\tau s = Tn - (n \cdot Tn)n. \tag{6.126}$$

If τ vanishes on the tangent plane at the point in question, then

$$Tn = \sigma n, \tag{6.127}$$

and n is a principal vector of T associated with the *principal stress* σ. Because of the symmetry of T there are three real-valued principal stresses and an associated orthonormal triad of principal vectors. Thus, there exist three mutually orthogonal planes intersecting at each $x \in \kappa_t$ on which the shear tractions vanish. These are uniquely determined if all the $\sigma' s$ are distinct.

Examples of states of stress

1. Pure pressure

In a state of pure pressure the traction on every plane containing a point $\boldsymbol{x}$ is perpendicular to that plane. Thus, $\boldsymbol{t}(\boldsymbol{x}, t; \boldsymbol{n}) = -p(\boldsymbol{x}, t)\boldsymbol{n}$ for all unit vectors $\boldsymbol{n}$, where p is the pressure. Suppressing the passive arguments $\boldsymbol{x}$ and t, we have that $\boldsymbol{T}\boldsymbol{n} = -p\boldsymbol{n}$. On multiplying by an arbitrary scalar v, say, we conclude that $\boldsymbol{T}\boldsymbol{v} = -p\boldsymbol{v} = (-p\boldsymbol{I})\boldsymbol{v}$ in which $\boldsymbol{v} = v\boldsymbol{n}$ is an arbitrary vector. From (1.41), it then follows that the stress is

$$\boldsymbol{T} = -p\boldsymbol{I}, \tag{6.128}$$

which is clearly symmetric, as required.

Alternatively, $T_{ij} = \boldsymbol{e}_i \cdot \boldsymbol{T}\boldsymbol{e}_j = \boldsymbol{e}_i \cdot \boldsymbol{t}(\boldsymbol{e}_j) = -p\boldsymbol{e}_i \cdot \boldsymbol{e}_j = -p\delta_{ij}$, which of course is the component form of our result. We may also derive it by using (6.107) to obtain $\boldsymbol{T}(\boldsymbol{x}, t) = \boldsymbol{t}(\boldsymbol{x}, t; \boldsymbol{e}_i) \otimes \boldsymbol{e}_i = -p(\boldsymbol{x}, t)\boldsymbol{e}_i \otimes \boldsymbol{e}_i$ and recalling (1.54).

2. Uniaxial tension/compression

In this example, the traction is of the form $\boldsymbol{t}(\boldsymbol{x}, t; \boldsymbol{n}) = \alpha(\boldsymbol{x}, t; \boldsymbol{n})\boldsymbol{e}$ for all unit vectors $\boldsymbol{n}$, where α is a scalar and $\boldsymbol{e}$ is a fixed unit vector (Figure 6.12).

Thus, $\boldsymbol{T}(\boldsymbol{x}, t)\boldsymbol{n} = \alpha(\boldsymbol{x}, t; \boldsymbol{n})\boldsymbol{e}$. Let $\alpha^*(\boldsymbol{x}, t; \cdot)$ be an extension of $\alpha(\boldsymbol{x}, t; \cdot)$ from the set of unit vectors to the set of all vectors, i.e., $\alpha^*(\boldsymbol{x}, t; \boldsymbol{v}) = \alpha(\boldsymbol{x}, t; \boldsymbol{v})$ for all $\boldsymbol{v}$ with $|\boldsymbol{v}| = 1$. Then, $\boldsymbol{T}(\boldsymbol{x}, t)\boldsymbol{v} = \alpha^*(\boldsymbol{x}, t; \boldsymbol{v})\boldsymbol{e}$ for all $\boldsymbol{v}$. Clearly $\boldsymbol{T}(\boldsymbol{x}, t)\boldsymbol{v}$ is a linear function of $\boldsymbol{v}$. Thus $\alpha^*(\boldsymbol{x}, t; \boldsymbol{v})$ is also a linear function of $\boldsymbol{v}$, and it follows from (1.19) that there is a vector function $\boldsymbol{a}(\boldsymbol{x}, t)$ such that $\alpha^*(\boldsymbol{x}, t; \boldsymbol{v}) = \boldsymbol{a}(\boldsymbol{x}, t) \cdot \boldsymbol{v}$. We note that to reach this conclusion it is necessary to introduce the extension α^* because the set of unit vectors is not a linear space; that is, an arbitrary linear combination of unit vectors is not a unit vector, and, moreover, the zero vector is obviously not a unit vector.

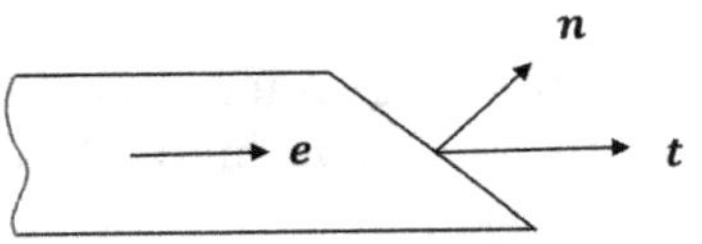

Figure 6.12. Uniaxial tension/compression.

We thus have $\boldsymbol{Tv} = (\boldsymbol{a} \cdot \boldsymbol{v})\boldsymbol{e} = (\boldsymbol{e} \otimes \boldsymbol{a})\boldsymbol{v}$ for all $\boldsymbol{v}$ and hence $\boldsymbol{T} = \boldsymbol{e} \otimes \boldsymbol{a}$. The symmetry condition (6.117) requires that $\boldsymbol{e} \otimes \boldsymbol{a} = \boldsymbol{a} \otimes \boldsymbol{e}$, and this in turn yields

$$\boldsymbol{a} = \boldsymbol{a}(\boldsymbol{e} \cdot \boldsymbol{e}) = (\boldsymbol{a} \otimes \boldsymbol{e})\boldsymbol{e} = (\boldsymbol{e} \otimes \boldsymbol{a})\boldsymbol{e} = \sigma\boldsymbol{e}, \quad \text{where } \sigma = \boldsymbol{a} \cdot \boldsymbol{e}. \tag{6.129}$$

Finally,

$$\boldsymbol{T}(\boldsymbol{x}, t) = \sigma(\boldsymbol{x}, t)\boldsymbol{e} \otimes \boldsymbol{e}, \tag{6.130}$$

where σ is the uniaxial tension or compression according as it is positive or negative, respectively.

6.5.6 *Referential equations of motion*

The reader will have noticed that we have obtained the local equations and associated jump conditions entirely in terms of the spatial description. In particular, we have not used the referential expressions $(6.25)_2$, $(6.27)_2$ or (6.29) in the course of deducing local equations and jump conditions from the linear momentum balance (6.34). We do so here in respect of the local equations and leave the derivation of the associated jump conditions as an exercise. On combining these equations we arrive at the linear momentum balance for an arbitrary part $P \subset \kappa$:

$$\int_{\partial P} \boldsymbol{p} dA = \int_P \rho_\kappa(\dot{\boldsymbol{v}} - \boldsymbol{b})dV, \tag{6.131}$$

where $\boldsymbol{p}$ is the Piola traction. We could use this in a version of Noll's theorem to conclude that $\boldsymbol{p}$ depends on ∂P through its local unit normal $\boldsymbol{N}$, and then adapt Cauchy's theorem to conclude that

$$\boldsymbol{p}(\boldsymbol{X}, t; \boldsymbol{N}) = \boldsymbol{P}(\boldsymbol{X}, t)\boldsymbol{N}, \tag{6.132}$$

where $\boldsymbol{P}$ — the *Piola Stress* — is a two-point tensor field; the Cartesian-component representation is

$$\boldsymbol{P} = P_{iA}\boldsymbol{e}_i \otimes \hat{\boldsymbol{E}}_A. \tag{6.133}$$

We can combine this with the referential form of the divergence theorem, for smooth (and hence continuous) fields, to obtain

$$\int_P \rho_\kappa(\dot{\boldsymbol{v}} - \boldsymbol{b})dV = \int_{\partial P} \boldsymbol{PN} dA = \int_P Div\boldsymbol{P} dV, \tag{6.134}$$

where, in terms of Cartesians,

$$Div\boldsymbol{P} = P_{iA,A}\boldsymbol{e}_i, \tag{6.135}$$

and then localize as usual to conclude that

$$\rho_\kappa \dot{\boldsymbol{v}} = \rho_\kappa \boldsymbol{b} + Div\boldsymbol{P}, \tag{6.136}$$

i.e.,

$$\rho_\kappa \dot{v}_i = \rho_\kappa b_i + P_{iA,A}, \tag{6.137}$$

at all points $\boldsymbol{X} \in \kappa$ that do not belong to a singular surface.

This formulation has the convenient feature that $\rho_\kappa(\boldsymbol{X})$ is a specified function, whereas the spatial mass density ρ appearing in the spatial equations must be such as to satisfy the differential equation (6.14), involving the material velocity field, and is not known in advance. Moreover, $\dot{\boldsymbol{v}} = \partial\hat{\boldsymbol{v}}(\boldsymbol{X},t)/\partial t$ is simply the partial time derivative in the referential description.

The connection (2.59) between the referential and spatial mass densities, combined with (6.109), suggests that if $\boldsymbol{P}$ is such that

$$Jdiv\boldsymbol{T} = Div\boldsymbol{P}, \tag{6.138}$$

then the referential and spatial forms of the balance of linear momentum are equivalent. To determine the requisite expression for $\boldsymbol{P}$ we combine (6.94) with (2.50) and the Piola–Nanson formula (2.54), obtaining

$$\begin{aligned}\int_P Jdiv\boldsymbol{T}dV = \int_{P_t} div\boldsymbol{T}dv = \int_{\partial P_t} \boldsymbol{Tn}da = \int_{\partial P} \boldsymbol{TF}^*\boldsymbol{N}dA \\ = \int_P Div(\boldsymbol{TF}^*)dV. \end{aligned} \tag{6.139}$$

Collecting the first and last integrals together and localizing then gives (6.138) with

$$\boldsymbol{P} = \boldsymbol{TF}^*, \tag{6.140}$$

or, terms of components,

$$P_{iA} = T_{ij}F^*_{jA}. \tag{6.141}$$

Indeed, the two expressions (6.28) and (6.29) for the contact force $\boldsymbol{f}_c$ imply that

$$\int_{\partial P} \boldsymbol{p} dA = \int_{\partial P_t} \boldsymbol{t} da = \int_{\partial P_t} \boldsymbol{T}\boldsymbol{n} da = \int_{\partial P} \boldsymbol{T}\boldsymbol{F}^* \boldsymbol{N} dA, \tag{6.142}$$

which is identically satisfied by (6.132) and (6.140).

The symmetry of the Cauchy stress imposes a restriction on the Piola stress and the deformation jointly. This follows easily on combining (6.140) with (2.55):

$$\boldsymbol{P}\boldsymbol{F}^t = \boldsymbol{F}\boldsymbol{P}^t, \tag{6.143}$$

i.e., $P_{iA}F_{jA} = F_{iA}P_{jA}$.

The purely referential *Piola–Kirchhoff Stress* $\boldsymbol{S}$, defined by

$$\boldsymbol{P} = \boldsymbol{F}\boldsymbol{S}, \tag{6.144}$$

is often useful in the formulation of constitutive equations, especially for solids. Its relationship to the Cauchy stress follows easily from (6.140) and (2.55). Thus,

$$J\boldsymbol{T} = \boldsymbol{F}\boldsymbol{S}\boldsymbol{F}^t. \tag{6.145}$$

It is clearly symmetric because $J\boldsymbol{T}^t = \boldsymbol{F}\boldsymbol{S}^t\boldsymbol{F}^t$ and $\boldsymbol{T}^t = \boldsymbol{T}$. We then have $\boldsymbol{F}\boldsymbol{S}^t\boldsymbol{F}^t = \boldsymbol{F}\boldsymbol{S}\boldsymbol{F}^t$, and pre- and post-multiplication by $\boldsymbol{F}^{-1}$ and $\boldsymbol{F}^{-t}$, respectively, yields $\boldsymbol{S}^t = \boldsymbol{S}$. In terms of components,

$$P_{iA} = F_{iB}S_{BA} \quad \text{and} \quad JT_{ij} = F_{iA}F_{jB}S_{AB}. \tag{6.146}$$

6.5.7 *Mechanical energy balance*

In particle mechanics the power generated by the force $\boldsymbol{f}$ acting on a particle is balanced by the rate of change of the kinetic energy of the particle. This follows easily from Newton's law $\boldsymbol{f} = m\dot{\boldsymbol{v}}$, where m and $\boldsymbol{v}$ respectively are, of course, the particle mass and velocity. The power is $\mathcal{P} = \boldsymbol{f} \cdot \boldsymbol{v} = m\boldsymbol{v} \cdot \dot{\boldsymbol{v}} = \frac{1}{2}m(\boldsymbol{v} \cdot \dot{\boldsymbol{v}} + \dot{\boldsymbol{v}} \cdot \boldsymbol{v}) = \frac{1}{2}m(\boldsymbol{v} \cdot \boldsymbol{v})^{\cdot}$. Thus, $\mathcal{P} = \dot{\mathcal{K}}$ as claimed, where $\mathcal{K} = \frac{1}{2}m\,|\boldsymbol{v}|^2$ is the kinetic energy. We derive an analogous result for continua.

To achieve this we form the inner product of (6.109) with the material velocity $\boldsymbol{v}$, obtaining

$$\frac{1}{2}\rho(|\boldsymbol{v}|^2)^{\cdot} = \rho\boldsymbol{b}\cdot\boldsymbol{v} + div\boldsymbol{T}\cdot\boldsymbol{v}$$
$$= \rho\boldsymbol{b}\cdot\boldsymbol{v} + div(\boldsymbol{T}^t\boldsymbol{v}) - \boldsymbol{T}\cdot\boldsymbol{L}, \tag{6.147}$$

where $\boldsymbol{L}$ is the spatial velocity gradient defined by $(4.1)_2$, the inner product is defined by (1.87), and use has been made of the identity (1.152) together with the symmetry $tr(\boldsymbol{AB}) = tr(\boldsymbol{BA})$. A reduction employing Cartesian coordinates is arguably more enlightening:

$$div\boldsymbol{T}\cdot\boldsymbol{v} = T_{ij,j}v_i = (T_{ij}v_i)_{,j} - T_{ij}v_{i,j} = div(\boldsymbol{T}^t\boldsymbol{v}) - \boldsymbol{T}\cdot\boldsymbol{L}. \tag{6.148}$$

Integrating over an arbitrary part $P_t = \boldsymbol{\chi}(S,t) \subset \kappa_t$ and invoking conservation of mass gives

$$\frac{d}{dt}\int_{P_t}\frac{1}{2}\rho\,|\boldsymbol{v}|^2\,dv + \int_{P_t}\boldsymbol{T}\cdot\boldsymbol{L}dv = \int_{P_t}\rho\boldsymbol{b}\cdot\boldsymbol{v}dv + \int_{\partial P_t}\boldsymbol{t}\cdot\boldsymbol{v}da, \tag{6.149}$$

where the time derivative pertains to the fixed set S of material points under consideration and we have used the divergence theorem for smooth fields to convert the integral of $div(\boldsymbol{T}^t\boldsymbol{v})$ over P_t to an integral over ∂P_t of $\boldsymbol{T}^t\boldsymbol{v}\cdot\boldsymbol{n} = \boldsymbol{v}\cdot\boldsymbol{Tn} = \boldsymbol{t}\cdot\boldsymbol{v}$.

Whence the *Mechanical Energy Balance*:

$$\frac{d}{dt}\mathcal{K}(S,t) + \mathcal{S}(S,t) = \mathcal{P}(S,t), \tag{6.150}$$

where

$$\mathcal{K}(S,t) = \frac{1}{2}\int_{P_t}\rho\,|\boldsymbol{v}|^2\,dv \tag{6.151}$$

is the kinetic energy of S,

$$\mathcal{P}(S,t) = \int_{P_t}\rho\boldsymbol{b}\cdot\boldsymbol{v}dv + \int_{\partial P_t}\boldsymbol{t}\cdot\boldsymbol{v}da \tag{6.152}$$

is the power of the forces acting on S, and

$$\mathcal{S}(S,t) = \int_{P_t} \boldsymbol{T} \cdot \boldsymbol{L} dv \tag{6.153}$$

is the *stress power* contained in S.

We observe that the stress power has no analog in particle mechanics. This is due, naturally, to the fact that ideal particles have no surface area. Thus, the concepts of traction and stress have no meaning for them.

The referential forms of the kinetic energy and power are

$$\mathcal{K}(S,t) = \frac{1}{2}\int_{P} \rho_\kappa |\boldsymbol{v}|^2 \, dV \quad \text{and}$$

$$\mathcal{P}(S,t) = \int_{P} \rho_\kappa \boldsymbol{b} \cdot \boldsymbol{v} dV + \int_{\partial P} \boldsymbol{p} \cdot \boldsymbol{v} dA, \tag{6.154}$$

respectively. As for the stress power, we use (6.140) to write

$$\begin{aligned}
\int_{P_t} \boldsymbol{T} \cdot \boldsymbol{L} dv &= \int_{P_t} J^{-1} \boldsymbol{P}\boldsymbol{F}^t \cdot \boldsymbol{L} dv \\
&= \int_{P} \boldsymbol{P}\boldsymbol{F}^t \cdot \boldsymbol{L} dV \\
&= \int_{P} tr(\boldsymbol{P}\boldsymbol{F}^t \boldsymbol{L}^t) dV.
\end{aligned} \tag{6.155}$$

Using $\boldsymbol{F}^t\boldsymbol{L}^t = (\boldsymbol{L}\boldsymbol{F})^t = \dot{\boldsymbol{F}}^t$ we then have $tr(\boldsymbol{P}\boldsymbol{F}^t\boldsymbol{L}^t) = tr(\boldsymbol{P}\dot{\boldsymbol{F}}^t) = \boldsymbol{P} \cdot \dot{\boldsymbol{F}}$. In terms of components, $\boldsymbol{P} \cdot \dot{\boldsymbol{F}} = P_{iA}\dot{F}_{iA}$. We note that whereas the trace of a two-point tensor is not defined, the product $\boldsymbol{P}\dot{\boldsymbol{F}}^t$ is a spatial tensor, and its trace furnishes a well-defined inner product of two-point tensors. Accordingly, the trace is meaningful and we reach

$$\mathcal{S}(S,t) = \int_{P} \boldsymbol{P} \cdot \dot{\boldsymbol{F}} dV. \tag{6.156}$$

Indeed the energy balance (6.150), with the kinetic energy, power and stress power given by (6.154) and (6.156), may be derived directly by repeating the foregoing procedure with (6.109) replaced by (6.136).

A stress measure that is juxtaposed with the material derivative of a deformation measure in an expression for the stress power is said to be *power conjugate* to that deformation measure. Thus, the Piola stress is power conjugate to the deformation gradient. We show that the Piola–Kirchhoff stress is power conjugate to the Lagrangian strain $\boldsymbol{E}$ defined by (2.97). This is easily proved by combining $(2.83)_2$ with (2.97) to obtain

$$\begin{aligned}\dot{\boldsymbol{E}} &= \frac{1}{2}(\boldsymbol{F}^t\boldsymbol{F} - \hat{\boldsymbol{I}})^{\cdot} = \frac{1}{2}(\boldsymbol{F}^t\dot{\boldsymbol{F}} + \dot{\boldsymbol{F}}^t\boldsymbol{F}) \\ &= \frac{1}{2}[\boldsymbol{F}^t\dot{\boldsymbol{F}} + (\boldsymbol{F}^t\dot{\boldsymbol{F}})^t] = Sym(\boldsymbol{F}^t\dot{\boldsymbol{F}}).\end{aligned} \tag{6.157}$$

Then,

$$\boldsymbol{P}\cdot\dot{\boldsymbol{F}} = \boldsymbol{F}\boldsymbol{S}\cdot\dot{\boldsymbol{F}} = tr(\boldsymbol{F}\boldsymbol{S}\dot{\boldsymbol{F}}^t) = tr(\boldsymbol{S}\dot{\boldsymbol{F}}^t\boldsymbol{F}), \tag{6.158}$$

in which the first trace pertains to a spatial tensor, yielding an inner product of two-point tensors, and the second pertains to a referential tensor. In detail, $tr(\boldsymbol{F}\boldsymbol{S}\dot{\boldsymbol{F}}^t) = (F_{iA}S_{AB})\dot{F}_{iB} = S_{AB}(\dot{F}_{iB}F_{iA}) = tr(\boldsymbol{S}\dot{\boldsymbol{F}}^t\boldsymbol{F})$. Thus, $\boldsymbol{P}\cdot\dot{\boldsymbol{F}} = \boldsymbol{S}\cdot\boldsymbol{F}^t\dot{\boldsymbol{F}}$, and the symmetry of the Piola–Kirchhoff stress reduces this to

$$\boldsymbol{P}\cdot\dot{\boldsymbol{F}} = \boldsymbol{S}\cdot Sym(\boldsymbol{F}^t\dot{\boldsymbol{F}}), \tag{6.159}$$

yielding the stress power in the form

$$\mathcal{S}(S,t) = \int_P \boldsymbol{S}\cdot\dot{\boldsymbol{E}}dV. \tag{6.160}$$

6.6 Energy balance

The mechanical energy balance (6.150) does not account for the interconvertibility of heat and mechanical power, a fundamental aspect of thermomechanical phenomena. To address this we introduce a new postulate, the *Balance of Energy.* Thus we assume the existence of scalars $\mathcal{U}(S,t)$ and $\mathcal{H}(S,t)$, the internal energy and heating of $S\subset B$, respectively, such that

$$\frac{d}{dt}[\mathcal{K}(S,t) + \mathcal{U}(S,t)] = \mathcal{P}(S,t) + \mathcal{H}(S,t). \tag{6.161}$$

We assume that $\mathcal{U}$ can be represented as an integral over P_t, i.e., that there exists a function $\varepsilon(\boldsymbol{x}, t)$ — the internal energy per unit mass — such that

$$\mathcal{U}(S, t) = \int_{P_t} \rho \varepsilon dv, \tag{6.162}$$

and that the heating is expressible in the form

$$\mathcal{H}(S, t) = \int_{P_t} \rho r dv + \int_{\partial P_t} h da, \tag{6.163}$$

where $r(\boldsymbol{x}, t)$ is the heating per unit mass — through radioactive decay, for example — and $h(\boldsymbol{x}, t)$ is the heating influx supplied via conduction; for example, by contact with material outside P_t.

We can combine the mechanical energy balance with (6.161) to arrive at the *Thermal Energy Balance*

$$\mathcal{H}(S, t) = \frac{d}{dt}\mathcal{U}(S, t) - \mathcal{S}(S, t). \tag{6.164}$$

This is coupled to the mechanical energy balance by the stress power. For example, if the stress is a pure pressure, then $\boldsymbol{T} \cdot \boldsymbol{L} = -p \, tr \boldsymbol{L} = -p\dot{J}/J$, and if the motion is isochoric ($\dot{J} = 0$), then the stress power vanishes and there is no thermomechanical coupling. The thermal and mechanical problems for inviscid incompressible fluids can thus be treated independently.

For rigid bodies the stress power is $\boldsymbol{T} \cdot \boldsymbol{L} = \boldsymbol{T} \cdot \boldsymbol{W}$, where $\boldsymbol{W}$ is the skew spin tensor, and this also vanishes because $\boldsymbol{T}$ is symmetric. Thus, the thermal and mechanical problems are decoupled in rigid bodies. This is why traditional courses on heat transfer treat bodies as rigid, with no consideration given to mechanical response. This can only be valid in an approximate sense, however, since real bodies are not perfectly rigid, just as real fluids are not perfectly inviscid.

Proceeding, we write the thermal energy balance, granted conservation of mass, as

$$\int_{\partial P_t} h da = \int_{P_t} [\rho(\dot{\varepsilon} - r) - \boldsymbol{T} \cdot \boldsymbol{L}] dv. \tag{6.165}$$

Assuming $\rho, \dot{\varepsilon}, r$ and $\boldsymbol{T} \cdot \boldsymbol{L}$ to be continuous, and hence bounded, we can adapt Noll's theorem to conclude that h depends on ∂P_t through its local unit normal $\boldsymbol{n}$, and then adapt Cauchy's theorem to conclude that there is a vector field $\boldsymbol{q}(\boldsymbol{x},t)$, the *heat flux vector*, such that

$$h(\boldsymbol{x},t;\boldsymbol{n}) = -\boldsymbol{q}(\boldsymbol{x},t)\cdot\boldsymbol{n}, \tag{6.166}$$

in which the minus sign is inserted to conform to the convention that $\boldsymbol{q}\cdot\boldsymbol{n} > 0$ corresponds to heat outflow ($h < 0$); and $\boldsymbol{q}\cdot\boldsymbol{n} < 0$, to heat inflow ($h > 0$). Then,

$$\int_{\partial P_t} h da = -\int_{\partial P_r} \boldsymbol{q}\cdot\boldsymbol{n} da = -\int_{P_t} div\boldsymbol{q} dv, \tag{6.167}$$

for smooth fields, and we can localize (6.165) to arrive at the local form

$$\rho\dot{\varepsilon} = \boldsymbol{T}\cdot\boldsymbol{L} - div\boldsymbol{q} + \rho r, \tag{6.168}$$

holding at points in κ_t that do not belong to a singular surface. Alternatively, this follows directly on making the appropriate identifications in (6.1) and (6.21), and (6.6) then delivers the associated jump condition

$$[\![\rho\varepsilon(\boldsymbol{v}-\boldsymbol{u}) + \boldsymbol{q}]\!]\cdot\boldsymbol{n} = 0, \tag{6.169}$$

holding at all points on a singular surface.

A referential form of the energy balance proceeds by re-writing the thermal energy balance as

$$\begin{aligned}\int_P [\rho_\kappa(\dot{\varepsilon} - r) - \boldsymbol{P}\cdot\dot{\boldsymbol{F}}]dV &= -\int_{\partial P_r} \boldsymbol{q}\cdot\boldsymbol{n} da = -\int_{\partial P} \boldsymbol{q}\cdot\boldsymbol{F}^*\boldsymbol{N} dA \\ &= -\int_{\partial P} \boldsymbol{q}_\kappa\cdot\boldsymbol{N} dA,\end{aligned} \tag{6.170}$$

where

$$\boldsymbol{q}_\kappa(\boldsymbol{X},t) = (\boldsymbol{F}^*)^t\boldsymbol{q} = J\boldsymbol{F}^{-1}\boldsymbol{q} \tag{6.171}$$

is the referential heat flux vector, with components

$$q_{(\kappa)A} = F^*_{iA}q_i, \tag{6.172}$$

in which q_i are those of the spatial vector $\boldsymbol{q}$. Then, for smooth fields,

$$\int_{\partial P} \boldsymbol{q}_\kappa\cdot\boldsymbol{N} dA = \int_P Div\boldsymbol{q}_\kappa dV, \quad \text{where } Div\boldsymbol{q}_\kappa = q_{(\kappa)A,A}, \tag{6.173}$$

and we localize to obtain

$$\rho_\kappa \dot{\varepsilon} = \boldsymbol{P} \cdot \dot{\boldsymbol{F}} - Div \boldsymbol{q}_\kappa + \rho_\kappa r \tag{6.174}$$

at points $\boldsymbol{X} \in \kappa$ removed from a singular surface. The derivation of the associated jump condition is again left to the reader.

This result may be derived directly if (6.168) is multiplied by J and use is made of the easily proved identity

$$J div \boldsymbol{q} = Div(J\boldsymbol{F}^{-1}\boldsymbol{q}). \tag{6.175}$$

6.7 Entropy and the Clausius–Duhem inequality

To account for the thermodynamical concept of dissipation attending any real physical process, we introduce a further hypothesis: the *Entropy Balance*. This is assumed to take the form

$$\frac{d}{dt}\int_{P_t} \rho\eta dv = \int_{P_t} \theta^{-1}(\rho r) dv + \int_{\partial P_t} \theta^{-1} h da + \int_{P_t} \rho s dv, \tag{6.176}$$

where $\eta(\boldsymbol{x}, t)$ is the entropy, per unit mass, contained in S, $\theta(\boldsymbol{x}, t)$ (> 0) is the absolute temperature, and s is the entropy production rate per unit mass. The first integral on the right-hand side represents the entropy supplied by sources in the bulk and the second, the flux of entropy through the surface. Their forms are suggested by the classical thermostatics of reversible processes, whereas the last integral on the right, which vanishes in a reversible process, represents the net entropy production attending an irreversible process. These are, by definition, such that this integral is positive and hence that $s > 0$. All admissible processes are then subject to the *Clausius–Duhem Inequality*

$$\frac{d}{dt}\int_{P_t} \rho\eta dv \geq \int_{P_t} \theta^{-1}(\rho r) dv + \int_{\partial P_t} \theta^{-1} h da. \tag{6.177}$$

This inequality may be motivated by an argument due to Truesdell [12]. Thus, we assume that there is a function $\mathcal{B}(S, t)$, depending on material constitution, that bounds the heating from above, i.e., $\mathcal{B}(S, t) \geq \mathcal{H}(S, t)$. From the energy balance (6.161), we then have that, in accordance with experience, there is a limit to the rate at which heat can be converted to energy in the absence of forces ($\mathcal{P} = 0$), i.e.,

$$\frac{d}{dt}[\mathcal{K}(S,t) + \mathcal{U}(S,t)] \leq \mathcal{B}(S,t). \tag{6.178}$$

For uniformly distributed temperatures ($grad\theta = \mathbf{0}$) we define the entropy

$$\mathcal{C}(S,t) = \int_{P_t} \rho\eta dv \tag{6.179}$$

by

$$\mathcal{C}(S,t) = \int_{t_0}^{t} \theta(\tau)^{-1}\mathcal{B}(S,\tau)d\tau, \tag{6.180}$$

where t_0 is a fixed time. Then,

$$\theta(t)\frac{d}{dt}\mathcal{C}(S,t) = \mathcal{B}(S,t) \geq \mathcal{H}(S,t), \tag{6.181}$$

which is just inequality (6.177) with the uniform reciprocal temperature factored outside the integrals on the right-hand side. Placing this inside the integrals and making the further assumption that the resulting inequality also applies to non-uniform temperature fields then yields the Clausius–Duhem inequality. Combining it with (6.166) and localizing as usual yields the inequality

$$\rho\dot{\eta} \geq \theta^{-1}(\rho r) - div(\theta^{-1}\boldsymbol{q}), \tag{6.182}$$

holding in regions of smoothness, whereas the jump condition

$$[\![\rho\eta(\boldsymbol{v} - \boldsymbol{u}) + \theta^{-1}\boldsymbol{q}]\!] \cdot \boldsymbol{n} \geq 0, \tag{6.183}$$

operative on singular surfaces, follows by making the appropriate identifications in (6.6) in which the equality is replaced by an inequality. Here we recall, with reference to Figure 5.2 in Chapter 5, that $\boldsymbol{n}$ is directed into the "+ region", and hence, with our definition (5.28) of the jump, that this inequality is independent of the orientation of $\boldsymbol{n}$.

The referential form of the Clausius–Duhem inequality and its associated local form and jump condition are easily derived and therefore not made explicit here. Further, the entropy supply and flux represented by the first two integrals on the right-hand side of the entropy balance (6.176) are not the most general forms that could be contemplated. The text by Liu [7] may be consulted for a general formulation of this balance law.

A more useful form of the local inequality (6.182) is obtained by substituting ρr from the thermal energy balance (6.168):

$$\rho\dot{\eta} \geq \theta^{-1}(\rho\dot{\varepsilon} - \boldsymbol{T}\cdot\boldsymbol{L} + div\boldsymbol{q}) - div(\theta^{-1}\boldsymbol{q}), \tag{6.184}$$

which, since $\theta > 0$ and recalling Problem 24(c) of Chapter 1, is equivalent to

$$\rho(\theta\dot{\eta} - \dot{\varepsilon}) + \boldsymbol{T}\cdot\boldsymbol{L} - \theta^{-1}\boldsymbol{q}\cdot grad\theta \geq 0. \tag{6.185}$$

On introducing the *Helmholtz free energy* per unit mass ψ, defined by

$$\psi = \varepsilon - \theta\eta, \tag{6.186}$$

we finally reduce the local Clausius–Duhem inequality to

$$-\rho(\dot{\psi} + \eta\dot{\theta}) + \boldsymbol{T}\cdot\boldsymbol{L} - \theta^{-1}\boldsymbol{q}\cdot grad\theta \geq 0. \tag{6.187}$$

This will play a central role in the discussion of constitutive equations for materials in subsequent chapters.

6.8 Problems

1. Derive the referential forms of the local balance laws and jump conditions discussed in Section 6.1. [*Hint*: Recall Problem 3 in Chapter 5].
2. Verify that $\bar{\boldsymbol{X}}$ appearing in (6.53) is, in fact, the position of the center of mass of $P \subset \kappa$.
3. Show that the co-rotational derivative defined by (6.68) is expressible in the form $\mathring{\boldsymbol{J}} = \dot{J}_{ij}\boldsymbol{e}_i^* \otimes \boldsymbol{e}_j^*$.
4. Recall the inertia tensor $\boldsymbol{J}(S,t) = \int_{P_t}\rho[(\boldsymbol{\pi}\cdot\boldsymbol{\pi})\boldsymbol{I} - \boldsymbol{\pi}\otimes\boldsymbol{\pi}]dv$ associated with an arbitrary part S of a body, where $\boldsymbol{\pi}$ is the position of a point in the region P_t, currently occupied by S, relative to the center of mass of P_t.

(a) Show that the integrand is positive semi-definite (A symmetric tensor $\boldsymbol{A}$ is positive semi-definite if and only if $\boldsymbol{a} \cdot \boldsymbol{A}\boldsymbol{a} \geq 0$ for all $\boldsymbol{a} \in E^3$).

(b) Show that $\boldsymbol{J}$ is *positive definite* ($\boldsymbol{a} \cdot \boldsymbol{J}\boldsymbol{a} > 0$ for all $\boldsymbol{a} \neq \boldsymbol{0}$).

5. Let $\{\sigma_i\}$ and $\{\boldsymbol{n}_i\}$ be the principal Cauchy stresses and associated (normalized) principal stress axes at a given position $\boldsymbol{x}$ in a body. We can use these axes to define a unit vector $\boldsymbol{n}(\theta, \phi) = \cos\phi\, \boldsymbol{e}_r(\theta) + \sin\phi\, \boldsymbol{n}_3$, where $\boldsymbol{e}_r(\theta) = \cos\theta\, \boldsymbol{n}_1 + \sin\theta\, \boldsymbol{n}_2$ and $0 \leq \theta < 2\pi$, $-\pi \leq \phi \leq \pi$. This is essentially the radial unit vector with origin located at $\boldsymbol{x}$.

   (a) Compute the shear stress $\tau(\theta, \phi)$ on a plane with unit normal $\boldsymbol{n}(\theta, \phi)$.

   (b) Compute the *octahedral* shear stress, i.e., the shear stress on a plane with normal inclined at equal angles to each of the principal axes.

6. Consider a body undergoing a homogeneous deformation, such that the components of the deformation gradient at a given time t are

$$(F_{iA}) = \begin{pmatrix} 2 & 3 & 0 \\ 1 & 2 & 0 \\ 0 & 0 & 1 \end{pmatrix}.$$

   All components are with respect to the fixed basis $\{\boldsymbol{e}_i \otimes \hat{\boldsymbol{E}}_A\}$. Assume that the Piola tractions $\boldsymbol{p}(\boldsymbol{N})$ acting on deformed surfaces that have unit normals $\hat{\boldsymbol{E}}_A$ in the reference configuration are

$$\boldsymbol{p}(\hat{\boldsymbol{E}}_1) = a\boldsymbol{e}_1 + 10\boldsymbol{e}_2,$$

$$\boldsymbol{p}(\hat{\boldsymbol{E}}_2) = (b-6)\boldsymbol{e}_3 + 3\boldsymbol{e}_2,$$

$$\boldsymbol{p}(\hat{\boldsymbol{E}}_3) = c\boldsymbol{e}_2 + 3\boldsymbol{e}_3,$$

   where a, b and c are constants.

   (a) Use $\boldsymbol{p}(\boldsymbol{N}) = \boldsymbol{P}\boldsymbol{N}$, where $\boldsymbol{P}$ is the Piola stress tensor, together with $\hat{\boldsymbol{I}} = \hat{\boldsymbol{E}}_A \otimes \hat{\boldsymbol{E}}_A$, to show that $\boldsymbol{P} = \boldsymbol{p}_A \otimes \hat{\boldsymbol{E}}_A$, where $\boldsymbol{p}_A = \boldsymbol{p}(\hat{\boldsymbol{E}}_A)$.

(b) Find the components P_{iA} of the Piola stress tensor as functions of a, b, c.
(c) Apply the rotational momentum balance $\boldsymbol{T} = \boldsymbol{T}^t$ to determine the values of a, b, c.
(d) Calculate the values of the components T_{ij} of the Cauchy stress tensor $\boldsymbol{T}$.

7. Consider the deformation

$$x_1 = a_1(X_1 + bX_2), \quad x_2 = a_2X_2, \quad x_3 = a_3X_3.$$

Suppose the Cauchy traction is given by $\boldsymbol{t} = \tau\boldsymbol{e}_1, \tau\boldsymbol{e}_2, \boldsymbol{0}$ on planes with unit normals $\boldsymbol{n} = \boldsymbol{e}_2, \boldsymbol{e}_1$ and $\boldsymbol{e}_3$, respectively (carefully note the correspondences). Find the stress tensors $\boldsymbol{T}$, $\boldsymbol{P}$ and $\boldsymbol{S}$. Express your answers in terms of τ, a_1, a_2, a_3 and b.
8. Establish the formulas (6.138) and (6.175) by direct calculation with the aid of the Piola identity.
9. Suppose a body is in equilibrium in configuration κ_t. Derive the *virtual power* equation

$$\int_{P_t} T_{ij}u_{i,j}dv = \int_{P_t} \rho b_i u_i dv + \int_{\partial P_t} t_i u_i da,$$

where $P_t \subset \kappa_t$ is any part of κ_t and $\boldsymbol{u}$ is any differentiable vector field whatsoever. This is also called the *weak form* of the equilibrium equation, because it is well defined even if the T_{ij} are not differentiable. The local form derived in the text requires differentiability — a stronger requirement — and is accordingly called the *strong form.* Note: This formula is the basis of the Finite Element Method.
10. Let $V(\kappa_t)$ be the volume of a configuration κ_t of a body. Define the mean Cauchy stress

$$\bar{\boldsymbol{T}} = [V(\kappa_t)]^{-1} \int_{\kappa_t} \boldsymbol{T} dv.$$

(a) If the body is in *equilibrium* in this configuration, show that

$$[V(\kappa_t)]\bar{T}_{ij} = \int_{\partial\kappa_t} t_i x_j da + \int_{\kappa_t} \rho b_i x_j dv.$$

(b) Apply the global form of the rotational momentum balance to show that $e_{kij}\bar{T}_{ij} = 0$ in equilibrium.

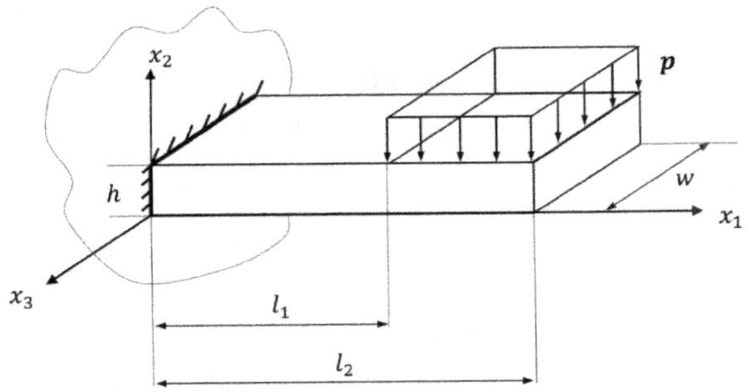

Figure 6.13. A clamped block (width w and height h) subjected to a uniform pressure on part of its top surface.

(c) Suppose the body lies on a rigid plane and is subjected to the gravity force $\boldsymbol{b} = -g\boldsymbol{e}_3$. Suppose there are no tractions applied to its boundary (except for those due to contact). Show that $\bar{T}_{33} = -\bar{\rho} g \bar{x}_3$, where $\bar{\rho}$ is the mean density in the configuration in question and $\bar{x}_3 = \bar{\boldsymbol{x}} \cdot \boldsymbol{e}_3$, where $\bar{\boldsymbol{x}}$ is the position of the center of mass relative to an origin lying on the plane. Note that the result involves the distance from the plane to the center of mass, and thus appears to depend on the choice of origin. Show that in fact this is *not* the case.

(d) Consider a block clamped at $x_1 = 0$, subjected to a uniform pressure acting in the interval $l_1 < x_1 < l_2$ (Figure 6.13). Evaluate $\bar{T}_{12}$ in terms of the dimensions shown in the figure.

11. All scalar balance laws that we have considered have the generic form

$$\frac{d}{dt}\int_{P_t} \rho\phi dv = \int_{P_t} \rho\sigma dv + \int_{\partial P_t} \boldsymbol{\pi} \cdot \boldsymbol{n} da,$$

for all $P_t \subset \kappa_t$, where ρ is the density, ϕ is the amount of the considered quantity per unit mass, σ is the rate, per unit mass, at which the quantity is supplied in the interior of P_t, and $\boldsymbol{\pi}$ is the flux of the quantity through the boundary.

If ϕ is the total energy per unit mass ($\phi = \varepsilon + \frac{1}{2}|\boldsymbol{v}|^2$), then what is the energy flux $\boldsymbol{\pi}$? Obtain the relevant jump condition, and compare it to (6.169). Thus, derive the jump condition for (6.150).

12. Let $\boldsymbol{A}$ and $\boldsymbol{B}$ be invertible tensors.
    (a) Show that $(\boldsymbol{AB})^* = \boldsymbol{A}^*\boldsymbol{B}^*$.
    (b) For invertible $\boldsymbol{Q}$, show that $\boldsymbol{Q}^* = \boldsymbol{Q}$ if, and only if, $\boldsymbol{Q}$ is a rotation.
    (c) Consider the referential *Biot* stress tensor $\boldsymbol{\sigma}$ (named after M. Biot) defined by $\boldsymbol{\sigma} = Sym(\boldsymbol{P}^t\boldsymbol{R})$, where $\boldsymbol{R}$ is the rotation in the polar decomposition of $\boldsymbol{F}$ and $Sym(\cdot)$ is the symmetric part of the indicated tensor. Show that $\boldsymbol{\sigma}$ is power-conjugate to the right-stretch tensor $\boldsymbol{U}$, i.e., that the stress power per unit reference volume is expressible as $\boldsymbol{\sigma} \cdot \dot{\boldsymbol{U}}$.
    (d) The spatial *Bell* stress tensor $\boldsymbol{\Sigma}$ (named after J.F. Bell) is defined by $\boldsymbol{\Sigma} = Sym(\boldsymbol{PR}^t)$. Derive an expression for the tensor $\boldsymbol{A}$ if $\boldsymbol{\Sigma} \cdot \boldsymbol{A}$ is the stress power per unit reference volume.

The Biot and Bell stress tensors are sometimes used in constitutive equations for solids. For applications involving the Biot stress, see [4].

Chapter 7

Observers and Frames of Reference

We have alluded to the notion of a frame of reference in connection with Euler's laws of motion. We explore the concept in detail in the present chapter.

7.1 Observers

An observer $\mathcal{O}$, perceiving a configuration κ_t of a body, is equipped for the purpose with the point space $\mathcal{E}^3$ and associated translation space E^3. These constitute the frame of reference adopted by $\mathcal{O}$. We are concerned with the relationship between this observer's perception and that of an alternative observer $\mathcal{O}^+$, say, who perceives the same configuration as $\kappa^+_{t^+}$, relative to a frame of reference consisting of the point space $\mathcal{E}^{3+}$ and translation space E^{3+}.

In classical non-relativistic physics, including classical continuum mechanics, all observers are presumed to agree on two things: the time intervals between successive events, and the distance between arbitrary pairs of material points. Thus, if t and t^+ are the times on the watches worn by $\mathcal{O}$ and $\mathcal{O}^+$, respectively, then

$$t^+ = t + a, \tag{7.1}$$

where a is a constant; and, if $\boldsymbol{x}, \boldsymbol{y} \in \mathcal{E}^3$ and $\boldsymbol{x}^+, \boldsymbol{y}^+ \in \mathcal{E}^{3+}$, respectively, are the positions of material points p and q as perceived by $\mathcal{O}$ and $\mathcal{O}^+$, then (Figure 7.1)

$$\left|\boldsymbol{x}^+ - \boldsymbol{y}^+\right| = |\boldsymbol{x} - \boldsymbol{y}|\,. \tag{7.2}$$

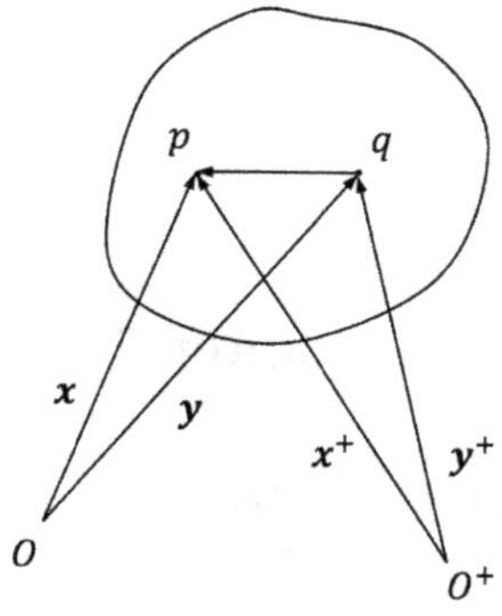

Figure 7.1. A body as perceived by two observers.

Assuming the two point spaces to be related by a differentiable mapping, we may proceed, as in the analysis of rigid-body motion (cf. Section 3.7), to conclude that the latter requirement is met if and only if

$$\boldsymbol{x}^+ = \boldsymbol{Q}(t)\boldsymbol{x} + \boldsymbol{c}(t), \tag{7.3}$$

where $\boldsymbol{c}(t) \in \mathcal{E}^{3+}$ is arbitrary and $\boldsymbol{Q}(t)$ is an arbitrary spatially uniform two-point orthogonal tensor mapping E^3 to E^{3+}. Because tensors operate on vectors rather than positions, this should be interpreted to mean [13]

$$\boldsymbol{x}^+ - \boldsymbol{y}^+ = \boldsymbol{Q}(t)(\boldsymbol{x} - \boldsymbol{y}). \tag{7.4}$$

In fact, as suggested by the figure, (7.3) also holds if the positions therein are replaced by position vectors [4].

7.2 Transformation of the motion

Following the development of Section 2.1, the motions of $p \in B$, as perceived by $\mathcal{O}$ and $\mathcal{O}^+$, are expressible as

$$\boldsymbol{x} = \boldsymbol{\chi}(p,t) \quad \text{and} \quad \boldsymbol{x}^+ = \boldsymbol{\chi}^+(p,t^+), \tag{7.5}$$

respectively. Accordingly,

$$\boldsymbol{\chi}^+(p,t^+) = \boldsymbol{Q}(t)\boldsymbol{\chi}(p,t) + \boldsymbol{c}(t). \tag{7.6}$$

The velocity of p in the frame E^{3+} is

$$\boldsymbol{v}^+ = \frac{\partial}{\partial t^+}\chi^+ = \frac{\partial}{\partial t}\chi^+ = \dot{\boldsymbol{Q}}\boldsymbol{x} + \boldsymbol{Q}\boldsymbol{v} + \dot{\boldsymbol{c}}, \tag{7.7}$$

where $\boldsymbol{v} = \partial\chi/\partial t$ is its velocity in the frame E^3. Thus, the velocities in the two frames are related by

$$\boldsymbol{v}^+ - \boldsymbol{Q}\boldsymbol{v} = \boldsymbol{\Omega}(\boldsymbol{x}^+ - \boldsymbol{c}) + \dot{\boldsymbol{c}}, \quad \text{where } \boldsymbol{\Omega}(t) = \dot{\boldsymbol{Q}}\boldsymbol{Q}^t \tag{7.8}$$

is the spin of E^{3+}. Using this result it is straightforward to show that the accelerations $\boldsymbol{a}^+ \in E^{3+}$ and $\boldsymbol{a} \in E^3$ of p are related by

$$\boldsymbol{a}^+ - \boldsymbol{Q}\boldsymbol{a} = \ddot{\boldsymbol{c}} + 2\boldsymbol{\Omega}(\boldsymbol{v}^+ - \dot{\boldsymbol{c}}) + (\dot{\boldsymbol{\Omega}} - \boldsymbol{\Omega}^2)(\boldsymbol{x}^+ - \boldsymbol{c}). \tag{7.9}$$

We observe that if $\ddot{\boldsymbol{c}} = \boldsymbol{0}$, i.e., if $\boldsymbol{c} = \boldsymbol{V}t + \boldsymbol{c}_0$ for constant $\boldsymbol{V}$ and $\boldsymbol{c}_0$, and, if $\boldsymbol{Q}$ is independent of t, then

$$\boldsymbol{x}^+ = \boldsymbol{Q}\boldsymbol{x} + \boldsymbol{V}t + \boldsymbol{c}_0 \quad \text{and} \quad \boldsymbol{a}^+ = \boldsymbol{Q}\boldsymbol{a}. \tag{7.10}$$

This is called a *Galilean transformation.* Its significance will be discussed later, in connection with the equations of motion.

7.3 Transformation of kinematic variables

To analyze the motion, observers $\mathcal{O}$ and $\mathcal{O}^+$ select reference configurations κ and κ^+, respectively, in which the positions of p are $\boldsymbol{X} = \boldsymbol{\kappa}(p)$ and $\boldsymbol{X}^+ = \boldsymbol{\kappa}^+(p)$. The motions of p in the two frames are then given by

$$\boldsymbol{x} = \chi_\kappa(\boldsymbol{X}, t) \quad \text{and} \quad \boldsymbol{x}^+ = \chi^+_{\kappa^+}(\boldsymbol{X}^+, t^+), \tag{7.11}$$

where $\chi_\kappa(\boldsymbol{X}, t) = \chi(\boldsymbol{\kappa}^{-1}(\boldsymbol{X}), t)$ and $\chi^+_{\kappa^+}(\boldsymbol{X}^+, t^+) = \chi^+(\boldsymbol{\kappa}^{+-1}(\boldsymbol{X}^+), t^+)$. Thus,

$$\chi^+_{\kappa^+}(\boldsymbol{X}^+, t^+) = \boldsymbol{Q}(t)\chi_\kappa(\boldsymbol{X}, t) + \boldsymbol{c}(t). \tag{7.12}$$

Suppose the observers $\mathcal{O}$ and $\mathcal{O}^+$ adopt κ_{t_0} and $\kappa^+_{t_0^+}$ as their respective reference configurations. Following the development of Section 2.6, we interpret this to mean

$$\chi_\kappa(\boldsymbol{X}, t_0) = \mathbf{1}\boldsymbol{X} \quad \text{and} \quad \chi^+_{\kappa^+}(\boldsymbol{X}^+, t_0^+) = \mathbf{1}^+\boldsymbol{X}^+, \tag{7.13}$$

with $\boldsymbol{X}$ and $\boldsymbol{X}^+$ regarded as position vectors in $\hat{E}^3$ and $\hat{E}^{3+}$, respectively, and where $\mathbf{1}$: $\hat{E}^3 \to E^3$ and $\mathbf{1}^+$: $\hat{E}^{3+} \to E^{3+}$ are the (uniform and orthogonal) shifters in the two frames. Thus,

$$\boldsymbol{X}^+ = \boldsymbol{K}\boldsymbol{X} + \boldsymbol{c}^+, \tag{7.14}$$

where

$$\boldsymbol{K} = (\mathbf{1}^+)^t \boldsymbol{Q}(t_0)\mathbf{1} \quad \text{and} \quad \boldsymbol{c}^+ = (\mathbf{1}^+)^t \boldsymbol{c}(t_0), \tag{7.15}$$

with the shifters evaluated at the times t_0 and t_0^+. Here, $\boldsymbol{K}$: $\hat{E}^3 \to \hat{E}^{3+}$ is a fixed orthogonal tensor and $\boldsymbol{c}^+ \in \hat{E}^{3+}$ is a fixed vector.

Fixing t (and hence t^+) and combining (2.34) with (7.3) and $d\boldsymbol{X}^+ = \boldsymbol{K}d\boldsymbol{X}$, which follows from (7.14), we find that the deformation gradients in the two frames are related by

$$\boldsymbol{F}^+ d\boldsymbol{X}^+ = d\boldsymbol{x}^+ = \boldsymbol{Q}(t)d\boldsymbol{x} = \boldsymbol{Q}(t)\boldsymbol{F}d\boldsymbol{X} = \boldsymbol{Q}(t)\boldsymbol{F}\boldsymbol{K}^t d\boldsymbol{X}^+, \tag{7.16}$$

i.e., that

$$\boldsymbol{F}^+ = \boldsymbol{Q}(t)\boldsymbol{F}\boldsymbol{K}^t. \tag{7.17}$$

This is a linear map from $\hat{E}^{3+}$ to E^{3+}. Because the shifter $\mathbf{1}^+$ is also such a map, it is natural to assume that

$$\mathbf{1}^+ = \boldsymbol{Q}(t)\mathbf{1}\boldsymbol{K}^t. \tag{7.18}$$

This recovers $(7.15)_1$ at time t_0 and ensures that E^{3+} and $\mathcal{O}^+$ evolve together as time passes. It also yields $\boldsymbol{F}^+(\boldsymbol{X}, t_0^+) = \mathbf{1}^+$ in accordance with (2.79) and the chosen reference configurations.

An inertial frame is a frame that neither accelerates nor spins. In this book, we always regard the frame E^3 of $\mathcal{O}$ as being inertial. Accordingly the shifter $\mathbf{1}$ is fixed, independent of time. It then follows that the shifter is fixed in any frame related to $\mathcal{O}$ by a Galilean transformation, i.e., in all inertial frames.

Because both observers have selected occupiable reference configurations, we have that $0 < J^+ = \det \boldsymbol{F}^+ = J \det \boldsymbol{Q} \det \boldsymbol{K}$ in which $J = \det \boldsymbol{F} > 0$. Then, because $\boldsymbol{Q}$ and $\boldsymbol{K}$ are orthogonal, $\det \boldsymbol{Q} \det \boldsymbol{K} = 1$, and hence

$$J^+ = J. \tag{7.19}$$

It is then a simple matter to show that the cofactor of $\boldsymbol{F}$ transforms to

$$(\boldsymbol{F}^+)^* = \boldsymbol{Q}\boldsymbol{F}^*\boldsymbol{K}^t. \tag{7.20}$$

With reference to (2.77), the displacement of p in the frame of $\mathcal{O}^+$ is

$$\begin{aligned} \boldsymbol{u}^+(\boldsymbol{X}^+, t^+) &= \chi^+_{\kappa^+}(\boldsymbol{X}^+, t^+) - \mathbf{1}^+\boldsymbol{X}^+ \\ &= \boldsymbol{Q}(t)[\chi_\kappa(\boldsymbol{X}, t) - \mathbf{1}\boldsymbol{K}^t(\boldsymbol{K}\boldsymbol{X} + \boldsymbol{c}^+)] + \boldsymbol{c}(t) \\ &= \boldsymbol{Q}(t)[\chi_\kappa(\boldsymbol{X}, t) - \mathbf{1}\boldsymbol{X}] + \boldsymbol{c}(t) - \boldsymbol{Q}(t)\mathbf{1}\boldsymbol{K}^t\boldsymbol{c}^+ \\ &= \boldsymbol{Q}(t)\boldsymbol{u}(\boldsymbol{X}, t) + \boldsymbol{d}(t), \end{aligned} \tag{7.21}$$

where $\boldsymbol{u}$ is the displacement in the frame of $\mathcal{O}$ and $\boldsymbol{d}(t) = \boldsymbol{c}(t) - \boldsymbol{Q}(t)\mathbf{1}\boldsymbol{K}^t\boldsymbol{c}^+$ vanishes at time t_0. Here, we have used the fact that $\boldsymbol{K}^t\boldsymbol{K} = \hat{\boldsymbol{I}}$, the identity for $\hat{E}^3$.

Let $\boldsymbol{H}$ be the gradient of $\boldsymbol{u}$ with respect to $\boldsymbol{X}$, i.e., $\boldsymbol{H}d\boldsymbol{X} = d\boldsymbol{u}$ at fixed t. It follows from (2.77) that

$$\boldsymbol{H} = \boldsymbol{F} - \mathbf{1}. \tag{7.22}$$

In terms of Cartesian components (cf. (2.78)),

$$\boldsymbol{H} = H_{iA}\boldsymbol{e}_i \otimes \hat{\boldsymbol{E}}_A, \quad \text{where } H_{iA} = \hat{u}_{i,A} = F_{iA} - \delta_{iA}. \tag{7.23}$$

This is a two-point tensor, mapping $\hat{E}^3$ to E^3.

Let $\boldsymbol{H}^+$ be the gradient of $\boldsymbol{u}^+$ with respect to $\boldsymbol{X}^+$. Then, from (7.21),

$$\boldsymbol{H}^+d\boldsymbol{X}^+ = d\boldsymbol{u}^+ = \boldsymbol{Q}(t)d\boldsymbol{u} = \boldsymbol{Q}(t)\boldsymbol{H}d\boldsymbol{X} = \boldsymbol{Q}(t)\boldsymbol{H}\boldsymbol{K}^td\boldsymbol{X}^+, \tag{7.24}$$

yielding

$$\boldsymbol{H}^+ = \boldsymbol{Q}(t)\boldsymbol{H}\boldsymbol{K}^t = \boldsymbol{F}^+ - \mathbf{1}^+. \tag{7.25}$$

This, of course, is also a two-point tensor, mapping $\hat{E}^{3+}$ to E^{3+}.

It is at this stage that the significance of the shifter becomes apparent. In particular, it ensures that every tensor in (7.22) is of the same type, and hence that it remains invariant under basis transformations in a given frame. It also ensures that the relationship between the deformation and displacement gradients is the same in all frames. In contrast, in the extant literature the shifter in (7.22) is typically replaced by the referential identity $\hat{\boldsymbol{I}}$. This transforms to $\hat{\boldsymbol{I}}^+$, where, according to (7.14),

$$\hat{\boldsymbol{I}}^+ d\boldsymbol{X}^+ = d\boldsymbol{X}^+ = \boldsymbol{K} d\boldsymbol{X} = \boldsymbol{K}\hat{\boldsymbol{I}} d\boldsymbol{X} = \boldsymbol{K}\hat{\boldsymbol{I}}\boldsymbol{K}^t d\boldsymbol{X}^+;$$

hence,

$$\hat{\boldsymbol{I}}^+ = \boldsymbol{K}\hat{\boldsymbol{I}}\boldsymbol{K}^t. \tag{7.26}$$

Now, if (7.22) is replaced by $\boldsymbol{H} = \boldsymbol{F} - \hat{\boldsymbol{I}}$, then $(7.25)_1$ yields $\boldsymbol{H}^+ = \boldsymbol{F}^+ - \boldsymbol{Q}(t)\hat{\boldsymbol{I}}\boldsymbol{K}^t \neq \boldsymbol{F}^+ - \hat{\boldsymbol{I}}^+$, so that the proposed relationship, if deemed valid by $\mathcal{O}$, must be regarded by $\mathcal{O}^+$ as being invalid. This would imply that $\mathcal{O}$ enjoys a privileged status, not shared by other observers. Such a state of affairs runs counter to the philosophy underlying modern physics. We will consider the displacement gradient again later, in the context of linear elasticity theory.

Continuing, we have the polar decompositions (3.32) of $\boldsymbol{F}$ and the polar decompositions

$$\boldsymbol{F}^+ = \boldsymbol{R}^+\boldsymbol{U}^+ = \boldsymbol{V}^+\boldsymbol{R}^+, \tag{7.27}$$

of $\boldsymbol{F}^+$ in terms of the symmetric, positive-definite stretch tensors $\boldsymbol{U}^+$ and $\boldsymbol{V}^+$ and the two-point rotation $\boldsymbol{R}^+$. Then,

$$\begin{aligned} \boldsymbol{R}^+\boldsymbol{U}^+ &= \boldsymbol{QRUK}^t = \boldsymbol{QRU}\hat{\boldsymbol{I}}\boldsymbol{K}^t = \boldsymbol{QRK}^t(\boldsymbol{KUK}^t) \quad \text{and} \\ \boldsymbol{V}^+\boldsymbol{R}^+ &= \boldsymbol{QVRK}^t = \boldsymbol{QVIRK}^t = (\boldsymbol{QVQ}^t)\boldsymbol{QRK}^t, \end{aligned} \tag{7.28}$$

and the uniqueness of the polar decompositions furnishes

$$\boldsymbol{U}^+ = \boldsymbol{KUK}^t, \quad \boldsymbol{V}^+ = \boldsymbol{QVQ}^t \quad \text{and} \quad \boldsymbol{R}^+ = \boldsymbol{QRK}^t. \tag{7.29}$$

The right- and left- Cauchy–Green tensors in the frame of $\mathcal{O}^+$ are thus given by

$$\boldsymbol{C}^+ = (\boldsymbol{U}^+)^2 = \boldsymbol{KUK}^t\boldsymbol{KUK}^t = \boldsymbol{KU}\hat{\boldsymbol{I}}\boldsymbol{UK}^t = \boldsymbol{KU}^2\boldsymbol{K}^t = \boldsymbol{KCK}^t,$$

and

$$\boldsymbol{B}^+ = (\boldsymbol{V}^+)^2 = \boldsymbol{QVQ}^t\boldsymbol{QVQ}^t = \boldsymbol{QVIVQ}^t = \boldsymbol{QV}^2\boldsymbol{Q}^t = \boldsymbol{QBQ}^t, \tag{7.30}$$

respectively. These also follow directly from (7.17) with $\boldsymbol{C}^+ = (\boldsymbol{F}^+)^t\boldsymbol{F}^+$ and $\boldsymbol{B}^+ = \boldsymbol{F}^+(\boldsymbol{F}^+)^t$.

Equation (7.17) also furnishes

$$\dot{\boldsymbol{F}}^+ = \boldsymbol{Q}\dot{\boldsymbol{F}}\boldsymbol{K}^t + \dot{\boldsymbol{Q}}\boldsymbol{F}\boldsymbol{K}^t. \tag{7.31}$$

Using $\dot{\boldsymbol{F}}^+ = \boldsymbol{L}^+\boldsymbol{F}^+$ and $\dot{\boldsymbol{F}} = \boldsymbol{LF}$, where $\boldsymbol{L}^+$ and $\boldsymbol{L}$, respectively, are the spatial velocity gradients as perceived by $\mathcal{O}^+$ and $\mathcal{O}$, we conclude that

$$\boldsymbol{L}^+\boldsymbol{F}^+ = \boldsymbol{QLQ}^t\boldsymbol{F}^+ + \dot{\boldsymbol{Q}}\boldsymbol{Q}^t\boldsymbol{F}^+, \tag{7.32}$$

and right-multiplication by $(\boldsymbol{F}^+)^{-1}$ gives

$$\boldsymbol{L}^+ = \boldsymbol{QLQ}^t + \boldsymbol{\Omega}, \tag{7.33}$$

where $\boldsymbol{\Omega}$ is the skew tensor defined in $(7.8)_2$. The stretching and spin tensors thus transform as

$$\boldsymbol{D}^+ = \boldsymbol{QDQ}^t \quad \text{and} \quad \boldsymbol{W}^+ = \boldsymbol{QWQ}^t + \boldsymbol{\Omega}. \tag{7.34}$$

To any unit vector $\boldsymbol{n} \in E^3$ we can associate two positions $\boldsymbol{x}, \boldsymbol{y} \in \mathcal{E}^3$ by $\boldsymbol{n} = (\boldsymbol{y} - \boldsymbol{x})/\,|\boldsymbol{y} - \boldsymbol{x}|$ (Figure 7.2).

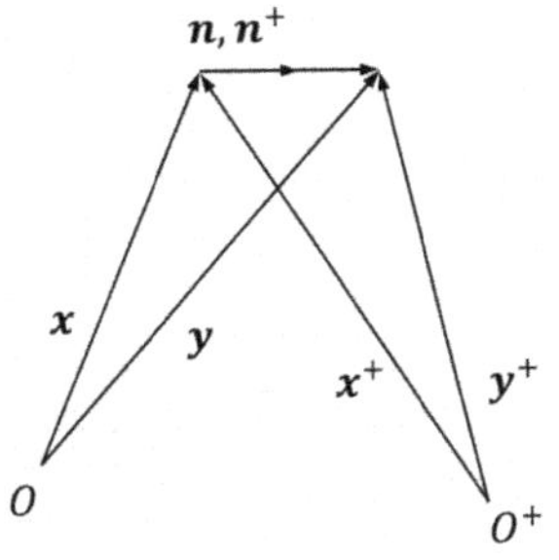

Figure 7.2. A unit vector as perceived by two observers.

Its counterpart in E^{3+} is

$$\begin{aligned} \boldsymbol{n}^+ &= (\boldsymbol{y}^+ - \boldsymbol{x}^+)/\left|\boldsymbol{y}^+ - \boldsymbol{x}^+\right| \\ &= (\boldsymbol{y}^+ - \boldsymbol{x}^+)/\left|\boldsymbol{y} - \boldsymbol{x}\right| \\ &= \boldsymbol{Q}(\boldsymbol{y} - \boldsymbol{x})/\left|\boldsymbol{y} - \boldsymbol{x}\right| \\ &= \boldsymbol{Q}\boldsymbol{n}. \end{aligned} \tag{7.35}$$

Similarly, for a unit vector $\boldsymbol{N} \in \hat{E}^3$ we have its counterpart in $\hat{E}^{3+}$:

$$\begin{aligned} \boldsymbol{N}^+ &= (\boldsymbol{Y}^+ - \boldsymbol{X}^+)/\left|\boldsymbol{Y}^+ - \boldsymbol{X}^+\right| \\ &= (\boldsymbol{Y}^+ - \boldsymbol{X}^+)/\left|\boldsymbol{Y} - \boldsymbol{X}\right| \\ &= \boldsymbol{K}(\boldsymbol{Y} - \boldsymbol{X})/\left|\boldsymbol{Y} - \boldsymbol{X}\right| \\ &= \boldsymbol{K}\boldsymbol{N}. \end{aligned} \tag{7.36}$$

7.4 Further transformations

We assume the density, energy, heating supply, temperature and entropy to be absolute scalars, and hence the same for all observers. Thus,

$$\rho^+ = \rho, \quad \varepsilon^+ = \varepsilon, \quad r^+ = r, \quad h^+ = h, \quad \theta^+ = \theta \quad \text{and} \quad \eta^+ = \eta. \tag{7.37}$$

The first of these is a natural consequence of the fact that neither the mass of a body, nor its volume, are affected by a change from one observer to another. It combines with (7.19) to yield the equality of the referential mass densities: $\rho^+_{\kappa^+} = \rho_\kappa$.

Further, we assume that the Cauchy and Piola tractions $\boldsymbol{t}, \boldsymbol{p} \in E^3$ and body force density $\boldsymbol{b} \in E^3$ transform as spatial vectors, i.e.,

$$\boldsymbol{t}^+ = \boldsymbol{Q}\boldsymbol{t}, \quad \boldsymbol{p}^+ = \boldsymbol{Q}\boldsymbol{p} \quad \text{and} \quad \boldsymbol{b}^+ = \boldsymbol{Q}\boldsymbol{b}. \tag{7.38}$$

The Cauchy stress $\boldsymbol{T}^+$ in the frame of $\mathcal{O}^+$ is then such that

$$\boldsymbol{T}^+\boldsymbol{n}^+ = \boldsymbol{t}^+ = \boldsymbol{Q}\boldsymbol{t} = \boldsymbol{Q}\boldsymbol{T}\boldsymbol{n} = \boldsymbol{Q}\boldsymbol{T}\boldsymbol{Q}^t\boldsymbol{n}^+ \tag{7.39}$$

for all unit vectors $\boldsymbol{n}^+$; thus,

$$\boldsymbol{T}^+ = \boldsymbol{Q}\boldsymbol{T}\boldsymbol{Q}^t, \tag{7.40}$$

and the Piola stress $\boldsymbol{P}^+$ is

$$\boldsymbol{P}^+ = \boldsymbol{T}^+(\boldsymbol{F}^+)^* = \boldsymbol{Q}\boldsymbol{T}\boldsymbol{Q}^t\boldsymbol{Q}\boldsymbol{F}^*\boldsymbol{K}^t = \boldsymbol{Q}\boldsymbol{P}\boldsymbol{K}^t. \tag{7.41}$$

This furnishes

$$\boldsymbol{p}^+ = \boldsymbol{P}^+\boldsymbol{N}^+ = \boldsymbol{Q}\boldsymbol{P}\boldsymbol{K}^t\boldsymbol{K}\boldsymbol{N} = \boldsymbol{Q}\boldsymbol{P}\boldsymbol{N} = \boldsymbol{Q}\boldsymbol{p}, \tag{7.42}$$

in accordance with $(7.38)_2$. Lastly, the Piola–Kirchhoff stress $\boldsymbol{S}^+$ follows from $\boldsymbol{P}^+ = \boldsymbol{F}^+\boldsymbol{S}^+$. We obtain

$$\boldsymbol{S}^+ = \boldsymbol{K}\boldsymbol{S}\boldsymbol{K}^t. \tag{7.43}$$

The relationship between the heat flux h through a surface with unit normal $\boldsymbol{n}$ and the heat flux vector $\boldsymbol{q}$ is $h = -\boldsymbol{q}\cdot\boldsymbol{n}$ (cf. (6.166)). The corresponding relationship in the frame of $\mathcal{O}^+$ is $h^+ = -\boldsymbol{q}^+\cdot\boldsymbol{n}^+$. We thus have, from (7.35) and $(7.37)_4$, that $\boldsymbol{q}^+\cdot\boldsymbol{n}^+ = \boldsymbol{q}\cdot\boldsymbol{Q}^t\boldsymbol{n}^+ = \boldsymbol{Q}\boldsymbol{q}\cdot\boldsymbol{n}^+$ for all unit vectors $\boldsymbol{n}^+$. Hence,

$$\boldsymbol{q}^+ = \boldsymbol{Q}\boldsymbol{q}, \tag{7.44}$$

and the referential heat flux vectors in the two frames are related by

$$\begin{aligned}\boldsymbol{q}^+_{\kappa^+} &= [(\boldsymbol{F}^+)^*]^t\boldsymbol{q}^+ = (\boldsymbol{Q}\boldsymbol{F}^*\boldsymbol{K}^t)^t\boldsymbol{Q}\boldsymbol{q}\\ &= \boldsymbol{K}(\boldsymbol{F}^*)^t\boldsymbol{Q}^t\boldsymbol{Q}\boldsymbol{q} = \boldsymbol{K}(\boldsymbol{F}^*)^t\boldsymbol{I}\boldsymbol{q}\\ &= \boldsymbol{K}\boldsymbol{q}_\kappa.\end{aligned} \tag{7.45}$$

7.5 Transformation of the balance laws

Let $\boldsymbol{w}(\boldsymbol{x},t)$ be an arbitrary spatial vector in the frame of $\mathcal{O}$, and suppose $\boldsymbol{w}^+(\boldsymbol{x}^+,t^+)$ is its counterpart in the frame of $\mathcal{O}^+$, i.e.,

$$\boldsymbol{w}^+(\boldsymbol{x}^+,t^+) = \boldsymbol{Q}(t)\boldsymbol{w}(\boldsymbol{x},t). \tag{7.46}$$

Then,

$$\boldsymbol{Q}(grad\boldsymbol{w})d\boldsymbol{x} = \boldsymbol{Q}d\boldsymbol{w} = d\boldsymbol{w}^+ = (grad^+\boldsymbol{w}^+)d\boldsymbol{x}^+ = (grad^+\boldsymbol{w}^+)\boldsymbol{Q}d\boldsymbol{x}, \tag{7.47}$$

and therefore

$$grad^+\boldsymbol{w}^+ = \boldsymbol{Q}(grad\boldsymbol{w})\boldsymbol{Q}^t. \tag{7.48}$$

This implies that the divergence is the same in both frames:

$$\begin{aligned} div^+\boldsymbol{w}^+ &= tr(grad^+\boldsymbol{w}^+) = tr[\boldsymbol{Q}(grad\boldsymbol{w})\boldsymbol{Q}^t] = tr[\boldsymbol{Q}^t\boldsymbol{Q}(grad\boldsymbol{w})] \\ &= tr(grad\boldsymbol{w}) = div\boldsymbol{w}. \end{aligned} \tag{7.49}$$

From (7.40),

$$\boldsymbol{T}^+\boldsymbol{w}^+ = \boldsymbol{Q}\boldsymbol{T}\boldsymbol{Q}^t\boldsymbol{Q}\boldsymbol{w} = \boldsymbol{Q}(\boldsymbol{T}\boldsymbol{w}). \tag{7.50}$$

If $\boldsymbol{w}$ is independent of $\boldsymbol{x}$ (i.e., $grad\boldsymbol{w} = \boldsymbol{0}$), then $\boldsymbol{w}^+$ is independent of $\boldsymbol{x}^+$, and the definition (1.148) of the divergence of a tensor yields

$$\begin{aligned} \boldsymbol{Q}^t(div^+\boldsymbol{T}^+)\cdot\boldsymbol{w} &= div^+\boldsymbol{T}^+\cdot\boldsymbol{w}^+ = div^+[(\boldsymbol{T}^+)^t\boldsymbol{w}^+] = div(\boldsymbol{T}^t\boldsymbol{w}) \\ &= div\boldsymbol{T}\cdot\boldsymbol{w}, \end{aligned} \tag{7.51}$$

leading, as $\boldsymbol{w}$ is arbitrary, to the conclusion that

$$div^+\boldsymbol{T}^+ = \boldsymbol{Q}(div\boldsymbol{T}). \tag{7.52}$$

Suppose now that $\mathcal{O}$ is an inertial observer. The laws holding in the frame of this observer are the conservation of mass, the balance of energy, the balances of linear and rotational momenta, and the Clausius–Duhem inequality:

$$\begin{gathered} \dot{\rho} + \rho div\boldsymbol{v} = 0, \\ \rho\dot{\varepsilon} + div\boldsymbol{q} - \boldsymbol{T}\cdot\boldsymbol{L} = \rho r, \\ \rho\boldsymbol{a} - div\boldsymbol{T} = \rho\boldsymbol{b}, \\ \boldsymbol{T}^t = \boldsymbol{T}, \quad \text{and} \\ -\rho(\dot{\psi} + \eta\dot{\theta}) + \boldsymbol{T}\cdot\boldsymbol{L} - \theta^{-1}\boldsymbol{q}\cdot grad\theta \geq 0, \end{gathered} \tag{7.53}$$

where $\boldsymbol{a}$ is the material acceleration. We seek the corresponding laws pertaining to an arbitrary observer $\mathcal{O}^+$. To this end we use (7.44), (7.46) and (7.49) to infer that

$$div^+\boldsymbol{q}^+ = div\boldsymbol{q}. \tag{7.54}$$

Also, from $grad^+\theta^+ \cdot d\boldsymbol{x}^+ = d\theta^+ = d\theta = grad\theta \cdot d\boldsymbol{x} = grad\theta \cdot \boldsymbol{Q}^t d\boldsymbol{x}^+$ we have that

$$grad^+\theta^+ = \boldsymbol{Q}(grad\theta), \quad \text{and therefore} \quad \boldsymbol{q}^+ \cdot grad^+\theta^+ = \boldsymbol{q} \cdot grad\theta. \tag{7.55}$$

The transformation formula (7.8) for the material velocity is not of the form (7.46). Nevertheless,

$$div^+\boldsymbol{v}^+ = div\boldsymbol{v}. \tag{7.56}$$

This follows from (7.33), i.e.,

$$\begin{aligned} div^+\boldsymbol{v}^+ &= tr\boldsymbol{L}^+ = tr(\boldsymbol{QLQ}^t + \boldsymbol{\Omega}) = tr(\boldsymbol{QLQ}^t) = tr(\boldsymbol{Q}^t\boldsymbol{QL}) \\ &= tr\boldsymbol{L} = div\boldsymbol{v}. \end{aligned} \tag{7.57}$$

Further, the symmetry of $\boldsymbol{T}$, and hence that of $\boldsymbol{QTQ}^t$, yields

$$\begin{aligned} \boldsymbol{T}^+ \cdot \boldsymbol{L}^+ &= \boldsymbol{QTQ}^t \cdot (\boldsymbol{QLQ}^t + \boldsymbol{\Omega}) = \boldsymbol{QTQ}^t \cdot \boldsymbol{QLQ}^t \\ &= tr[\boldsymbol{QTQ}^t(\boldsymbol{QL}^t\boldsymbol{Q}^t)] = tr(\boldsymbol{QTL}^t\boldsymbol{Q}^t) \\ &= tr(\boldsymbol{Q}^t\boldsymbol{QTL}^t) = tr(\boldsymbol{TL}^t) = \boldsymbol{T} \cdot \boldsymbol{L}. \end{aligned} \tag{7.58}$$

With these results and (7.37), together with the invariance of the Helmholtz free energy (cf. (6.186)) and the material derivative, we conclude that the balances of mass and energy, and the Clausius–Duhem inequality, are the same in all frames:

$$\begin{aligned} &\dot{\rho}^+ + \rho^+ div^+\boldsymbol{v}^+ = 0, \\ &\rho^+\dot{\varepsilon}^+ + div^+\boldsymbol{q}^+ - \boldsymbol{T}^+ \cdot \boldsymbol{L}^+ = \rho^+ r^+, \quad \text{and} \\ &-\rho^+(\dot{\psi}^+ + \eta^+\dot{\theta}^+) + \boldsymbol{T}^+ \cdot \boldsymbol{L}^+ - (\theta^+)^{-1}\boldsymbol{q}^+ \cdot grad^+\theta^+ \geq 0, \end{aligned} \tag{7.59}$$

and, of course, (7.40) and $(7.53)_4$ ensure that

$$(\boldsymbol{T}^+)^t = \boldsymbol{T}^+. \tag{7.60}$$

The situation for the linear momentum balance is different. To investigate this we combine $(7.38)_3$ and $(7.53)_3$, obtaining

$$\begin{aligned} \rho^+\boldsymbol{b}^+ &= \rho^+\boldsymbol{Qb} = \rho\boldsymbol{Qb} = \rho\boldsymbol{Qa} - \boldsymbol{Q}(div\boldsymbol{T}) \\ &= \rho^+\boldsymbol{Qa} - div^+\boldsymbol{T}^+. \end{aligned} \tag{7.61}$$

Substituting from (7.9), we arrive at the linear momentum balance

$$div^{+}\boldsymbol{T}^{+} + \rho^{+}(\boldsymbol{b}^{+} + \boldsymbol{i}^{+}) = \rho^{+}\boldsymbol{a}^{+}, \tag{7.62}$$

where

$$\boldsymbol{i}^{+} = \ddot{\boldsymbol{c}} + 2\boldsymbol{\Omega}(\boldsymbol{v}^{+} - \dot{\boldsymbol{c}}) + (\dot{\boldsymbol{\Omega}} - \boldsymbol{\Omega}^{2})(\boldsymbol{x}^{+} - \boldsymbol{c}) \tag{7.63}$$

is the contribution to the net body force arising from the spin and translation of the frame of $\mathcal{O}^{+}$. This is typically referred to as the *inertial force*. However, if the frame of $\mathcal{O}^{+}$ is related to that of $\mathcal{O}$ by a Galilean transformation, then $\boldsymbol{i}^{+}$ vanishes and the balance of linear momentum is identical in both frames, i.e.,

$$div^{+}\boldsymbol{T}^{+} + \rho^{+}\boldsymbol{b}^{+} = \rho^{+}\boldsymbol{a}^{+}. \tag{7.64}$$

Thus, if an inertial frame exists (cf. Euler's postulates (6.34) and (6.35)), then there are infinitely many, simply because there are infinitely many Galilean transformations.

It is possible to show [7] that

$$\boldsymbol{a}^{++} - \boldsymbol{i}^{++} = \boldsymbol{Q}^{+}(t^{+})(\boldsymbol{a}^{+} - \boldsymbol{i}^{+}) \tag{7.65}$$

under a further transformation

$$\boldsymbol{x}^{++} = \boldsymbol{Q}^{+}(t^{+})\boldsymbol{x}^{+} + \boldsymbol{c}^{+}(t^{+}), \quad t^{++} = t^{+} + a^{+} \tag{7.66}$$

from the frame of $\mathcal{O}^{+}$ to that of another observer $\mathcal{O}^{++}$, where $\boldsymbol{i}^{++}$ is the inertial force in the latter frame, and hence that

$$div^{++}\boldsymbol{T}^{++} + \rho^{++}(\boldsymbol{b}^{++} + \boldsymbol{i}^{++}) = \rho^{++}\boldsymbol{a}^{++}. \tag{7.67}$$

Thus, the linear momentum balance, with the inertial forces included, is valid in all frames. Moreover, the jump conditions derived in Chapter 6 are identical in all frames [7].

7.6 Problems

1. Derive (7.9).
2. Show that $(\mathbf{1}^{+})^{\cdot} = \boldsymbol{\Omega}\mathbf{1}^{+}$, where $\boldsymbol{\Omega} = \dot{\boldsymbol{Q}}\boldsymbol{Q}^{t}$.

3. Observer $\mathcal{O}^+$ adopts a frame of reference attached to the surface of a planet of radius R, with origin located at azimuth θ (degrees longitude) and elevation ϕ (degrees latitude). The planet spins about its axis (unit vector $\mathbf{e}_3$) at the constant rate ω, i.e., $\theta = \omega t + \theta_0$, where $\theta_0 = \theta(0)$. What are the equations of motion for this observer?

Chapter 8

Constitutive Equations

To make progress in the formulation of predictive theories in continuum mechanics it is necessary to connect the variables appearing in the balance laws, such as stress and heat flux, for example, to the processes experienced by the material at hand. These *constitutive relations* are mathematical idealizations of material behavior. They are judged to be useful models if they deliver predictions that are found to be in accord with experimental observation. In general, for any particular substance, such fidelity is to be expected only under a limited set of conditions. Moreover, given the vast range of behavior of actual materials, constitutive relations necessarily reflect both the properties of the material and the conditions to which it is subjected.

In consideration of the introductory nature of this book, we consider only the classical constitutive theories pertaining to viscous and inviscid fluids and elastic solids. These represent the most highly developed and widely applied classes of constitutive relations. They are far from comprehensive models of material behavior, however. Nevertheless, mastery of these models is a firm pre-requisite to the study of more complex theories of material behavior. Indeed the subject of constitutive theory is a principal aspect of ongoing research.

8.1 Elastic materials with heat conduction and viscosity

For our purposes the dependent variables for which constitutive equations are sought are the present values of the stress $\boldsymbol{T}$, the heat flux $\boldsymbol{q}$, the energy density ε and the entropy density η. These all pertain to a material point $p \in B$, with position $\boldsymbol{X} \in \kappa$. We will develop these relations from the perspective of an observer $\mathcal{O}$. Later, we will impose the requirement of *material frame indifference*, which stipulates that constitutive relations pertaining to another observer, $\mathcal{O}^+$ say, should reflect the agreement of the two observers regarding the nature of material response. Were it otherwise, then the characterization of material behavior would depend on who is determining it, whereas experience supports the intuitive notion that such behavior should instead be intrinsic to the material. We will see that such observer agreement imposes certain restrictions on the constitutive relations pertaining to any one observer.

In the course of constructing models of material behavior, certain other restrictions suggest themselves. Among these is the notion of *locality*. This is the idea that the response of a material at a point p should depend on the values of the variables constituting a process in a neighborhood of p, i.e., in a neighborhood of $\boldsymbol{X}$. In particular, we limit attention here to the special class of material models in which the processes to which the material is deemed to be sensitive consist of the values of the current deformation $\chi_\kappa(\boldsymbol{X}', t)$, the current material velocity $\hat{\boldsymbol{v}}(\boldsymbol{X}', t)$ and the current temperature $\hat{\theta}(\boldsymbol{X}', t)$ for all $\boldsymbol{X}' \in \kappa$ contained in an arbitrarily small neighborhood of $\boldsymbol{X}$. Assuming these to be smooth functions of position in κ, we may approximate them by (see (3.102) and (4.2))

$$\begin{aligned} \chi_\kappa(\boldsymbol{X}', t) &= \chi_\kappa(\boldsymbol{X}, t) + \boldsymbol{F}(\boldsymbol{X}, t)(\boldsymbol{X}' - \boldsymbol{X}) + \boldsymbol{o}(|\boldsymbol{X}' - \boldsymbol{X}|), \\ \hat{\boldsymbol{v}}(\boldsymbol{X}', t) &= \hat{\boldsymbol{v}}(\boldsymbol{X}, t) + \dot{\boldsymbol{F}}(\boldsymbol{X}, t)(\boldsymbol{X}' - \boldsymbol{X}) + \boldsymbol{o}(|\boldsymbol{X}' - \boldsymbol{X}|), \quad \text{and} \\ \hat{\theta}(\boldsymbol{X}', t) &= \hat{\theta}(\boldsymbol{X}, t) + (Grad\hat{\theta}(\boldsymbol{X}, t)) \cdot (\boldsymbol{X}' - \boldsymbol{X}) + o(|\boldsymbol{X}' - \boldsymbol{X}|). \end{aligned} \tag{8.1}$$

For $|\boldsymbol{X}' - \boldsymbol{X}|$ sufficiently small these are all dominated by the terms shown explicitly; for example, $\chi_\kappa(\boldsymbol{X}', t)$ is determined to linear-order accuracy by $\chi_\kappa(\boldsymbol{X}, t)$ and $\boldsymbol{F}(\boldsymbol{X}, t)$. Accordingly, if we seek to account for effects at this order then we would assume the constitutive response of the material, consisting of the set

$$C' = \{\boldsymbol{T}, \boldsymbol{q}, \varepsilon, \eta\}, \tag{8.2}$$

evaluated at $\boldsymbol{X}$ and t, to be determined by the list

$$L' = \{\boldsymbol{x}, \boldsymbol{v}, \theta, \boldsymbol{F}, \dot{\boldsymbol{F}}, Grad\theta\}, \tag{8.3}$$

also evaluated at $\boldsymbol{X}$ and t. Here, $\boldsymbol{x} = \chi_\kappa(\boldsymbol{X}, t)$, $\boldsymbol{v} = \hat{\boldsymbol{v}}(\boldsymbol{X}, t)$ and $\theta = \hat{\theta}(\boldsymbol{X}, t)$. We will show later that considerations having to do with frame indifference imply that the constitutive equations pertaining to the point p cannot, in fact, depend on $\boldsymbol{x}$ or $\boldsymbol{v}$.

In place of C', it will prove to be more convenient to work with the equivalent set

$$C = \{\boldsymbol{T}, \boldsymbol{q}, \psi, \eta\}, \tag{8.4}$$

where ψ is the Helmholtz free energy. Further, from

$$Grad\hat{\theta} \cdot d\boldsymbol{X} = d\theta = grad\tilde{\theta} \cdot d\boldsymbol{x} = grad\tilde{\theta} \cdot \boldsymbol{F} d\boldsymbol{X}, \tag{8.5}$$

it follows that

$$Grad\hat{\theta} = \boldsymbol{F}^t grad\tilde{\theta} \tag{8.6}$$

and hence that L' is equivalent to the list

$$L = \{\boldsymbol{x}, \boldsymbol{v}, \theta, \boldsymbol{F}_\kappa, \boldsymbol{L}, grad\theta\}, \tag{8.7}$$

where $\boldsymbol{L}$ is the spatial velocity gradient (cf. (4.3)) and the subscript κ has been attached to the deformation gradient to identify the reference configuration relative to which it is computed. Of course none of the other variables in this list depends on a reference configuration.

The special constitutive assumption adopted here may be motivated by an analogy with a spring-damper system, the response of which is determined by the present values of deformation and velocity. Accordingly we study a particular class of material models that exhibit both elasticity and viscosity. We emphasize that this class is not sufficient to capture general viscoelastic behavior [14, 15], the consideration of which lies outside the scope of this book. It does, however, suffice for our limited purposes.

8.2 Change of reference configuration and material symmetry

The list L of variables, presumed to constitute the relevant process, depends on the choice of reference configuration due to the presence of the deformation gradient. To see how this choice affects the theory, consider a change of reference configuration from κ to μ, say (Figure 8.1). Because the body and any reference configuration are presumed to be in one-to-one correspondence, we assume that there exists an invertible map $\boldsymbol{\mu}(p)$ taking $p \in B$ to $\boldsymbol{Y} \in \mu$. Further, as B and κ are also in one-to-one correspondence, the same is true of κ and μ. Thus we assume the existence of a smooth, invertible function $\boldsymbol{\lambda}$ such that

$$\boldsymbol{Y} = \boldsymbol{\lambda}(\boldsymbol{X}) \quad \text{and} \quad d\boldsymbol{Y} = \boldsymbol{R}(\boldsymbol{X})d\boldsymbol{X}, \quad \text{where } \boldsymbol{R}(\boldsymbol{X}) = Grad\boldsymbol{\lambda}. \tag{8.8}$$

The deformation gradients relative to κ and μ are related by

$$\boldsymbol{F}_\kappa d\boldsymbol{X} = d\boldsymbol{x} = \boldsymbol{F}_\mu d\boldsymbol{Y} = \boldsymbol{F}_\mu \boldsymbol{R} d\boldsymbol{X}; \quad \text{hence } \boldsymbol{F}_\kappa = \boldsymbol{F}_\mu \boldsymbol{R}. \tag{8.9}$$

We emphasize that this $\boldsymbol{R}$ must not be confused with the two-point rotation tensor occurring in the polar decomposition of the deformation gradient.

Let $F \in C$ be an entry in the list of variables characterizing the response of the material. Its value is determined by the entries of list L, which, as we have noted, depends on the choice of reference

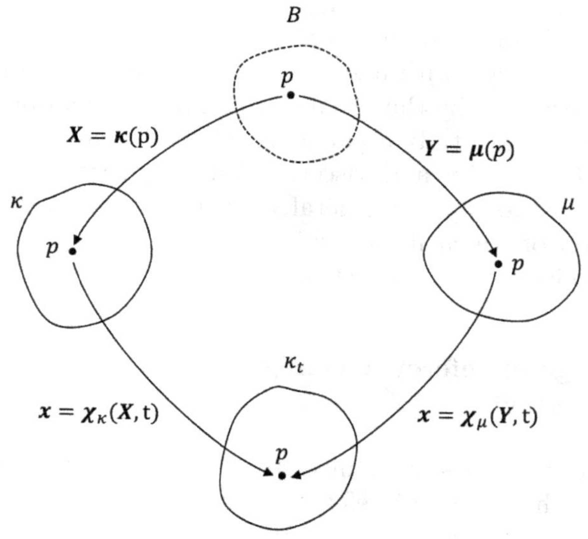

Figure 8.1. Change of reference configuration.

configuration. For example, choosing κ as reference, we have

$$F(p,t) = F_{\kappa}(\boldsymbol{x}, \boldsymbol{v}, \theta, \boldsymbol{F}_{\kappa}, \boldsymbol{L}, grad\theta; \boldsymbol{X}), \tag{8.10}$$

where the right-hand side is the constitutive function delivering the values of F. This *function* necessarily depends on κ because it involves the deformation gradient, whereas it values do not. The function itself must therefore depend on the reference configuration adopted. We have also included a parametric dependence on $\boldsymbol{X}$, to identify the material point in question. Using (8.8) and (8.9), we then have

$$F(p,t) = F_{\mu}(\boldsymbol{x}, \boldsymbol{v}, \theta, \boldsymbol{F}_{\mu}, \boldsymbol{L}, grad\theta; \boldsymbol{Y}), \tag{8.11}$$

where

$$\begin{aligned} &F_{\mu}(\boldsymbol{x}, \boldsymbol{v}, \theta, \boldsymbol{F}_{\mu}, \boldsymbol{L}, grad\theta; \boldsymbol{Y}) \\ &\quad = F_{\kappa}(\boldsymbol{x}, \boldsymbol{v}, \theta, \boldsymbol{F}_{\mu}\boldsymbol{R}(\boldsymbol{\lambda}^{-1}(\boldsymbol{Y})), \boldsymbol{L}, grad\theta; \boldsymbol{\lambda}^{-1}(\boldsymbol{Y})), \end{aligned} \tag{8.12}$$

yielding the constitutive function pertaining to any choice of reference configuration when that pertaining to one is given. Thus, the list of variables deemed to be relevant to material response is the same for all reference configurations.

Since constitutive response is defined pointwise, it suffices for our purposes to confine attention to local changes of reference configuration in the neighborhood of a material point p_0, say, rather than for the entire body (Figure 8.2). Thus consider a smooth invertible function $\boldsymbol{\lambda}(\boldsymbol{X})$, defined on an open neighborhood $N_{\kappa}(p_0) \subset \kappa$ and

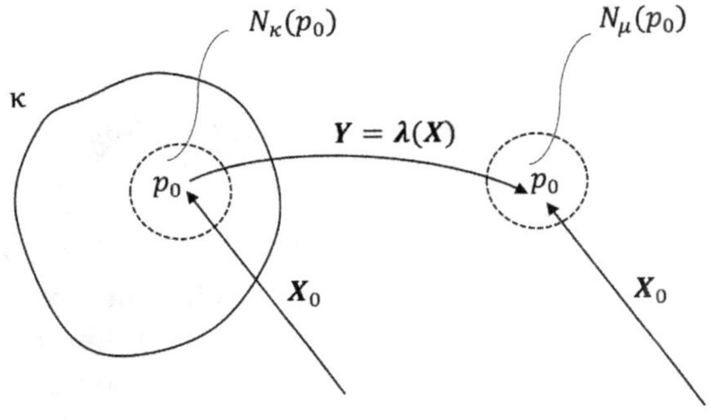

Figure 8.2. Local change of reference.

mapping it to another open neighborhood $N_\mu(p_0)$. Let $\boldsymbol{X}_0 = \boldsymbol{\kappa}(p_0)$ be a pivot point of the map, i.e., $\boldsymbol{Y}_0 = \boldsymbol{\lambda}(\boldsymbol{X}_0) = \boldsymbol{X}_0$.

Then,

$$\begin{aligned} F(p_0, t) &= F_\kappa(\boldsymbol{x}, \boldsymbol{v}, \theta, \boldsymbol{F}_\kappa, \boldsymbol{L}, grad\theta; \boldsymbol{X}_0) \\ &= F_\mu(\boldsymbol{x}, \boldsymbol{v}, \theta, \boldsymbol{F}_\mu, \boldsymbol{L}, grad\theta; \boldsymbol{X}_0), \quad \text{with } \boldsymbol{F}_\kappa = \boldsymbol{F}_\mu \boldsymbol{R}(\boldsymbol{X}_0), \end{aligned} \tag{8.13}$$

in which all arguments are evaluated at $\boldsymbol{X}_0$.

Suppose now that both $N_\kappa(p_0)$ and $N_\mu(p_0)$ are subjected to the same experiment, consisting of a given deformation $\chi(\cdot, t)$ and temperature distribution $\theta(\cdot, t)$. This experiment acts on $N_\kappa(p_0)$ to produce the fields

$$\begin{aligned} \chi(\boldsymbol{X}, t) &= \chi(\boldsymbol{X}_0, t) + \boldsymbol{F}(\boldsymbol{X}_0, t)(\boldsymbol{X} - \boldsymbol{X}_0) + \boldsymbol{o}(|\boldsymbol{X} - \boldsymbol{X}_0|), \\ \hat{\boldsymbol{v}}(\boldsymbol{X}, t) &= \hat{\boldsymbol{v}}(\boldsymbol{X}_0, t) + \dot{\boldsymbol{F}}(\boldsymbol{X}_0, t)(\boldsymbol{X} - \boldsymbol{X}_0) + \boldsymbol{o}(|\boldsymbol{X} - \boldsymbol{X}_0|) \quad \text{and} \\ \theta(\boldsymbol{X}, t) &= \theta(\boldsymbol{X}_0, t) + Grad\theta(\boldsymbol{X}_0, t) \cdot (\boldsymbol{X} - \boldsymbol{X}_0) + o(|\boldsymbol{X} - \boldsymbol{X}_0|), \end{aligned} \tag{8.14}$$

where $\boldsymbol{X} \in N_\kappa(p_0)$, with $\hat{\boldsymbol{v}}(\boldsymbol{X}_0, t) = \dot{\chi}(\boldsymbol{X}_0, t)$ and $\boldsymbol{F}(\boldsymbol{X}_0, t) = Grad\chi_{|\boldsymbol{X}_0}$; and on $N_\mu(p_0)$ to produce

$$\begin{aligned} \chi(\boldsymbol{Y}, t) &= \chi(\boldsymbol{X}_0, t) + \boldsymbol{F}(\boldsymbol{X}_0, t)(\boldsymbol{Y} - \boldsymbol{X}_0) + \boldsymbol{o}(|\boldsymbol{Y} - \boldsymbol{X}_0|), \\ \hat{\boldsymbol{v}}(\boldsymbol{Y}, t) &= \hat{\boldsymbol{v}}(\boldsymbol{X}_0, t) + \dot{\boldsymbol{F}}(\boldsymbol{X}_0, t)(\boldsymbol{Y} - \boldsymbol{X}_0) + \boldsymbol{o}(|\boldsymbol{Y} - \boldsymbol{X}_0|) \quad \text{and} \\ \theta(\boldsymbol{Y}, t) &= \theta(\boldsymbol{X}_0, t) + Grad\theta(\boldsymbol{X}_0, t) \cdot (\boldsymbol{Y} - \boldsymbol{X}_0) + o(|\boldsymbol{Y} - \boldsymbol{X}_0|), \end{aligned} \tag{8.15}$$

where $\boldsymbol{Y} \in N_\mu(p_0)$, the point being that the values of these variables, and their gradients, are identical at the point $\boldsymbol{X}_0$.

The response of the material at p_0, using $N_\kappa(p_0)$ as reference, is given by $F_\kappa(\boldsymbol{x}, \boldsymbol{v}, \theta, \boldsymbol{F}, \boldsymbol{L}, grad\theta; \boldsymbol{X}_0)$, with $\boldsymbol{x} = \chi(\boldsymbol{X}_0, t)$, $\boldsymbol{v} = \dot{\chi}(\boldsymbol{X}_0, t)$, $\theta = \theta(\boldsymbol{X}_0, t)$, $\boldsymbol{F} = \boldsymbol{F}(\boldsymbol{X}_0, t)$, $\boldsymbol{L} = \dot{\boldsymbol{F}}(\boldsymbol{X}_0, t)\boldsymbol{F}(\boldsymbol{X}_0, t)^{-1}$ and $grad\theta = \boldsymbol{F}(\boldsymbol{X}_0, t)^{-t} Grad\theta(\boldsymbol{X}_0, t)$. On the other hand, the response elicited at p_0 by using $N_\mu(p_0)$ as reference is $F_\mu(\boldsymbol{x}, \boldsymbol{v}, \theta, \boldsymbol{F}, \boldsymbol{L}, grad\theta; \boldsymbol{X}_0)$, in which the arguments of the two functions are identical. Having subjected two different local neighborhoods of p_0 to the same experiment, we would expect the two response functions, insofar as they are evaluated at the *same* arguments, to yield *different* values. This conclusion follows from (8.13). However, there may be certain distinguished maps $\boldsymbol{\lambda}$ for which the responses of the two neighborhoods, evaluated at the point p_0, are identical. In this case, no experiment can distinguish between local neighborhoods that are related by such a map. These special maps are called *material symmetry transformations.*

Thus, if $N_\kappa(p_0)$ and $N_\mu(p_0)$ are related by a symmetry transformation, then $F_\kappa(\boldsymbol{x}, \boldsymbol{v}, \theta, \boldsymbol{F}, \boldsymbol{L}, grad\theta; \boldsymbol{X}_0) = F_\mu(\boldsymbol{x}, \boldsymbol{v}, \theta, \boldsymbol{F}, \boldsymbol{L}, grad\theta; \boldsymbol{X}_0)$, and it then follows from (8.13) that

$$F_\kappa(\boldsymbol{x}, \boldsymbol{v}, \theta, \boldsymbol{F}, \boldsymbol{L}, grad\theta; \boldsymbol{X}_0) = F_\kappa(\boldsymbol{x}, \boldsymbol{v}, \theta, \boldsymbol{F}\boldsymbol{R}, \boldsymbol{L}, grad\theta; \boldsymbol{X}_0),$$
$$\text{where } \boldsymbol{R} = Grad\boldsymbol{\lambda}(\boldsymbol{X})_{|\boldsymbol{X}_0}. \qquad (8.16)$$

We recall that the map $\boldsymbol{\lambda}$ in this discussion is defined on a neighborhood of the material point in question. In particular, $\boldsymbol{R}$ need not be the gradient of a single function defined on the body. Thus, the notion of material symmetry is purely local. This allows for different symmetries at different points. The concept of material symmetry thus accommodates composites, for example, in which the type of material varies from one point of a body to another.

It follows from (8.16) that $\boldsymbol{FR}$ is the value of an admissible deformation gradient at the point $\boldsymbol{X}_0$. If κ is an occupiable reference configuration, this requires that $0 < \det(\boldsymbol{FR}) = (\det \boldsymbol{F})(\det \boldsymbol{R})$ with $\det \boldsymbol{F} > 0$. Thus, $\det \boldsymbol{R} > 0$. We denote by $\mathcal{G}_{\kappa(p_0)}$ the set of $\boldsymbol{R}'s$ for which (8.16) holds for all admissible values of $\boldsymbol{F}$. Thus,

$$\mathcal{G}_{\kappa(p_0)} = \{\boldsymbol{R} : \det \boldsymbol{R} > 0 \text{ and (8.16) holds}\}. \qquad (8.17)$$

This set is a *group*, i.e., it has the following three properties:

(1) $\hat{\boldsymbol{I}} \in \mathcal{G}_{\kappa(p_0)}$
(2) If $\boldsymbol{R}_1 \in \mathcal{G}_{\kappa(p_0)}$ and $\boldsymbol{R}_2 \in \mathcal{G}_{\kappa(p_0)}$, then $\boldsymbol{R}_1\boldsymbol{R}_2 \in \mathcal{G}_{\kappa(p_0)}$.
(3) If $\boldsymbol{R} \in \mathcal{G}_{\kappa(p_0)}$, then $\boldsymbol{R}^{-1} \in \mathcal{G}_{\kappa(p_0)}$.

The first of these follows trivially from the definition of $\mathcal{G}_{\kappa(p_0)}$. The second property follows, on suppressing all arguments that do not involve $\boldsymbol{R}$, from $F_\kappa(\boldsymbol{F}(\boldsymbol{R}_1\boldsymbol{R}_2)) = F_\kappa((\boldsymbol{FR}_1)\boldsymbol{R}_2) = F_\kappa(\boldsymbol{FR}_1) = F_\kappa(\boldsymbol{F})$; and the third from $F_\kappa(\boldsymbol{FR}^{-1}) = F_\kappa((\boldsymbol{FR}^{-1})\boldsymbol{R}) = F_\kappa(\boldsymbol{F}\hat{\boldsymbol{I}}) = F_\kappa(\boldsymbol{F})$.

It follows from the second property that if $\boldsymbol{R} \in \mathcal{G}_{\kappa(p_0)}$, then $\boldsymbol{R}^n \in \mathcal{G}_{\kappa(p_0)}$ for any positive integer n. Thus, $F_\kappa(\boldsymbol{F}) = F_\kappa(\boldsymbol{F}\boldsymbol{R}^n)$, where $\det(\boldsymbol{F}\boldsymbol{R}^n) = (\det \boldsymbol{F})(\det \boldsymbol{R}^n) = (\det \boldsymbol{F})(\det \boldsymbol{R})^n$. Let $n \to \infty$. If $\det \boldsymbol{R} > 1$ we have $\det(\boldsymbol{F}\boldsymbol{R}^n) \to \infty$, corresponding to unbounded dilation of the material, whereas if $\det \boldsymbol{R} < 1$ then $\det(\boldsymbol{F}\boldsymbol{R}^n) \to 0$, corresponding to unbounded compaction, neither of which affects the response of the material. To avoid this untenable conclusion, we must restrict $\boldsymbol{R}$ such that $\det \boldsymbol{R} = 1$. Therefore,

$$\mathcal{G}_{\kappa(p_0)} \subset U = \{\boldsymbol{R} : \det \boldsymbol{R} = 1\}. \tag{8.18}$$

This is called the *unimodular group*.

It was Noll's elegant idea [16] to classify materials as being either solid or fluid on the basis of the symmetry group. For example, the stress in a compressible inviscid fluid under isothermal conditions is $\boldsymbol{T} = -p(\rho)\boldsymbol{I}$, where $p(\rho)$ is the constitutive function for the density-dependent pressure. From (2.50) and (2.59) we have $\rho = \rho_\kappa(\boldsymbol{X})/\det \boldsymbol{F} = \rho_\kappa(\boldsymbol{X})/\det(\boldsymbol{F}\boldsymbol{R})$, provided that $\boldsymbol{R} \in U$. Therefore $\mathcal{G}_{\kappa(p_0)} = U$. This motivates the definition of a fluid as a material of the considered class for which there exists a local reference configuration $\kappa^*(p_0)$, say, such that $\mathcal{G}_{\kappa^*(p_0)} = U$.

The situation for solids is different. For these, experience indicates that a strain induced by the map $\boldsymbol{\lambda}(\boldsymbol{X})$ is experimentally detectable. Thus, if $\boldsymbol{\lambda}$ is to be a symmetry transformation, and hence undetectable, then it should be strain free; that is, the right Cauchy–Green tensor associated with $\boldsymbol{R}$ should be such that $\boldsymbol{R}^t\boldsymbol{R} = \hat{\boldsymbol{I}}$. Combining this with (8.18) we conclude that $\boldsymbol{R}$ is a rotation. Accordingly, we define a solid to be a material of the considered class for which there exists a local reference $\kappa^*(p_0)$ such that $\mathcal{G}_{\kappa^*(p_0)}$ is contained in $Orth^+$, the set of *proper-orthogonal* tensors, or simply the rotations, i.e., the orthogonal tensors having determinant equal to +1. Thus,

$$\mathcal{G}_{\kappa^*(p_0)} \subset Orth^+ = \{\boldsymbol{R} : \boldsymbol{R} \in U \text{ and } \boldsymbol{R}^t\boldsymbol{R} = \hat{\boldsymbol{I}}\}. \tag{8.19}$$

We note that $\kappa^*(p_0)$ need not bear any relationship to any particular reference configuration one may use for the purposes of analysis. In solids, for example, $\kappa^*(p_0)$ would correspond to a unit cell of an undistorted lattice of a crystalline material. These are mapped to themselves by certain discrete rotations that characterize the particular crystalline symmetry at hand. For this reason, $\kappa^*(p_0)$ is often called an *undistorted* local configuration of the solid.

8.3 Noll's rule

We have seen that material symmetry is a characteristic of the nature of the material as well as the adopted reference configuration. It is therefore of interest to examine the extent to which material symmetry is influenced by the choice of reference configuration. To investigate this issue consider two reference configurations κ_1 and κ_2 in which the positions of $p_0 \in B$ are $\boldsymbol{X}_1$ and $\boldsymbol{X}_2$, respectively. We assume the existence of a smooth invertible map $\boldsymbol{\pi}$, defined on a neighborhood of $\boldsymbol{X}_1$, such that $\boldsymbol{X}_2 = \boldsymbol{\pi}(\boldsymbol{X}_1)$. Suppressing the variables that do not involve a reference configuration, we have, from (8.12), that

$$F_{\kappa_2}(\boldsymbol{F}; \boldsymbol{X}_2) = F_{\kappa_1}(\boldsymbol{F}\boldsymbol{K}; \boldsymbol{X}_1), \quad \text{where } \boldsymbol{K} = Grad\boldsymbol{\pi}_{|\boldsymbol{X}_1}. \tag{8.20}$$

This in turn implies that

$$F_{\kappa_2}(\boldsymbol{F}\boldsymbol{K}^{-1}; \boldsymbol{X}_2) = F_{\kappa_1}(\boldsymbol{F}; \boldsymbol{X}_1). \tag{8.21}$$

Suppose $\boldsymbol{R} \in \mathcal{G}_{\kappa_1(p_0)}$, so that

$$F_{\kappa_1}(\boldsymbol{F}; \boldsymbol{X}_1) = F_{\kappa_1}(\boldsymbol{F}\boldsymbol{R}; \boldsymbol{X}_1). \tag{8.22}$$

Then, on replacing $\boldsymbol{F}$ by $\boldsymbol{F}\boldsymbol{R}$ in (8.21),

$$F_{\kappa_1}(\boldsymbol{F}\boldsymbol{R}; \boldsymbol{X}_1) = F_{\kappa_2}(\boldsymbol{F}\boldsymbol{R}\boldsymbol{K}^{-1}; \boldsymbol{X}_2). \tag{8.23}$$

Thus,

$$F_{\kappa_2}(\boldsymbol{F}\boldsymbol{R}\boldsymbol{K}^{-1}; \boldsymbol{X}_2) = F_{\kappa_1}(\boldsymbol{F}; \boldsymbol{X}_1) = F_{\kappa_2}(\boldsymbol{F}\boldsymbol{K}^{-1}; \boldsymbol{X}_2). \tag{8.24}$$

Let $\hat{\boldsymbol{F}} = \boldsymbol{F}\boldsymbol{K}^{-1}$. Then, $\boldsymbol{F}\boldsymbol{R}\boldsymbol{K}^{-1} = \hat{\boldsymbol{F}}\boldsymbol{K}\boldsymbol{R}\boldsymbol{K}^{-1}$ and

$$F_{\kappa_2}(\hat{\boldsymbol{F}}; \boldsymbol{X}_2) = F_{\kappa_2}(\hat{\boldsymbol{F}}\boldsymbol{K}\boldsymbol{R}\boldsymbol{K}^{-1}; \boldsymbol{X}_2). \tag{8.25}$$

We thus arrive at Noll's rule:

$$\mathcal{G}_{\kappa_2(p_0)} = \{\boldsymbol{K}\boldsymbol{R}\boldsymbol{K}^{-1} : \boldsymbol{R} \in \mathcal{G}_{\kappa_1(p_0)}\}. \tag{8.26}$$

Here, again $\boldsymbol{K}$ is the gradient, at $\boldsymbol{X}_1$, of a map defined on a neighborhood of $\boldsymbol{X}_1$. It need not be the gradient of a map defined on the entire body. This fact has major implications for plasticity theory, for example [17–19].

We observe that if $\mathcal{G}_{\kappa_1(p_0)}$ is in $Orth^+$ then $\mathcal{G}_{\kappa_2(p_0)}$ is not, in general. Thus, if $\kappa_1(p_0)$ is an undistorted state of a crystalline lattice, for example, then the lattice is distorted in $\kappa_2(p_0)$. For this reason the existence of a (local) configuration such that (8.19) holds is an essential part of the definition of a solid. However, (8.18) is satisfied in all configurations. Thus, the fact of a substance being fluid, i.e., the identification of the symmetry group with the unimodular group, is independent of a reference configuration. This is in accord with the intuitive notion of fluidity.

8.4 Frame invariance

In the course of formulating the basic structure of our restricted class of constitutive equations we have adopted the perspective of a particular observer, $\mathcal{O}$ say. We have already mentioned the concept of material frame indifference, according to which all observers should agree on the nature of the material at hand, and hence on the class of processes to which the material responds. Accordingly, with reference to (8.4) and (8.7), the response functions in the frame of observer $\mathcal{O}^+$ should deliver the values of the elements of the set

$$C^+ = \{\boldsymbol{T}^+, \boldsymbol{q}^+, \psi^+, \eta^+\} \tag{8.27}$$

in terms of the elements of the list

$$L^+ = \{\boldsymbol{x}^+, \boldsymbol{v}^+, \theta^+, \boldsymbol{F}^+_{\kappa^+}, \boldsymbol{L}^+, grad^+\theta^+\}, \tag{8.28}$$

all evaluated at the material point $p \in B$ and the time t^+. Thus, if $\mathcal{O}$ adopts the constitutive function $F_\kappa(\boldsymbol{x}, \boldsymbol{v}, \theta, \boldsymbol{F}_\kappa, \boldsymbol{L}, grad\theta; \boldsymbol{X})$ to determine the values of $F(p,t) \in C$, then $\mathcal{O}^+$ should adopt the function $F^+_{\kappa^+}(\boldsymbol{x}^+, \boldsymbol{v}^+, \theta^+, \boldsymbol{F}^+_{\kappa^+}, \boldsymbol{L}^+, grad^+\theta^+; \boldsymbol{X}^+)$ to evaluate its counterpart $F^+(p,t^+) \in C^+$. Of course all variables pertaining to $\mathcal{O}^+$ are given in terms of their counterparts in the frame of $\mathcal{O}$ by the formulas developed in Chapter 7.

For example, if the free energy and entropy are determined by $\mathcal{O}$ to be

$$\psi = \hat{\psi}_\kappa(\boldsymbol{x}, \boldsymbol{v}, \theta, \boldsymbol{F}_\kappa, \boldsymbol{L}, grad\theta; \boldsymbol{X}) \quad \text{and}$$
$$\eta = \hat{\eta}_\kappa(\boldsymbol{x}, \boldsymbol{v}, \theta, \boldsymbol{F}_\kappa, \boldsymbol{L}, grad\theta; \boldsymbol{X}), \tag{8.29}$$

in terms of constitutive functions $\hat{\psi}_\kappa$ and $\hat{\eta}_\kappa$, respectively, then, with reference to (6.186) and (7.37), the corresponding constitutive equations in the frame of observer $\mathcal{O}^+$ are

$$\begin{aligned}
&\hat{\psi}^+_{\kappa^+}(\boldsymbol{x}^+, \boldsymbol{v}^+, \theta^+, \boldsymbol{F}^+_{\kappa^+}, \boldsymbol{L}^+, grad^+\theta^+; \boldsymbol{X}^+)\\
&\quad = \hat{\psi}_\kappa(\boldsymbol{x}, \boldsymbol{v}, \theta, \boldsymbol{F}_\kappa, \boldsymbol{L}, grad\theta; \boldsymbol{X}) \quad \text{and}\\
&\hat{\eta}^+_{\kappa^+}(\boldsymbol{x}^+, \boldsymbol{v}^+, \theta^+, \boldsymbol{F}^+_{\kappa^+}, \boldsymbol{L}^+, grad^+\theta^+; \boldsymbol{X}^+)\\
&\quad = \hat{\eta}_\kappa(\boldsymbol{x}, \boldsymbol{v}, \theta, \boldsymbol{F}_\kappa, \boldsymbol{L}, grad\theta; \boldsymbol{X}),
\end{aligned} \tag{8.30}$$

respectively.

Similarly, if the heat flux is determined by $\mathcal{O}$ to be

$$\boldsymbol{q} = \hat{\boldsymbol{q}}_\kappa(\boldsymbol{x}, \boldsymbol{v}, \theta, \boldsymbol{F}_\kappa, \boldsymbol{L}, grad\theta; \boldsymbol{X}) \tag{8.31}$$

in terms of a constitutive function $\hat{\boldsymbol{q}}_\kappa$ (not to be confused with the values of the referential heat flux — see (6.171)), then, with reference to (7.44), the corresponding constitutive equation in the frame of observer $\mathcal{O}^+$ is

$$\begin{aligned}
&\hat{\boldsymbol{q}}^+_{\kappa^+}(\boldsymbol{x}^+, \boldsymbol{v}^+, \theta^+, \boldsymbol{F}^+_{\kappa^+}, \boldsymbol{L}^+, grad^+\theta^+; \boldsymbol{X}^+)\\
&\quad = \boldsymbol{Q}\hat{\boldsymbol{q}}_\kappa(\boldsymbol{x}, \boldsymbol{v}, \theta, \boldsymbol{F}_\kappa, \boldsymbol{L}, grad\theta; \boldsymbol{X}),
\end{aligned} \tag{8.32}$$

where $\boldsymbol{Q}$ is the orthogonal transformation of (7.3) that characterizes the change of frame.

Lastly, if $\mathcal{O}$ finds the stress to be determined constitutively by

$$\boldsymbol{T} = \hat{\boldsymbol{T}}_\kappa(\boldsymbol{x}, \boldsymbol{v}, \theta, \boldsymbol{F}_\kappa, \boldsymbol{L}; \boldsymbol{X}), \tag{8.33}$$

then the stress is determined by $\mathcal{O}^+$ to be

$$\boldsymbol{T}^+ = \hat{\boldsymbol{T}}^+_{\kappa^+}(\boldsymbol{x}^+, \boldsymbol{v}^+, \theta^+, \boldsymbol{F}^+_{\kappa^+}, \boldsymbol{L}^+; \boldsymbol{X}^+), \tag{8.34}$$

where (see (7.40))

$$\hat{\boldsymbol{T}}^+_{\kappa^+}(\boldsymbol{x}^+, \boldsymbol{v}^+, \theta^+, \boldsymbol{F}^+_{\kappa^+}, \boldsymbol{L}^+; \boldsymbol{X}^+) = \boldsymbol{Q}\hat{\boldsymbol{T}}_\kappa(\boldsymbol{x}, \boldsymbol{v}, \theta, \boldsymbol{F}_\kappa, \boldsymbol{L}; \boldsymbol{X})\boldsymbol{Q}^t. \tag{8.35}$$

Here, we have suppressed the dependence of the stress on the temperature gradient. This is in accord with classical treatments of the present class of constitutive equations. See [16, 20], for example. We will return to this issue in subsequent chapters.

Frame invariance alone does not suffice to restrict the class of admissible constitutive functions. We must also ensure that they are compatible with the balances of mass, energy, momentum and the Clausius–Duhem inequality. These matters are explored in depth in the following chapters, in the contexts of distinct sub-models for viscous fluids and elastic solids.

Chapter 9

Viscous Fluids

For fluids the constitutive equation (8.33) for the stress, in the frame of $\mathcal{O}$, satisfies the material symmetry condition

$$\hat{\boldsymbol{T}}_\kappa(\boldsymbol{x}, \boldsymbol{v}, \theta, \boldsymbol{F}, \boldsymbol{L}; \boldsymbol{X}) = \hat{\boldsymbol{T}}_\kappa(\boldsymbol{x}, \boldsymbol{v}, \theta, \boldsymbol{F}\boldsymbol{R}, \boldsymbol{L}; \boldsymbol{X}) \tag{9.1}$$

for all fixed $\boldsymbol{R} \in U$, the unimodular group. To derive a necessary condition for this we consider a fixed instant t_*, say, and select $\boldsymbol{R} = J^{1/3}(\boldsymbol{1}^t\boldsymbol{F})^{-1}$ in which $\boldsymbol{1}$ is the fixed shifter (see (2.68)) in the frame of $\mathcal{O}$ and $\boldsymbol{F}$ is fixed at the value $\boldsymbol{F}(\boldsymbol{X}, t_*)$, with determinant J. In terms of components, $\boldsymbol{1}^t\boldsymbol{F} = \delta_{iA}F_{iB}\hat{\boldsymbol{E}}_A \otimes \hat{\boldsymbol{E}}_B$. The shifter is needed here to ensure that $\boldsymbol{R}$ is a referential tensor, in accordance with (8.8). With reference to Problem 21(a) of Chapter 1, this choice furnishes a fixed unimodular $\boldsymbol{R}$, and implies that

$$\hat{\boldsymbol{T}}_\kappa(\boldsymbol{x}, \boldsymbol{v}, \theta, \boldsymbol{F}_\kappa, \boldsymbol{L}; \boldsymbol{X}) = \hat{\boldsymbol{T}}_\kappa(\boldsymbol{x}, \boldsymbol{v}, \theta, J^{1/3}\boldsymbol{1}, \boldsymbol{L}; \boldsymbol{X}). \tag{9.2}$$

Conversely, this satisfies (9.1) for *all* unimodular $\boldsymbol{R}$ and is thus also sufficient for fluidity at the instant t_*. This instant is arbitrary, however, and so this holds at all instants. Further, from (2.59) and the conservation of mass, we have that $J = \rho_\kappa(\boldsymbol{X})/\rho$. Thus, the stress at $\boldsymbol{X}$ is determined by the list $\{\boldsymbol{x}, \boldsymbol{v}, \theta, \rho, \boldsymbol{L}\}$. Then, using $\boldsymbol{X} = \boldsymbol{\kappa}(p)$ — see (2.3) — we may eliminate the reference configuration from the constitutive function altogether and write

$$\boldsymbol{T} = \mathcal{G}(\boldsymbol{x}, \boldsymbol{v}, \theta, \rho, \boldsymbol{L}; p). \tag{9.3}$$

9.1 Material frame indifference

Of course any other observer, $\mathcal{O}^+$ say, can follow the same line of reasoning to conclude that the stress is expressible in the form

$$\boldsymbol{T}^+ = \mathcal{G}^+(\boldsymbol{x}^+, \boldsymbol{v}^+, \theta^+, \rho^+, \boldsymbol{L}^+; p), \tag{9.4}$$

for some suitable constitutive function $\mathcal{G}^+$. This function is related to $\mathcal{G}$ by

$$\mathcal{G}^+(\boldsymbol{x}^+, \boldsymbol{v}^+, \theta^+, \rho^+, \boldsymbol{L}^+; p) = \boldsymbol{Q}\mathcal{G}(\boldsymbol{x}, \boldsymbol{v}, \theta, \rho, \boldsymbol{L}; p)\boldsymbol{Q}^t, \tag{9.5}$$

where $\boldsymbol{Q}(t)$: $E^3 \to E^{3+}$ is the orthogonal tensor associated with the change of frame. We wish to use this statement to derive restrictions on the constitutive function $\mathcal{G}$ pertaining to $\mathcal{O}$. That pertaining to $\mathcal{O}^+$, if desired, can then be obtained *a posteriori*.

Following [13], consider two relative motions of the observers, characterized by orthogonal tensors $\boldsymbol{Q}_1(t)$ and $\boldsymbol{Q}_2(t)$, and vectors $\boldsymbol{c}_1(t)$ and $\boldsymbol{c}_2(t)$, respectively. From the perspective of $\mathcal{O}^+$, the positions $\boldsymbol{x}_1$ and $\boldsymbol{x}_2$ of p, as perceived by $\mathcal{O}$, are related by (cf. (7.3))

$$\boldsymbol{Q}_1(t)\boldsymbol{x}_1 + \boldsymbol{c}_1(t) = \boldsymbol{x}^+ = \boldsymbol{Q}_2(t)\boldsymbol{x}_2 + \boldsymbol{c}_2(t), \tag{9.6}$$

and from (7.8) and (7.33), the associated velocities and velocity gradients are related by

$$\boldsymbol{Q}_1\boldsymbol{v}_1 + \dot{\boldsymbol{Q}}_1\boldsymbol{Q}_1^t(\boldsymbol{x}^+ - \boldsymbol{c}_1) + \dot{\boldsymbol{c}}_1 = \boldsymbol{v}^+ = \boldsymbol{Q}_2\boldsymbol{v}_2 + \dot{\boldsymbol{Q}}_2\boldsymbol{Q}_2^t(\boldsymbol{x}^+ - \boldsymbol{c}_2) + \dot{\boldsymbol{c}}_2 \tag{9.7}$$

and

$$\boldsymbol{Q}_1\boldsymbol{L}_1\boldsymbol{Q}_1^t + \dot{\boldsymbol{Q}}_1\boldsymbol{Q}_1^t = \boldsymbol{L}^+ = \boldsymbol{Q}_2\boldsymbol{L}_2\boldsymbol{Q}_2^t + \dot{\boldsymbol{Q}}_2\boldsymbol{Q}_2^t, \tag{9.8}$$

respectively, whereas (7.37) implies that $\rho_1 = \rho^+ = \rho_2$ and $\theta_1 = \theta^+ = \theta_2$. Denoting these simply by ρ and θ, we then have

$$\begin{aligned}\boldsymbol{Q}_1\mathcal{G}(\boldsymbol{x}_1, \boldsymbol{v}_1, \theta, \rho, \boldsymbol{L}_1; p)\boldsymbol{Q}_1^t &= \mathcal{G}^+(\boldsymbol{x}^+, \boldsymbol{v}^+, \theta^+, \rho^+, \boldsymbol{L}^+; p)\\ &= \boldsymbol{Q}_2\mathcal{G}(\boldsymbol{x}_2, \boldsymbol{v}_2, \theta, \rho, \boldsymbol{L}_2; p)\boldsymbol{Q}_2^t,\end{aligned} \tag{9.9}$$

and hence the restriction

$$\mathcal{G}(\boldsymbol{x}_2, \boldsymbol{v}_2, \theta, \rho, \boldsymbol{L}_2; p) = \boldsymbol{Q}\mathcal{G}(\boldsymbol{x}_1, \boldsymbol{v}_1, \theta, \rho, \boldsymbol{L}_1; p)\boldsymbol{Q}^t \tag{9.10}$$

on the constitutive equation used by $\mathcal{O}$, where

$$\boldsymbol{Q} = \boldsymbol{Q}_2^t \boldsymbol{Q}_1 \colon E^3 \to E^3 \tag{9.11}$$

is an arbitrary orthogonal tensor mapping E^3 to itself.

Straightforward manipulations making use of (9.6)–(9.8) yield

$$\boldsymbol{x}_2 = \boldsymbol{Q}\boldsymbol{x}_1 + \boldsymbol{d}(t), \quad \text{where } \boldsymbol{d}(t) = \boldsymbol{Q}_2^t(\boldsymbol{c}_1 - \boldsymbol{c}_2), \quad \text{and} \quad \boldsymbol{v}_2 = \boldsymbol{Q}\boldsymbol{v}_1 + \boldsymbol{e}, \tag{9.12}$$

where $\boldsymbol{e}$ is a function of $\boldsymbol{x}^+$ and t, and

$$\boldsymbol{L}_2 = \boldsymbol{Q}\boldsymbol{L}_1\boldsymbol{Q}^t + \dot{\boldsymbol{Q}}\boldsymbol{Q}^t; \quad \text{hence } \boldsymbol{D}_2 = \boldsymbol{Q}\boldsymbol{D}_1\boldsymbol{Q}^t \quad \text{and}$$
$$\boldsymbol{W}_2 = \boldsymbol{Q}\boldsymbol{W}_1\boldsymbol{Q}^t + \dot{\boldsymbol{Q}}\boldsymbol{Q}^t, \tag{9.13}$$

in which $\boldsymbol{Q}$ is given by (9.11).

To derive a necessary condition for (9.10) we exploit the arbitrariness of $\boldsymbol{Q}$ and select $\boldsymbol{Q}(t) = \boldsymbol{I}$, the identity on E^3. Inserting this into (9.11) and left-multiplying by $\boldsymbol{Q}_2$ gives $\boldsymbol{Q}_2\boldsymbol{I} = \boldsymbol{I}^+\boldsymbol{Q}_1$, where $\boldsymbol{I}^+$ is the identity on E^{3+}. The left- and right-hand sides of this are simply $\boldsymbol{Q}_2$ and $\boldsymbol{Q}_1$, respectively, and our choice thus yields $\boldsymbol{Q}_2 = \boldsymbol{Q}_1$. We find that $\boldsymbol{e}$ then reduces to a function of t alone, given by

$$\boldsymbol{e}(t) = \boldsymbol{Q}_1^t(\dot{\boldsymbol{c}}_1 - \dot{\boldsymbol{c}}_2) + \dot{\boldsymbol{Q}}_1\boldsymbol{Q}_1^t(\boldsymbol{c}_1 - \boldsymbol{c}_2), \tag{9.14}$$

and that (9.10) reduces, on dropping the subscript $_1$, to

$$\mathcal{G}(\boldsymbol{x} + \boldsymbol{d}, \boldsymbol{v} + \boldsymbol{e}, \theta, \rho, \boldsymbol{L}; p) = \mathcal{G}(\boldsymbol{x}, \boldsymbol{v}, \theta, \rho, \boldsymbol{L}; p), \tag{9.15}$$

for arbitrary $\boldsymbol{d}$ and $\boldsymbol{e}$. It follows that the constitutive function is insensitive to variations of its first two arguments, and hence that it is independent of those arguments. We thus arrive at the major simplification

$$\mathcal{G}(\boldsymbol{x}, \boldsymbol{v}, \theta, \rho, \boldsymbol{L}; p) = \boldsymbol{G}(\theta, \rho, \boldsymbol{L}; p) \tag{9.16}$$

for some function $\boldsymbol{G}$. Substituting back into (9.10), we find that this function must be such that

$$\boldsymbol{G}(\theta, \rho, \boldsymbol{Q}\boldsymbol{L}\boldsymbol{Q}^t + \boldsymbol{\Omega}; p) = \boldsymbol{Q}\boldsymbol{G}(\theta, \rho, \boldsymbol{L}; p)\boldsymbol{Q}^t \tag{9.17}$$

for arbitrary orthogonal $\boldsymbol{Q}(t) \colon E^3 \to E^3$, with $\boldsymbol{\Omega} = \dot{\boldsymbol{Q}}\boldsymbol{Q}^t$.

At any fixed instant t_*, say, we may choose $\boldsymbol{Q}$ and $\boldsymbol{\Omega}$ independently. For example, with reference to (C.24) of Appendix C, consider the orthogonal tensor

$$\boldsymbol{Q}(t) = [\exp(t - t_*)\hat{\boldsymbol{\Omega}}]\hat{\boldsymbol{Q}}, \tag{9.18}$$

in which $\hat{\boldsymbol{\Omega}}$ and $\hat{\boldsymbol{Q}}$, respectively, are arbitrary fixed skew and orthogonal tensors. Then, $\boldsymbol{Q}(t_*) = \hat{\boldsymbol{Q}}$ and $\dot{\boldsymbol{Q}}\boldsymbol{Q}^t_{|t*} = \hat{\boldsymbol{\Omega}}$, and from (9.17), evaluated at $t = t_*$,

$$\boldsymbol{G}(\theta, \rho, \hat{\boldsymbol{Q}}(\boldsymbol{D}_* + \boldsymbol{W}_*)\hat{\boldsymbol{Q}}^t + \hat{\boldsymbol{\Omega}}; p) = \hat{\boldsymbol{Q}}\boldsymbol{G}(\theta, \rho, \boldsymbol{L}_*; p)\hat{\boldsymbol{Q}}^t, \tag{9.19}$$

where $\boldsymbol{L}_*$ is the velocity gradient at the instant t_*, and $\boldsymbol{D}_* = Sym\boldsymbol{L}_*$ and $\boldsymbol{W}_* = Skw\boldsymbol{L}_*$ are the associated stretching and spin tensors. Now, since the material point p is fixed in this discussion, all arguments of the constitutive function depend on time alone. We are thus free to choose $\hat{\boldsymbol{\Omega}} = -\boldsymbol{W}_*$. Setting $\hat{\boldsymbol{Q}} = \boldsymbol{I}$, we arrive at

$$\boldsymbol{G}(\theta, \rho, \boldsymbol{L}_*; p) = \boldsymbol{G}(\theta, \rho, \boldsymbol{D}_*; p). \tag{9.20}$$

Since t_* is arbitrary, we conclude that, for (9.17) to hold, it is necessary that

$$\boldsymbol{G}(\theta, \rho, \boldsymbol{L}; p) = \boldsymbol{G}(\theta, \rho, \boldsymbol{D}; p) \tag{9.21}$$

at all t, and hence that the stress can depend on the velocity gradient only through the stretching tensor. On substituting back into (9.17) and recalling (9.13), we find that this dependence must be such that

$$\boldsymbol{G}(\theta, \rho, \boldsymbol{Q}\boldsymbol{D}\boldsymbol{Q}^t; p) = \boldsymbol{Q}\boldsymbol{G}(\theta, \rho, \boldsymbol{D}; p)\boldsymbol{Q}^t \tag{9.22}$$

for *all* orthogonal $\boldsymbol{Q}(t)\colon E^3 \to E^3$. Since this function delivers the values of the Cauchy stress, it must also be symmetric.

9.1.1 *Isotropic symmetric tensor-valued functions of a symmetric tensor*

Dropping the passive variables θ, ρ and p for convenience, we seek a general representation of a symmetric tensor-valued function $\boldsymbol{G}$, of an arbitrary symmetric tensor $\boldsymbol{D}$, satisfying

$$\boldsymbol{G}(\boldsymbol{Q}\boldsymbol{D}\boldsymbol{Q}^t) = \boldsymbol{Q}\boldsymbol{G}(\boldsymbol{D})\boldsymbol{Q}^t \tag{9.23}$$

for all orthogonal $\boldsymbol{Q}(t)$: $E^3 \to E^3$. Such a function is said to be *isotropic*, despite the fact that (9.22) pertains to frame invariance rather than to the notion of isotropic material symmetry, the latter to be discussed in Chapter 11.

We will prove that the most general function of this type is of the form

$$\boldsymbol{G}(\boldsymbol{D}) = \varphi_0(I_1, I_2, I_3)\boldsymbol{I} + \varphi_1(I_1, I_2, I_3)\boldsymbol{D} + \varphi_2(I_1, I_2, I_3)\boldsymbol{D}^2, \quad (9.24)$$

where $\varphi_{0,1,2}$ are scalar-valued functions of the principal invariants of $\boldsymbol{D}$ (see (3.3)):

$$I_1(\boldsymbol{D}) = tr\boldsymbol{D}, \quad I_2(\boldsymbol{D}) = \frac{1}{2}[I_1^2 - tr(\boldsymbol{D}^2)] \quad \text{and} \quad I_3(\boldsymbol{D}) = \det \boldsymbol{D}. \quad (9.25)$$

The proof of this important result is in four parts.

Part 1

Suppose the scalar-valued function $\varphi(\boldsymbol{D})$ is isotropic, i.e., $\varphi(\boldsymbol{QDQ}^t) = \varphi(\boldsymbol{D})$. We show that it is necessary and sufficient that $\varphi(\boldsymbol{D}) = \bar{\varphi}(I_1, I_2, I_3)$ for some function $\bar{\varphi}$. To prove necessity, it suffices to show that φ is determined by the invariants, i.e., that $\varphi(\boldsymbol{A}) = \varphi(\boldsymbol{B})$ whenever $I_k(\boldsymbol{A}) = I_k(\boldsymbol{B})$; $k = 1, 2, 3$. Here, $\boldsymbol{A}$ and $\boldsymbol{B}$ are arbitrary symmetric tensors. Thus, suppose that $I_k(\boldsymbol{A}) = I_k(\boldsymbol{B})$. From (3.2), we have that $\boldsymbol{A}$ and $\boldsymbol{B}$ share the same principal values. Their spectral representations are

$$\boldsymbol{A} = \sum \lambda_i \boldsymbol{u}_i \otimes \boldsymbol{u}_i \quad \text{and} \quad \boldsymbol{B} = \sum \lambda_i \boldsymbol{v}_i \otimes \boldsymbol{v}_i, \quad (9.26)$$

where $\{\boldsymbol{u}_i\}$ and $\{\boldsymbol{v}_i\}$ are orthonormal sets of principal vectors. We use them to construct the orthogonal tensor $\tilde{\boldsymbol{Q}} = \boldsymbol{u}_i \otimes \boldsymbol{v}_i$. Thus, $\boldsymbol{A} = \tilde{\boldsymbol{Q}}\boldsymbol{B}\tilde{\boldsymbol{Q}}^t$. Recalling that $\varphi(\tilde{\boldsymbol{Q}}\boldsymbol{B}\tilde{\boldsymbol{Q}}^t) = \varphi(\boldsymbol{B})$ by hypothesis, we arrive at $\varphi(\boldsymbol{A}) = \varphi(\boldsymbol{B})$, and so φ is determined by the invariants, as claimed.

To prove sufficiency, suppose $\varphi(\boldsymbol{A}) = \bar{\varphi}(I_1(\boldsymbol{A}), I_2(\boldsymbol{A}), I_3(\boldsymbol{A}))$. Then for any orthogonal $\boldsymbol{Q}$, $I_1(\boldsymbol{QAQ}^t) = tr(\boldsymbol{QAQ}^t) = tr(\boldsymbol{Q}^t\boldsymbol{QA}) = I_1(\boldsymbol{A})$, $I_2(\boldsymbol{QAQ}^t) = \frac{1}{2}[I_1^2 - tr(\boldsymbol{QAQ}^t\boldsymbol{QAQ}^t)] = \frac{1}{2}[I_1^2 - tr(\boldsymbol{QA}^2\boldsymbol{Q}^t)] = I_2(\boldsymbol{A})$, and $I_3(\boldsymbol{QAQ}^t) = \det(\boldsymbol{QAQ}^t) = (\det \boldsymbol{Q})^2 \det \boldsymbol{A} = I_3(\boldsymbol{A})$, so that $\varphi(\boldsymbol{A}) = \varphi(\boldsymbol{QAQ}^t)$.

Part 2

Let η_i be principal values of $\boldsymbol{D}$, and let $\boldsymbol{\nu}_i$ be the associated principal vectors, so that $\boldsymbol{D} = \sum \eta_i \boldsymbol{\nu}_i \otimes \boldsymbol{\nu}_i$. We show that $\boldsymbol{\nu}_i$ are also principal vectors of $\boldsymbol{G}(\boldsymbol{D})$. To prove this claim consider the orthogonal tensor $\bar{\boldsymbol{Q}} = 2\boldsymbol{\nu}_1 \otimes \boldsymbol{\nu}_1 - \boldsymbol{I}$, representing a 180° rotation about the axis $\boldsymbol{\nu}_1$, i.e., $\bar{\boldsymbol{Q}}\boldsymbol{\nu}_1 = \boldsymbol{\nu}_1$, $\bar{\boldsymbol{Q}}\boldsymbol{\nu}_2 = -\boldsymbol{\nu}_2$ and $\bar{\boldsymbol{Q}}\boldsymbol{\nu}_3 = -\boldsymbol{\nu}_3$. From (9.23),

$$\bar{\boldsymbol{Q}}\boldsymbol{G}(\boldsymbol{D})\bar{\boldsymbol{Q}}^t = \boldsymbol{G}\left(\sum \eta_i \bar{\boldsymbol{Q}}\boldsymbol{\nu}_i \otimes \bar{\boldsymbol{Q}}\boldsymbol{\nu}_i\right) = \boldsymbol{G}\left(\sum \eta_i \boldsymbol{\nu}_i \otimes \boldsymbol{\nu}_i\right) = \boldsymbol{G}(\boldsymbol{D}). \tag{9.27}$$

Thus, $\bar{\boldsymbol{Q}}\boldsymbol{G}(\boldsymbol{D}) = \boldsymbol{G}(\boldsymbol{D})\bar{\boldsymbol{Q}}$ and

$$\bar{\boldsymbol{Q}}\boldsymbol{G}(\boldsymbol{D})\boldsymbol{\nu}_1 = \boldsymbol{G}(\boldsymbol{D})\bar{\boldsymbol{Q}}\boldsymbol{\nu}_1 = \boldsymbol{G}(\boldsymbol{D})\boldsymbol{\nu}_1. \tag{9.28}$$

Therefore, $\boldsymbol{G}(\boldsymbol{D})\boldsymbol{\nu}_1$ is also an axis of $\bar{\boldsymbol{Q}}$. But this axis is unique (Appendix C); therefore, $\boldsymbol{G}(\boldsymbol{D})\boldsymbol{\nu}_1 = \beta_1(\boldsymbol{D})\boldsymbol{\nu}_1$ for some scalar function β_1, and $\boldsymbol{\nu}_1$ is a principal vector of $\boldsymbol{G}(\boldsymbol{D})$. Repeating this argument with $\boldsymbol{\nu}_1$ replaced by $\boldsymbol{\nu}_2$ or $\boldsymbol{\nu}_3$, we conclude that $\{\boldsymbol{\nu}_i\}$ is a principal basis for $\boldsymbol{G}(\boldsymbol{D})$, as claimed, and hence that

$$\boldsymbol{G}(\boldsymbol{D}) = \sum \beta_i(\boldsymbol{D})\boldsymbol{\nu}_i \otimes \boldsymbol{\nu}_i, \tag{9.29}$$

for some scalar-valued functions β_i.

Part 3

We show that the set $\{\boldsymbol{I}, \boldsymbol{D}, \boldsymbol{D}^2\}$ is linearly independent. To prove this, we first note that since the constitutive function $\boldsymbol{G}$ is defined for all symmetric $\boldsymbol{D}$, it is defined for $\boldsymbol{D}$ having distinct principal values η_i. For such $\boldsymbol{D}$, we will assume that $\boldsymbol{I}, \boldsymbol{D}$ and $\boldsymbol{D}^2$ are linearly dependent and reach a contradiction. Thus, we assume that

$$\boldsymbol{O} = a\boldsymbol{I} + b\boldsymbol{D} + c\boldsymbol{D}^2 = \sum (a + b\eta_i + c\eta_i^2)\boldsymbol{\nu}_i \otimes \boldsymbol{\nu}_i, \tag{9.30}$$

with at least one element of $\{a, b, c\}$ unequal to zero. This implies that $a + b\eta_i + c\eta_i^2 = 0$ for $i = 1, 2, 3$, and hence that the η_i are all the roots of the same quadratic equation. But such equations have at most two distinct roots, contrary to our choice of $\boldsymbol{D}$. This contradiction implies that the premise is false, and hence that the set $\{\boldsymbol{I}, \boldsymbol{D}, \boldsymbol{D}^2\}$ is linearly independent, as claimed.

Note that

$$\boldsymbol{I}, \boldsymbol{D}, \boldsymbol{D}^2 \in S = Span\{\boldsymbol{\nu}_1 \otimes \boldsymbol{\nu}_1, \boldsymbol{\nu}_2 \otimes \boldsymbol{\nu}_2, \boldsymbol{\nu}_3 \otimes \boldsymbol{\nu}_3\}, \tag{9.31}$$

a three-dimensional vector space. Since n linearly independent elements of an n-dimensional vector space constitute a basis for that space, we have

$$S = Span\{\boldsymbol{I}, \boldsymbol{D}, \boldsymbol{D}^2\}. \tag{9.32}$$

Part 4

It follows from (9.29) that $\boldsymbol{G}(\boldsymbol{D}) \in S$. Thus, (9.32) implies that there are unique scalars $\psi_0(\boldsymbol{D}), \psi_1(\boldsymbol{D})$ and $\psi_2(\boldsymbol{D})$ such that

$$\boldsymbol{G}(\boldsymbol{D}) = \psi_0(\boldsymbol{D})\boldsymbol{I} + \psi_1(\boldsymbol{D})\boldsymbol{D} + \psi_2(\boldsymbol{D})\boldsymbol{D}^2. \tag{9.33}$$

To see this explicitly, we combine it with (9.29) to get

$$\beta_i(\boldsymbol{D}) = \psi_0 + \psi_1\eta_i + \psi_2\eta_i^2; \quad i = 1, 2, 3, \tag{9.34}$$

i.e.,

$$\begin{pmatrix} 1 & \eta_1 & \eta_1^2 \\ 1 & \eta_2 & \eta_2^2 \\ 1 & \eta_3 & \eta_3^2 \end{pmatrix} \begin{Bmatrix} \psi_0 \\ \psi_1 \\ \psi_2 \end{Bmatrix} = \begin{Bmatrix} \beta_1 \\ \beta_2 \\ \beta_3 \end{Bmatrix}, \tag{9.35}$$

in which the coefficient matrix has determinant $(\eta_1 - \eta_1)$ $(\eta_2 - \eta_3)$ $(\eta_3 - \eta_1)$. This is non-zero for the family of $\boldsymbol{D}'s$ considered. Then, since the η_i are determined by the $I_k(\boldsymbol{D})$, it follows that the $\psi's$ are determined by $\boldsymbol{D}$.

Now, from (9.33) we have

$$\begin{aligned} &\boldsymbol{G}(\boldsymbol{Q}\boldsymbol{D}\boldsymbol{Q}^t) - \boldsymbol{Q}\boldsymbol{G}(\boldsymbol{D})\boldsymbol{Q}^t \\ &\quad = \boldsymbol{Q}\{[\psi_0(\boldsymbol{Q}\boldsymbol{D}\boldsymbol{Q}^t) - \psi_0(\boldsymbol{D})]\boldsymbol{I} + [\psi_1(\boldsymbol{Q}\boldsymbol{D}\boldsymbol{Q}^t) - \psi_1(\boldsymbol{D})]\boldsymbol{D} \\ &\qquad + [\psi_2(\boldsymbol{Q}\boldsymbol{D}\boldsymbol{Q}^t) - \psi_2(\boldsymbol{D})]\boldsymbol{D}^2\}\boldsymbol{Q}^t, \end{aligned} \tag{9.36}$$

where we have used $\boldsymbol{Q}\boldsymbol{Q}^t = \boldsymbol{I}$ and $(\boldsymbol{Q}\boldsymbol{D}\boldsymbol{Q}^t)^2 = \boldsymbol{Q}\boldsymbol{D}^2\boldsymbol{Q}^t$. We impose (9.23) and left- and right-multiply the resulting equation by $\boldsymbol{Q}^t$ and $\boldsymbol{Q}$ respectively. Using the result of Part 3, we obtain

$\psi_i(\boldsymbol{QDQ}^t) = \psi_i(\boldsymbol{D})$. Finally, we apply the result of Part 1 to conclude that $\psi_i(\boldsymbol{D}) = \bar{\psi}_i(I_1, I_2, I_3)$ and substitute into (9.33) to arrive at (9.24). Conversely, it is a simple matter to confirm that (9.24) ensures the isotropy of $\boldsymbol{G}(\boldsymbol{D})$.

This result subsumes the case in which $\boldsymbol{D}$ has only two distinct principal values, or just one. The first of these is covered by taking $\varphi_2 = 0$, and the third by taking $\varphi_1 = \varphi_2 = 0$ [3].

Reinstating the passive variables, we conclude that the stress at a material point p in a viscous fluid is of the form

$$\begin{aligned}\boldsymbol{T} = {} & \varphi_0(I_1, I_2, I_3, \theta, \rho; p)\boldsymbol{I} + \varphi_1(I_1, I_2, I_3, \theta, \rho; p)\boldsymbol{D} \\ & + \varphi_2(I_1, I_2, I_3, \theta, \rho; p)\boldsymbol{D}^2, \end{aligned} \tag{9.37}$$

where the I_k are the principal invariants of $\boldsymbol{D}$. It is a matter of experiment to determine the three functions φ_0, φ_1 and φ_2 in terms of the five variables I_1, I_2, I_3, θ and ρ. While this task presents a formidable challenge, it is far more tractable than that presented by the original constitutive hypothesis embodied in (9.3).

Assuming the $\varphi' s$ to be continuous functions of the invariants, we observe that the stress reduces to

$$\boldsymbol{T}_{|\boldsymbol{D}=\boldsymbol{O}} = -\tilde{p}(\theta, \rho; p)\boldsymbol{I} \tag{9.38}$$

in the absence of stretching, where $\tilde{p} = -\varphi_0(0, 0, 0, \theta, \rho; p)$ is the pressure. If this holds in the presence of stretching, then the fluid is *inviscid*. The latter model is descriptive in regions exterior to "boundary layers" in which the effects of viscosity are important. The inviscid fluid model therefore plays a major role in the theories of aero- and hydro-dynamics.

9.2 The heat flux

Beginning with (8.31), we may proceed, as in the passage from (9.1) to (9.22), to conclude that the heat flux is given constitutively in the form

$$\boldsymbol{q} = \boldsymbol{g}(\theta, \rho, \boldsymbol{D}, grad\theta; p), \tag{9.39}$$

in the frame of $\mathcal{O}$, where the function $\boldsymbol{g}$ is such that

$$\boldsymbol{g}(\theta, \rho, \boldsymbol{QDQ}^t, \boldsymbol{Q}grad\theta; p) = \boldsymbol{Qg}(\theta, \rho, \boldsymbol{D}, grad\theta; p) \tag{9.40}$$

for all orthogonal $\boldsymbol{Q}$: $E^3 \to E^3$. We will show that the general form of such a function is

$$\boldsymbol{g}(\theta, \rho, \boldsymbol{D}, grad\theta; p) = \boldsymbol{K} grad\theta, \tag{9.41}$$

where

$$\boldsymbol{K}(\theta, \rho, \boldsymbol{D}, grad\theta; p) = \phi_0(I_1, \ldots, I_6, \theta, \rho; p)\boldsymbol{I} + \phi_1(I_1, \ldots, I_6, \theta, \rho; p)\boldsymbol{D} + \phi_2(I_1, \ldots, I_6, \theta, \rho; p)\boldsymbol{D}^2, \tag{9.42}$$

with

$$I_4(\boldsymbol{D}, grad\theta) = grad\theta \cdot \boldsymbol{D}(grad\theta), \quad I_5(\boldsymbol{D}, grad\theta) = grad\theta \cdot \boldsymbol{D}^2(grad\theta)$$
$$\text{and } I_6(grad\theta) = |grad\theta|. \tag{9.43}$$

To simplify the notation we again suppress the passive variables and write (9.40) as

$$\boldsymbol{g}(\boldsymbol{QDQ}^t, \boldsymbol{Qd}) = \boldsymbol{Qg}(\boldsymbol{D}, \boldsymbol{d}), \tag{9.44}$$

for all symmetric $\boldsymbol{D}$: $E^3 \to E^3$ and all $\boldsymbol{d} = grad\theta \in E^3$, where $\boldsymbol{Q}$: $E^3 \to E^3$ is an arbitrary orthogonal tensor. For $\boldsymbol{Q} = -\boldsymbol{I}$ this yields

$$\boldsymbol{g}(\boldsymbol{D}, -\boldsymbol{d}) = -\boldsymbol{g}(\boldsymbol{D}, \boldsymbol{d}). \tag{9.45}$$

Our proof is again in four parts.

Part 1

Suppose $\varphi(\boldsymbol{D}, \boldsymbol{d}) = \varphi(\boldsymbol{D}, -\boldsymbol{d})$. We show that φ then depends on $\boldsymbol{d}$ via $\boldsymbol{d} \otimes \boldsymbol{d}$. Dropping the passive variable $\boldsymbol{D}$, our task is to show that if $\psi(\boldsymbol{d}) = \psi(-\boldsymbol{d})$, then $\psi(\boldsymbol{d}) = \bar{\psi}(\boldsymbol{d} \otimes \boldsymbol{d})$, i.e., that $\psi(\boldsymbol{a}) = \psi(\boldsymbol{b})$ whenever $\boldsymbol{a} \otimes \boldsymbol{a} = \boldsymbol{b} \otimes \boldsymbol{b}$. Suppose, then, that the latter is true. It follows that

$$|\boldsymbol{a}|^2 \boldsymbol{a} = (\boldsymbol{a} \cdot \boldsymbol{b})\boldsymbol{b} \quad \text{and} \quad |\boldsymbol{b}|^2 \boldsymbol{b} = (\boldsymbol{a} \cdot \boldsymbol{b})\boldsymbol{a}, \tag{9.46}$$

which imply that $|\boldsymbol{a}| = |\boldsymbol{b}|$ and $(\boldsymbol{a} \cdot \boldsymbol{b})^2 = |\boldsymbol{a}|^2 |\boldsymbol{b}|^2$. But from (1.5) there is $\alpha \in \mathbb{R}$ such that $\boldsymbol{a} \cdot \boldsymbol{b} = |\boldsymbol{a}| |\boldsymbol{b}| \cos \alpha$. Then, $\cos \alpha = \pm 1$ and either of (9.46) gives $\boldsymbol{b} = \pm \boldsymbol{a}$. The solution $\boldsymbol{b} = \boldsymbol{a}$ trivially yields $\psi(\boldsymbol{b}) = \psi(\boldsymbol{a})$, and the solution $\boldsymbol{b} = -\boldsymbol{a}$ gives $\psi(\boldsymbol{b}) = \psi(-\boldsymbol{a})$. But $\psi(\boldsymbol{a}) = \psi(-\boldsymbol{a})$ by hypothesis, and therefore $\psi(\boldsymbol{a}) = \psi(\boldsymbol{b})$ whenever $\boldsymbol{a} \otimes \boldsymbol{a} = \boldsymbol{b} \otimes \boldsymbol{b}$. Thus, $\psi(\boldsymbol{d}) = \bar{\psi}(\boldsymbol{d} \otimes \boldsymbol{d})$ for some function $\bar{\psi}$, as claimed. Conversely, if $\psi(\boldsymbol{d}) = \bar{\psi}(\boldsymbol{d} \otimes \boldsymbol{d})$, then it follows immediately that $\psi(\boldsymbol{d}) = \psi(-\boldsymbol{d})$.

Part 2

We show that if $\boldsymbol{d}$ is arbitrary and $\boldsymbol{D}$ is an arbitrary symmetric tensor, then $\{\boldsymbol{d}, \boldsymbol{D}\boldsymbol{d}, \boldsymbol{D}^2\boldsymbol{d}\}$ is a basis for E^3. We again discuss only the case in which the principal values η_i of $\boldsymbol{D}$ are distinct. Thus, we assume these vectors to be linearly dependent, i.e., that

$$a\boldsymbol{d} + b\boldsymbol{D}\boldsymbol{d} + c\boldsymbol{D}^2\boldsymbol{d} = \boldsymbol{0}, \tag{9.47}$$

with at least one element of $\{a, b, c\}$ unequal to zero. Using the spectral decomposition of $\boldsymbol{D}$, we write this as

$$\sum (a + b\eta_i + c\eta_i^2) d_i \boldsymbol{\nu}_i = \boldsymbol{0}, \quad \text{where } d_i = \boldsymbol{\nu}_i \cdot \boldsymbol{d}. \tag{9.48}$$

As $\boldsymbol{d}$ is arbitrary, this requires that $a + b\eta_i + c\eta_i^2 = 0; i = 1, 2, 3$. This is the contradiction encountered previously. It follows that a, b, c all vanish and hence that the set $\{\boldsymbol{d}, \boldsymbol{D}\boldsymbol{d}, \boldsymbol{D}^2\boldsymbol{d}\}$ is linearly independent.

Thus,

$$\boldsymbol{g}(\boldsymbol{D}, \boldsymbol{d}) = \psi_0(\boldsymbol{D}, \boldsymbol{d})\boldsymbol{d} + \psi_1(\boldsymbol{D}, \boldsymbol{d})\boldsymbol{D}\boldsymbol{d} + \psi_2(\boldsymbol{D}, \boldsymbol{d})\boldsymbol{D}^2\boldsymbol{d}. \tag{9.49}$$

Part 3

From the result of Part 2, we have

$$\boldsymbol{g}(\boldsymbol{D}, -\boldsymbol{d}) = -\psi_0(\boldsymbol{D}, -\boldsymbol{d})\boldsymbol{d} - \psi_1(\boldsymbol{D}, -\boldsymbol{d})\boldsymbol{D}\boldsymbol{d} - \psi_2(\boldsymbol{D}, -\boldsymbol{d})\boldsymbol{D}^2\boldsymbol{d}. \tag{9.50}$$

Imposing (9.45) yields

$$\begin{aligned}[]
[\psi_0(\boldsymbol{D}, \boldsymbol{d}) - \psi_0(\boldsymbol{D}, -\boldsymbol{d})]\boldsymbol{d} &+ [\psi_1(\boldsymbol{D}, \boldsymbol{d}) - \psi_1(\boldsymbol{D}, -\boldsymbol{d})]\boldsymbol{D}\boldsymbol{d} \\
&+ [\psi_2(\boldsymbol{D}, \boldsymbol{d}) - \psi_2(\boldsymbol{D}, -\boldsymbol{d})]\boldsymbol{D}^2\boldsymbol{d} = \boldsymbol{0},
\end{aligned} \tag{9.51}$$

and therefore $\psi_i(\boldsymbol{D}, \boldsymbol{d}) = \psi_i(\boldsymbol{D}, -\boldsymbol{d}); i = 0, 1, 2$. It then follows from Part 1 that

$$\psi_i(\boldsymbol{D}, \boldsymbol{d}) = \bar{\psi}_i(\boldsymbol{D}, \boldsymbol{d} \otimes \boldsymbol{d}); \quad i = 0, 1, 2. \tag{9.52}$$

Thus,

$$\begin{aligned}
&\boldsymbol{g}(\boldsymbol{QDQ}^t, \boldsymbol{Qd}) - \boldsymbol{Qg}(\boldsymbol{D}, \boldsymbol{d}) \\
&\quad = [\bar{\psi}_0(\boldsymbol{QDQ}^t, \boldsymbol{Qd} \otimes \boldsymbol{Qd}) - \bar{\psi}_0(\boldsymbol{D}, \boldsymbol{d} \otimes \boldsymbol{d})]\boldsymbol{Qd} \\
&\qquad + [\bar{\psi}_1(\boldsymbol{QDQ}^t, \boldsymbol{Qd} \otimes \boldsymbol{Qd}) - \bar{\psi}_1(\boldsymbol{D}, \boldsymbol{d} \otimes \boldsymbol{d})]\boldsymbol{QDd} \\
&\qquad + [\bar{\psi}_2(\boldsymbol{QDQ}^t, \boldsymbol{Qd} \otimes \boldsymbol{Qd}) - \bar{\psi}_2(\boldsymbol{D}, \boldsymbol{d} \otimes \boldsymbol{d})]\boldsymbol{QD}^2\boldsymbol{d}. \qquad (9.53)
\end{aligned}$$

Imposing (9.44), left-multiplying by $\boldsymbol{Q}^t$ and invoking the result of Part 2, we conclude that

$$\bar{\psi}_i(\boldsymbol{QDQ}^t, \boldsymbol{Qd} \otimes \boldsymbol{Qd}) = \bar{\psi}_i(\boldsymbol{D}, \boldsymbol{d} \otimes \boldsymbol{d}); \quad i = 0, 1, 2, \qquad (9.54)$$

for all orthogonal $\boldsymbol{Q}$.

Part 4

Suppose $\varphi(\boldsymbol{D}, \boldsymbol{d} \otimes \boldsymbol{d}) = \varphi(\boldsymbol{QDQ}^t, \boldsymbol{Qd} \otimes \boldsymbol{Qd})$ for all orthogonal $\boldsymbol{Q}$. We demonstrate that φ is then determined by $I_1 - I_6$, i.e., that $\varphi(\boldsymbol{A}, \boldsymbol{a} \otimes \boldsymbol{a}) = \varphi(\boldsymbol{B}, \boldsymbol{b} \otimes \boldsymbol{b})$ whenever

$$\begin{aligned}
&\{I_1(\boldsymbol{A}), I_2(\boldsymbol{A}), I_3(\boldsymbol{A}), I_4(\boldsymbol{A}, \boldsymbol{a}), I_5(\boldsymbol{A}, \boldsymbol{a}), I_6(\boldsymbol{a})\} \\
&\quad = \{I_1(\boldsymbol{B}), I_2(\boldsymbol{B}), I_3(\boldsymbol{B}), I_4(\boldsymbol{B}, \boldsymbol{b}), I_5(\boldsymbol{B}, \boldsymbol{b}), I_6(\boldsymbol{b})\}, \qquad (9.55)
\end{aligned}$$

where $I_1 - I_6$ are defined in (9.25) and (9.43). Thus, suppose these two sets are equal. Then the invariants I_1, I_2, I_3 of $\boldsymbol{A}$ and $\boldsymbol{B}$ coincide and (9.26) holds in particular, implying that $\boldsymbol{A} = \tilde{\boldsymbol{Q}}\boldsymbol{B}\tilde{\boldsymbol{Q}}^t$ and $\boldsymbol{A}^2 = \tilde{\boldsymbol{Q}}\boldsymbol{B}^2\tilde{\boldsymbol{Q}}^t$, where $\tilde{\boldsymbol{Q}} = \boldsymbol{u}_i \otimes \boldsymbol{v}_i$ is orthogonal.

Let

$$c_i = (\tilde{\boldsymbol{Q}}^t\boldsymbol{a} \cdot \boldsymbol{v}_i)^2 - (\boldsymbol{b} \cdot \boldsymbol{v}_i)^2. \qquad (9.56)$$

Then the three equations expressing the equality of I_4, I_5 and I_6 in (9.55) may be summarized in the matrix form

$$\begin{pmatrix} 1 & 1 & 1 \\ \lambda_1 & \lambda_2 & \lambda_3 \\ \lambda_1^2 & \lambda_2^2 & \lambda_2^2 \end{pmatrix} \begin{Bmatrix} c_1 \\ c_2 \\ c_3 \end{Bmatrix} = \begin{Bmatrix} 0 \\ 0 \\ 0 \end{Bmatrix}, \qquad (9.57)$$

where λ_i are the principal values of $\boldsymbol{A}$ and $\boldsymbol{B}$. As we are concerned with tensors having distinct principal values, the determinant of the coefficient matrix, $(\lambda_1 - \lambda_2)(\lambda_2 - \lambda_3)(\lambda_3 - \lambda_1)$, is nonzero, and therefore all the c_i vanish, i.e., $\tilde{\boldsymbol{Q}}^t \boldsymbol{a} \cdot \boldsymbol{v}_i = \pm \boldsymbol{b} \cdot \boldsymbol{v}_i; i = 1, 2, 3$. Since $\{\boldsymbol{v}_i\}$ is a basis for E^3 it follows that $\boldsymbol{a} = \pm \tilde{\boldsymbol{Q}} \boldsymbol{b}$. Thus, the equality of the two sets of scalars $I_1 - I_6$ implies that $\boldsymbol{a} \otimes \boldsymbol{a} = \tilde{\boldsymbol{Q}}\boldsymbol{b} \otimes \tilde{\boldsymbol{Q}}\boldsymbol{b}$. Finally, as $\varphi(\boldsymbol{B}, \boldsymbol{b} \otimes \boldsymbol{b}) = \varphi(\tilde{\boldsymbol{Q}}\boldsymbol{B}\tilde{\boldsymbol{Q}}^t, \tilde{\boldsymbol{Q}}\boldsymbol{b} \otimes \tilde{\boldsymbol{Q}}\boldsymbol{b})$ by hypothesis, we have $\varphi(\boldsymbol{A}, \boldsymbol{a} \otimes \boldsymbol{a}) = \varphi(\boldsymbol{B}, \boldsymbol{b} \otimes \boldsymbol{b})$, as claimed.

Collecting these results and reinstating the passive variables, we conclude that the heat flux is given by (9.41) and (9.42). This subsumes the cases in which the stretching tensor has only one or two distinct principal values. Conversely, it is a straightforward matter to show that (9.41) and (9.42) yield (9.40) for *all* orthogonal $\boldsymbol{Q}: E^3 \to E^3$, and hence that the two statements are equivalent. Further, assuming the $\phi's$ in (9.42) to be continuous functions of the $I_k; k = 1, \ldots, 6$, it follows from the continuity of this representation with respect to $grad\theta$ that

$$\boldsymbol{g}(\theta, \rho, \boldsymbol{D}, \boldsymbol{0}; p) = \boldsymbol{0}, \tag{9.58}$$

so that there is no heat conduction in the absence of a temperature gradient; and, from its continuity with respect to $\boldsymbol{D}$, that

$$\boldsymbol{g}(\theta, \rho, \boldsymbol{O}, grad\theta; p) = K(I_6, \theta, \rho; p) grad\theta, \tag{9.59}$$

for some scalar function K.

Empirical facts indicate that fluids of the present class, such as air and water, are well described by constitutive equations for the stress and heat flux that are linear in $\boldsymbol{D}$ and $grad\theta$, respectively, and that the heat flux is insensitive to $\boldsymbol{D}$. Recalling that I_2 and I_3 are nonlinear functions of $\boldsymbol{D}$, whereas $I_1 = tr\boldsymbol{D}$ is linear, the relevant expression for the stress is given by the *Newtonian fluid*

$$\boldsymbol{T} = [-\tilde{p}(\theta, \rho; p) + \lambda(\theta, \rho; p) tr\boldsymbol{D}]\boldsymbol{I} + 2\mu(\theta, \rho; p)\boldsymbol{D}, \tag{9.60}$$

where $\tilde{p}$ is the pressure in the absence of flow, and λ, μ are viscosity coefficients; and that for the heat flux, by the *Fourier law*

$$\boldsymbol{q} = K(\theta, \rho; p) grad\theta, \tag{9.61}$$

where K is the thermal conductivity.

Regarding the constitutive functions for the Helmholtz energy and the entropy, the only consequences of frame invariance and symmetry under the unimodular group that are required for our purposes are that these functions are determined by the list $\{\theta, \rho, \boldsymbol{D}, grad\theta\}$. We write these simply as $\psi(\theta, \rho, \boldsymbol{D}, grad\theta; p)$ and $\eta(\theta, \rho, \boldsymbol{D}, grad\theta; p)$, respectively.

9.3 Thermodynamical considerations

The stress, energy and entropy must be related if the balance laws and the Clausius–Duhem inequality are to be satisfied. To see this we invoke the latter inequality (see (6.187)) at the material point $p \in B$ for a process consisting of arbitrary values of $\theta, \rho, \boldsymbol{D}$ and $grad\theta$. In principle the balance laws for energy and momentum, together with the constitutive equations, furnish the energy supply r and body force $\boldsymbol{b}$ required to effect this process. Here, we note that θ and $grad\theta$ can be specified independently at any one point.

Proceeding, we use the chain rule to obtain

$$\dot{\psi} = \psi_\theta \dot{\theta} + \psi_\rho \dot{\rho} + \psi_{\boldsymbol{D}} \cdot \dot{\boldsymbol{D}} + \psi_{grad\theta} \cdot (grad\theta)^{\cdot}, \tag{9.62}$$

where subscripts are used to denote derivatives with respect to scalar, tensor or vector variables. Invoking $(6.17)_1$ in the form $\dot{\rho} = -\rho tr\boldsymbol{D} = -\rho\boldsymbol{I} \cdot \boldsymbol{D}$ and recalling that $\boldsymbol{T}$ is symmetric, we thus reduce the Clausius–Duhem inequality to the form

$$\begin{aligned} &- \rho(\psi_\theta + \eta)\dot{\theta} - \rho\psi_{\boldsymbol{D}} \cdot \dot{\boldsymbol{D}} - \rho\psi_{grad\theta} \cdot (grad\theta)^{\cdot} \\ &+ (\boldsymbol{T} + \rho^2\psi_\rho \boldsymbol{I}) \cdot \boldsymbol{D} - \theta^{-1}\boldsymbol{q} \cdot grad\theta \geq 0. \end{aligned} \tag{9.63}$$

At any fixed instant, t_* say, the lists $\{\theta, \boldsymbol{D}, grad\theta\}$ and $\{\dot{\theta}, \dot{\boldsymbol{D}}, (grad\theta)^{\cdot}\}$ can be assigned independently. Accordingly this inequality is of the form

$$\boldsymbol{a}(\boldsymbol{z}) \cdot \boldsymbol{w} + b(\boldsymbol{z}) \geq 0 \tag{9.64}$$

for arbitrary, independent values of the arrays

$$\boldsymbol{z} = (\theta, \rho, \boldsymbol{D}, grad\theta) \quad \text{and} \quad \boldsymbol{w} = (\dot{\theta}, \dot{\boldsymbol{D}}, (grad\theta)^{\cdot}), \tag{9.65}$$

where

$$\boldsymbol{a}(\boldsymbol{z}) = (-\rho(\psi_\theta + \eta), -\rho\psi_{\boldsymbol{D}}, -\rho\psi_{grad\theta}) \tag{9.66}$$

and

$$b(\boldsymbol{z}) = (\boldsymbol{T} + \rho^2\psi_\rho \boldsymbol{I}) \cdot \boldsymbol{D} - \theta^{-1}\boldsymbol{q} \cdot grad\theta. \tag{9.67}$$

To derive a necessary condition for (9.64), we choose $\boldsymbol{w} = \delta\boldsymbol{a}(\boldsymbol{z})$ with $\delta \in \mathbb{R}$ arbitrary. Then, $\delta |\boldsymbol{a}|^2 + b \geq 0$. If $|\boldsymbol{a}| \neq 0$, then this is violated for $\delta < -b/|\boldsymbol{a}|^2$, contrary to the requirement that it be satisfied for all δ. Hence, the necessary conditions

$$\boldsymbol{a}(\boldsymbol{z}) = (0, \boldsymbol{O}, \boldsymbol{0}) \quad \text{and} \quad b(\boldsymbol{z}) \geq 0 \tag{9.68}$$

for (9.64), which are obviously also sufficient. We conclude that $\psi_{\boldsymbol{D}} = \boldsymbol{O}$ and $\psi_{grad\theta} = \boldsymbol{0}$, and hence that that the free energy is a function $\psi(\theta, \rho; p)$ of θ and ρ, and that

$$\eta = -\psi_\theta, \tag{9.69}$$

so that the entropy is also a function of θ and ρ. This leaves the *residual dissipation inequality*

$$(\boldsymbol{T} + \rho^2\psi_\rho \boldsymbol{I}) \cdot \boldsymbol{D} - \theta^{-1}\boldsymbol{q} \cdot grad\theta \geq 0, \tag{9.70}$$

restricting the stress and heat flux jointly, for all values of $\theta, \rho, \boldsymbol{D}$ and $grad\theta$. Moreover, as the material point p and the instant t_* are arbitrary, these conclusions hold at all material points and at all instants.

Consider a process for which $grad\theta$ vanishes at a particular point and instant, and let

$$f(\boldsymbol{D}) = (\boldsymbol{T} + \rho^2\psi_\rho \boldsymbol{I}) \cdot \boldsymbol{D} \tag{9.71}$$

in which θ and ρ are fixed. Then, $f(\boldsymbol{O}) = 0$ and $f(\boldsymbol{D}) \geq 0$ for all symmetric $\boldsymbol{D}$, so that f is minimized — and hence stationary — at $\boldsymbol{D} = \boldsymbol{O}$, i.e.,

$$(\boldsymbol{T}_{|\boldsymbol{D}=\boldsymbol{O}} + \rho^2\psi_\rho \boldsymbol{I}) \cdot d\boldsymbol{D} = 0 \quad \text{for all symmetric } d\boldsymbol{D}. \tag{9.72}$$

This implies that

$$\boldsymbol{T}_{|\boldsymbol{D}=\boldsymbol{O}} = -\rho^2\psi_\rho \boldsymbol{I} \tag{9.73}$$

and hence, from (9.38), that

$$\tilde{p}(\theta, \rho; p) = \rho^2\psi_\rho. \tag{9.74}$$

This is called the *thermodynamic pressure.*

Invoking (9.70) for the Newtonian fluid model (9.60), again with $grad\theta = \mathbf{0}$, yields

$$\lambda(tr\boldsymbol{D})^2 + 2\mu\boldsymbol{D}\cdot\boldsymbol{D} \geq 0, \tag{9.75}$$

for all symmetric $\boldsymbol{D}$. To derive necessary conditions for this we invoke the orthogonal decomposition

$$\boldsymbol{D} = dev\boldsymbol{D} + \frac{1}{3}(tr\boldsymbol{D})\boldsymbol{I}, \tag{9.76}$$

where $dev\boldsymbol{D}$ is the deviatoric part of $\boldsymbol{D}$ (see Problem 15 of Chapter 1). Our inequality becomes

$$\kappa(tr\boldsymbol{D})^2 + 2\mu\,|dev\boldsymbol{D}|^2 \geq 0, \tag{9.77}$$

for all $\boldsymbol{D}$, where

$$\kappa = \lambda + \frac{2}{3}\mu. \tag{9.78}$$

Consider two flows, the first with $\boldsymbol{D} = \boldsymbol{I}$ ($dev\boldsymbol{D} = \boldsymbol{O}$) and the second with $\boldsymbol{D} = \frac{1}{2}(\boldsymbol{a}\otimes\boldsymbol{b} + \boldsymbol{b}\otimes\boldsymbol{a})$, where $\boldsymbol{a}\cdot\boldsymbol{b} = 0$, the latter yielding $tr\boldsymbol{D} = 0$ and hence $\boldsymbol{D} = dev\boldsymbol{D}$. These yield the necessary and sufficient conditions

$$\kappa \geq 0 \quad \text{and} \quad \mu \geq 0. \tag{9.79}$$

Because the second form of $\boldsymbol{D}$ represents a shear flow, we refer to μ as the *shear* viscosity, whereas κ is the *bulk* viscosity.

Invoking (9.70) for the classical Fourier law (9.61), with $\boldsymbol{D} = \boldsymbol{O}$, we conclude that $K\,|grad\theta|^2 \leq 0$ and hence that the thermal conductivity is non-positive:

$$K \leq 0. \tag{9.80}$$

In turn, these inequalities ensure that a model consisting of the Newtonian fluid and Fourier's law satisfy the Clausius–Duhem inequality without qualification, i.e., for all processes.

At this stage a remark is in order concerning our *a priori* omission of the temperature gradient from the list of independent variables in the constitutive function for the stress. If one does not omit the temperature gradient, then it is possible to show that the leading-order model emerging for small values of $|\boldsymbol{D}|$ and $|grad\theta|$ is precisely the Newtonian–Fourier model; and that, to satisfy the Clausius–Duhem inequality, the associated stress *must* be independent of the temperature gradient. The details of the fairly lengthy argument may be found in Chapter 4 of [21].

Many fluids are approximately incompressible under feasible conditions, including small deviations in temperature. Their motions are therefore approximately isochoric, satisfying $J = 1$ very nearly. It is thus appropriate to impose the constraint $\dot{J} = 0$ at a material point. This in turn is equivalent to $0 = tr\boldsymbol{D} = \boldsymbol{I}\cdot\boldsymbol{D}$, reducing the dissipation inequality (9.70) to

$$dev\boldsymbol{T} \cdot dev\boldsymbol{D} - \theta^{-1}\boldsymbol{q} \cdot grad\theta \geq 0, \tag{9.81}$$

where, for the incompressible Newtonian fluid,

$$dev\boldsymbol{T} = 2\mu dev\boldsymbol{D} = 2\mu\boldsymbol{D}. \tag{9.82}$$

The foregoing argument yields only $(9.79)_2$. Moreover, conservation of mass, i.e., $(\rho J)^{\cdot} = 0$, then implies that $\dot{\rho} = 0$, and hence that ρ is fixed at a material point. It therefore enters the functions ψ, η and μ as a fixed parameter.

Thermodynamic considerations do not restrict the spherical part of the stress, which is thus constitutively indeterminate and hence outside the purview of our earlier discussion about observer consensus concerning material response. The stress is thus of the form

$$\boldsymbol{T} = -p(\boldsymbol{x}, t)\boldsymbol{I} + 2\mu\boldsymbol{D}, \tag{9.83}$$

where $p(\boldsymbol{x}, t)$ is entirely arbitrary. With $\boldsymbol{D} = Sym(grad\boldsymbol{v})$, the linear momentum balance is a vector equation for the determination of $\boldsymbol{v}$, ρ and p, assuming the body force to be assigned. Conservation of mass provides the additional equation

$$\rho' + grad\rho \cdot \boldsymbol{v} = 0, \tag{9.84}$$

where $(\cdot)'$ is the time derivative in the spatial description, and the constraint furnishes the final equation

$$div\boldsymbol{v} = 0. \tag{9.85}$$

Thus, we have five equations for the five variables consisting of ρ, p and the components of $\boldsymbol{v}$. Clearly, this system would be overdetermined, and hence generally insoluble, if not for the presence of the additional unknown function p. Accordingly the latter takes whatever values that may be necessary to furnish a solution to the problem at hand.

9.4 Problems

1. Verify (9.12)–(9.14).
2. What are the counterparts of (9.37) and (9.41) in the frame of $\mathcal{O}^+$?
3. Show that (9.41) and (9.42) satisfy (9.40) for arbitrary orthogonal $\boldsymbol{Q}$: $E^3 \to E^3$.
4. Show that all observers agree on the value of p in (9.83).
5. Consider an incompressible viscous liquid described by (9.83), and take μ to be constant. Consider the steady flow (i.e, $\boldsymbol{v}' = \boldsymbol{0}$) defined by $\boldsymbol{v} = f(x_1, x_2)\boldsymbol{e}_3$.

   (a) Show that such a motion is isochoric (as required for incompressible fluids).
   (b) Assuming zero body force, show that the pressure depends on x_3 only and that the pressure gradient $dp/dx_3 = P$, a constant. Thus, reduce the problem to the equation $\Delta f = P/\mu$, where Δ is the two-dimensional Laplacian ($\Delta f = div(gradf)$).
   (c) Solve this problem for flow through a pipe with an elliptical cross-section in the x_1, x_2 plane with major and minor axes aligned with the x_1, x_2 axes, respectively. The semi-axis lengths are a and $b(< a)$. Assume the *no slip* condition ($\boldsymbol{v} = \boldsymbol{0}$ on the boundary). [*Hint*: If $\phi(x_1, x_2) = 0$ is the equation of the boundary, then $c\phi$, where c is an appropriate constant, also solves the governing differential equation (a lucky coincidence in this problem!)].

6. Use the same model to solve the following problems, in which the body force is zero.

(a) Steady flow of the form: $\boldsymbol{v} = v_1(x_2)\boldsymbol{e}_1$, with $v_1(0) = 0$ and $v_1(d) = V$. Obtain a restriction on p and use it to find the function $v_1(x_2)$.
(b) Steady pipe flow: $\boldsymbol{v} = w(r)\boldsymbol{e}_3$ in polar coordinates, inside a pipe of radius a. Assume $w(a) = 0$. Again, obtain a restriction on p and use it to find the function $w(r)$.

7. The same liquid experiences a steady velocity field of the form

$$\boldsymbol{v} = r\omega\boldsymbol{e}_\theta$$

in a polar coordinate system, where ω is a constant. Assume the density ρ of the fluid to be uniform and independent of time in the spatial description.

(a) Describe this motion in words. Is mass conservation satisfied?
(b) Evaluate the tensors $\boldsymbol{L}, \boldsymbol{D}, \boldsymbol{W}$.
(c) Suppose the fluid is subjected to the uniform gravitational body force $\boldsymbol{b} = -g\boldsymbol{e}_3$ ($g = const.$). Derive the equation $z(r)$ of the surface of the liquid if the surface is exposed to the air and the air exerts the uniform atmospheric pressure $p_a(= const.)$ on the entire surface. Ignore the viscosity of the air.

8. Suppose this liquid experiences a steady velocity field of the kind described in Problem 10 of Chapter 4, and suppose the body force vanishes.

(a) Find the function f in that velocity field for the problem of flow in an annulus bounded by cylinders of radii a and $b(> a)$. The inner and outer cylinders spin at the constant rates ω_a and ω_b, respectively, about the $\boldsymbol{e}_3$ axis. Assume the "no-slip" condition, i.e., the fluid velocity matches the cylindrical wall velocities at $r = a, b$.
(b) Calculate the torque per unit length of the $\boldsymbol{e}_3$ axis needed to drive each cylinder.
(c) Find the ratio ω_b/ω_a corresponding to irrotational flow, i.e., the ratio needed to nullify the vorticity everywhere in the fluid.

9. Suppose a configuration κ_t of a body is completely immersed in an incompressible fluid of constant density. The body and fluid are in *equilibrium* in a uniform gravitational field.

   (a) Show that the force on the body, due to the pressure distribution at its boundary, is equal to the weight of the fluid that is displaced by the body. [*Remark*: This is the famous Law of Buoyancy discovered by Archimedes in the year 246 BC!]

   (b) Show that the moment of this pressure distribution, relative to the centroid of the body, is zero. [*Hint*: The centroid is located at $\boldsymbol{x}_c = V^{-1} \int_{\kappa_t} \boldsymbol{x} dv$, where V = volume of κ_t.]

   (c) Revisit this problem for the case in which the body is only partly immersed in the fluid.

10. Consider an incompressible Newtonian fluid in which the absolute temperature field is $\theta(\boldsymbol{x}, t)$. Suppose the internal energy per unit mass ε is some function $\hat{\varepsilon}(\theta)$ of temperature, and suppose the internal heat supply vanishes.

    (a) Derive the thermal energy balance

    $$\boldsymbol{T} \cdot \boldsymbol{L} = div \boldsymbol{q},$$

    where $\boldsymbol{q}$ is the spatial heat flux, assuming that the temperature does not vary with time at a material point (thus, the material derivative $\dot{\theta}$ is zero). Under what restrictions does this remain valid for temperature fields that are steady in the sense of the spatial description ($\theta' = 0$)?

11. For the Newtonian fluid and the Fourier law, solve this thermal energy balance equation for the flow of Problem 8 above, assuming the temperature θ to be a function of r. Assume the viscosity μ and conductivity K to be constant (this is realistic if the temperature does not vary significantly at a material point). The inner and outer cylinders are held at the fixed temperatures θ_a and θ_b, respectively.

12. An incompressible Newtonian fluid with a constant shear viscosity experiences a spherically symmetric velocity field $\boldsymbol{v} = v(r)\boldsymbol{e}_\rho$ in the spatial description, where $r = |\boldsymbol{x}|$ and $\boldsymbol{e}_\rho = r^{-1}\boldsymbol{x}$. There are no body forces acting, and the problem is *steady* (in the sense that $\boldsymbol{v}' = \boldsymbol{0}$ and $\rho' = 0$).

(a) Obtain an expression for the spatial velocity gradient $\boldsymbol{L}$. Assuming $v(a) = v_a$ to be a known constant, find the function $v(r)$ consistent with the incompressibility constraint (i.e., consistent with the motion being isochoric).

(b) What is the material acceleration? Consider a spherical *material* surface of radius a at time $t = 0$. Assuming $v_a > 0$, what is the time at which the radius of the material surface doubles? [*Hint*: Switch to the referential description.]

(c) Solve the equation of motion to obtain the pressure field $p(\boldsymbol{x})$, assuming that $\boldsymbol{T} \to -p_\infty \boldsymbol{I}$ as $r \to \infty$, where p_∞ is a known constant. [*Hint*: Use the Cartesian-coordinate formula for the divergence of a tensor, together with $r_{,i} = r^{-1} x_i$.]

Chapter 10

Kinematic Constraints

Quite often the motions that a material can undergo are such as to satisfy one or more kinematic constraints. Guided by our experience with incompressible fluids, we consider local constraints of the form

$$g_{\kappa}(\boldsymbol{F}) = 0, \tag{10.1}$$

from the perspective of observer $\mathcal{O}$, say, where g_{κ} is a scalar-valued function, $\boldsymbol{F}$ is the deformation gradient and the subscript κ identifies the reference configuration relative to which it is defined. For the case of incompressibility, for example, $g_{\kappa}(\boldsymbol{F}) = \det \boldsymbol{F} - 1$.

10.1 Observer consensus

Recall that we require all observers to agree on material response. Because the constraints characterize the nature of such response, we require that all observers agree that a constraint of the form (10.1) is in force. Thus, the constraint, as perceived by observer $\mathcal{O}^{+}$, say, is

$$g^{+}_{\kappa^{+}}(\boldsymbol{F}^{+}) = 0, \tag{10.2}$$

where

$$g^{+}_{\kappa^{+}}(\boldsymbol{F}^{+}) = g_{\kappa}(\boldsymbol{F}). \tag{10.3}$$

From (7.17), we then have

$$g^{+}_{\kappa^{+}}(\boldsymbol{F}^{+}) = g_{\kappa}(\boldsymbol{Q}^{t}\boldsymbol{F}^{+}\boldsymbol{K}). \tag{10.4}$$

As in the previous chapter, we impose this requirement for two relative motions of the observers from the perspective of $\mathcal{O}^+$, concluding, in terms of the notation used there, that

$$g_\kappa(\boldsymbol{Q}_2^t\boldsymbol{F}^+\boldsymbol{K}) = g^+_{\kappa^+}(\boldsymbol{F}^+) = g_\kappa(\boldsymbol{Q}_1^t\boldsymbol{F}^+\boldsymbol{K}). \tag{10.5}$$

Here, we assume that $\boldsymbol{Q}_2(t_0) = \boldsymbol{Q}_1(t_0)$, so that $\boldsymbol{K}$ is the same for both relative motions (see (7.15)). Let $\boldsymbol{Q} = \boldsymbol{Q}_2^t\boldsymbol{Q}_1\colon E^3 \to E^3$, as in (9.11), and let

$$\boldsymbol{F} = \boldsymbol{Q}_1^t\boldsymbol{F}^+\boldsymbol{K}. \tag{10.6}$$

Then,

$$\boldsymbol{Q}_2^t\boldsymbol{F}^+\boldsymbol{K} = \boldsymbol{QF}; \quad \text{hence } (\det\boldsymbol{Q})J = (\det\boldsymbol{Q}_2)(\det\boldsymbol{K})J^+. \tag{10.7}$$

Recalling the discussion leading to (7.19), pertaining to the choice of occupiable reference configurations, we have $(\det\boldsymbol{Q}_2)(\det\boldsymbol{K}) = 1$ and $J^+ = J$. Thus, $\det\boldsymbol{Q} = 1$ and (10.5) yields

$$g_\kappa(\boldsymbol{QF}) = g_\kappa(\boldsymbol{F}), \tag{10.8}$$

for arbitrary *rotations* $\boldsymbol{Q}$.

To obtain a necessary condition we select $\boldsymbol{Q} = \boldsymbol{1}\boldsymbol{R}^t$, where $\boldsymbol{1}$ is the shifter and $\boldsymbol{R}$ is the two-point rotation in the polar factorization (3.32) of $\boldsymbol{F}$. This furnishes $g_\kappa(\boldsymbol{F}) = g_\kappa(\boldsymbol{1}\hat{\boldsymbol{I}}\boldsymbol{U}) = g_\kappa(\boldsymbol{1U})$, where $\boldsymbol{U}$ is the right stretch tensor. Because $\boldsymbol{1}$ is fixed in the frame of $\mathcal{O}$, this is a function of $\boldsymbol{U}$ alone, and since the latter is uniquely determined by the right Cauchy–Green tensor $\boldsymbol{C} = \boldsymbol{F}^t\boldsymbol{F}$ (see (3.38)), we conclude that

$$g_\kappa(\boldsymbol{F}) = \bar{g}_\kappa(\boldsymbol{C}) \tag{10.9}$$

for some function $\bar{g}_\kappa$. Conversely, if this holds, then

$$g_\kappa(\boldsymbol{QF}) = \bar{g}_\kappa((\boldsymbol{QF})^t\boldsymbol{QF}) = \bar{g}_\kappa(\boldsymbol{F}^t\boldsymbol{IF}) = \bar{g}_\kappa(\boldsymbol{F}^t\boldsymbol{F}) = g_\kappa(\boldsymbol{F})$$

for *all* rotations $\boldsymbol{Q}$, so that (10.9) is both necessary and sufficient for (10.8).

The constraint, from the perspective of $\mathcal{O}^+$, follows from (10.4):

$$\begin{aligned} g^+_{\kappa^+}(\boldsymbol{F}^+) &= g_\kappa(\boldsymbol{Q}^t\boldsymbol{F}^+\boldsymbol{K}) = \bar{g}_\kappa((\boldsymbol{Q}^t\boldsymbol{F}^+\boldsymbol{K})^t\boldsymbol{Q}^t\boldsymbol{F}^+\boldsymbol{K}) \\ &= \bar{g}_\kappa(\boldsymbol{K}^t(\boldsymbol{F}^+)^t\boldsymbol{Q}\boldsymbol{Q}^t\boldsymbol{F}^+\boldsymbol{K}) = \bar{g}_\kappa(\boldsymbol{K}^t(\boldsymbol{F}^+)^t\boldsymbol{I}^+\boldsymbol{F}^+\boldsymbol{K}) \\ &= \bar{g}_\kappa(\boldsymbol{K}^t(\boldsymbol{F}^+)^t\boldsymbol{F}^+\boldsymbol{K}) = \bar{g}_\kappa(\boldsymbol{K}^t\boldsymbol{C}^+\boldsymbol{K}) \\ &= \bar{g}^+_{\kappa^+}(\boldsymbol{C}^+), \end{aligned} \tag{10.10}$$

where $\boldsymbol{K}$ is a fixed parameter and $\boldsymbol{Q}\colon E^3 \to E^{3+}$ is now the orthogonal tensor characterizing the observer transformation.

Examples

1. Incompressibility

Because $J > 0$, the constraint $J = 1$ associated with incompressibility is equivalent to $1 = J^2 = \det(\boldsymbol{F}^t\boldsymbol{F}) = \det \boldsymbol{C}$; hence,

$$\bar{g}_\kappa(\boldsymbol{C}) = \det \boldsymbol{C} - 1. \tag{10.11}$$

2. Inextensibility

In the engineering theory of fiber-reinforced composites, the fibers are idealized as being continuously distributed material curves in κ having the unit-tangent field $\boldsymbol{M}(\boldsymbol{X})$, say. If the fibers are very stiff in comparison to the matrix material in which they are embedded, then their stretches μ remain close to unity in a typical deformation. In these circumstances it is appropriate to impose the constraint $\boldsymbol{M} \cdot \boldsymbol{C}\boldsymbol{M} = 1$ (see (2.83)). Accordingly,

$$\bar{g}_\kappa(\boldsymbol{C}) = \boldsymbol{M} \cdot \boldsymbol{C}\boldsymbol{M} - 1. \tag{10.12}$$

3. Rigidity

If the body is rigid then $\boldsymbol{F}$ is a spatially uniform rotation (see Section 3.7), and therefore $\boldsymbol{C} = \hat{\boldsymbol{I}}$. Because of the symmetry of $\boldsymbol{C}$, this is equivalent to the *six* constraints

$$\bar{g}_{(\kappa)AB} = C_{AB} - \delta_{AB}. \tag{10.13}$$

We note that, in general, there can be no more than six *independent* constraints because there at most six independent entries of the symmetric matrix (C_{AB}). Thus rigidity is the maximal constraint, in that all components of this matrix are fixed.

10.2 Modification of the stress to accommodate constraints

We have seen, in connection with the incompressible Newtonian fluid, that the stress decomposes into the sum of a *reactive part*, $-p\boldsymbol{I}$, and a constitutive part that depends on the properties of the material. The salient feature of the reactive part is that it makes no contribution to the stress power density whenever the constraint is operative; that is, $-p\boldsymbol{I} \cdot \boldsymbol{D} = -p\, tr\boldsymbol{D}$ vanishes whenever $tr\boldsymbol{D} = 0$. We extend this idea to general constraints by assuming that the net Cauchy stress $\boldsymbol{T}$ is of the form

$$\boldsymbol{T} = \boldsymbol{N} + \boldsymbol{G}, \tag{10.14}$$

where $\boldsymbol{G}$ is a symmetric-tensor-valued constitutive function appropriate to the material and conditions at hand, and $\boldsymbol{N}$ is a symmetric tensor such that $\boldsymbol{N} \cdot \boldsymbol{D}$ vanishes in any motion that is compatible with the constraint. Once the structure of this tensor is determined, any remaining indeterminacy must be such as to secure a solution, if indeed one exists, to the equations of motion and any boundary and initial conditions.

Suppose, then, that the constraint is operative at a particular material point in an interval I of time. We then require that

$$\boldsymbol{N} \cdot \boldsymbol{D} = 0 \quad \text{for all } \boldsymbol{D} \quad \text{such that} \quad \bar{g}_{\kappa}(\boldsymbol{C}(t)) = 0; \quad t \in I. \tag{10.15}$$

Because $\bar{g}_{\kappa}$ vanishes identically in this interval we may differentiate it to arrive at

$$0 = \dot{g} = g_{\boldsymbol{C}} \cdot \dot{\boldsymbol{C}} = tr((g_{\boldsymbol{C}})\dot{\boldsymbol{C}}), \tag{10.16}$$

where $g_{\boldsymbol{C}} = (g_{\boldsymbol{C}})^t$ is the symmetric referential tensor-valued gradient of g (see Sections B.2 and B.3 of Appendix B), and we have written g in place of $\bar{g}_{\kappa}$ in the interests of promoting a tractable notation. The symmetry of $g_{\boldsymbol{C}}$ follows from the fact that $\dot{\boldsymbol{C}}$ — the limit of the difference quotient of positive definite symmetric tensors — is a symmetric tensor. Accordingly the inner product picks out only the symmetric part of g_C. Because the latter is effectively defined by the second equality in (10.16), it follows that no generality is lost by taking g_C to be symmetric. We remark that $g(\boldsymbol{C})$ is well defined for all symmetric positive definite $\boldsymbol{C}$, not only those that satisfy the constraint. Accordingly g_C is similarly well defined for all such $\boldsymbol{C}$. The derivative g_C appearing in (10.16), however, is obtained by imposing the constraint *a posteriori*.

Recalling (4.11) and making use of the properties of the trace operation, (10.16) is seen to be equivalent to

$$0 = tr(g_C(\boldsymbol{F}^t\boldsymbol{D}\boldsymbol{F})) = tr((\boldsymbol{F}(g_C)\boldsymbol{F}^t)\boldsymbol{D}) = \boldsymbol{F}(g_C)\boldsymbol{F}^t \cdot \boldsymbol{D}, \quad (10.17)$$

in which the first factor on the right-hand side of the third equality is a symmetric spatial tensor. Thus, we require that $\boldsymbol{N}$ be perpendicular, in the vector space Sym of symmetric spatial tensors, to *all* $\boldsymbol{D}$ that are perpendicular to $\boldsymbol{F}(g_C)\boldsymbol{F}^t$ (Figure 10.1). Accordingly, $\boldsymbol{N}$ must be parallel to $\boldsymbol{F}(g_C)\boldsymbol{F}^t$, i.e.,

$$\boldsymbol{N} = \lambda(\boldsymbol{x}, t)\boldsymbol{F}(g_C)\boldsymbol{F}^t, \quad (10.18)$$

in which the multiplier $\lambda(\boldsymbol{x}, t)$ is determined solely by the equations of motion, with the stress given by (10.14), together with the constraint equation and initial/boundary conditions.

If n constraints $g^{(i)}(\boldsymbol{C}) = 0$; $i = 1, \ldots, n$ are in force during an interval of time, then $\boldsymbol{N} \cdot \boldsymbol{D}$ vanishes for all $\boldsymbol{D}$ such that the $\boldsymbol{F}(g_C^{(i)})\boldsymbol{F}^t \cdot \boldsymbol{D}$ vanish for each i. The n constraints are independent provided that the elements of the set $\{\boldsymbol{F}(g_C^{(i)})\boldsymbol{F}^t\}$ are linearly independent, in which the $g_C^{(i)}$ are computed for all positive definite

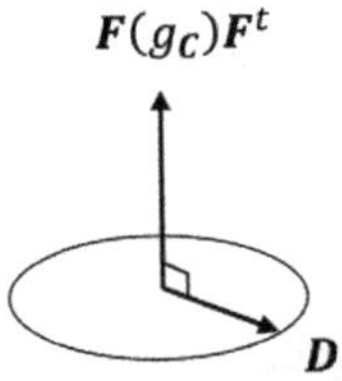

Figure 10.1. Orthogonality of $\boldsymbol{F}(g_C)\boldsymbol{F}^t$ and $\boldsymbol{D}$ in the space of symmetric tensors.

symmetric $\boldsymbol{C}$. In this case $n \leq 6$ because Sym is six-dimensional, and $Span\{\boldsymbol{F}(g_{\boldsymbol{C}}^{(i)})\boldsymbol{F}^t\}$ is then a subspace of Sym. Thus,

$$\boldsymbol{N} = \sum_{i=1}^{n\leq 6} \lambda_i(\boldsymbol{x}, t)\boldsymbol{F}(g_{\boldsymbol{C}}^{(i)})\boldsymbol{F}^t = \boldsymbol{F}\left[\sum_{i=1}^{n} \lambda_i g_{\boldsymbol{C}}^{(i)}\right]\boldsymbol{F}^t, \tag{10.19}$$

where the multipliers $\lambda_i(\boldsymbol{x}, t)$ are again such as to secure a solution to the equations of motion together with the n constraint equations and initial/boundary conditions; and the derivatives $g_{\boldsymbol{C}}^{(i)}$ are evaluated, post facto, at those $\boldsymbol{C}'s$ that satisfy *all* n constraints simultaneously.

Examples

We determine the forms of the reactive stresses for the constraints of incompressibility, inextensibility and rigidity.

(1) In the case of incompressibility we use (10.11), with (B.26) and (B.27) of Appendix B, to obtain

$$g_{\boldsymbol{C}} \cdot \dot{\boldsymbol{C}} = (\det \boldsymbol{C})^{\cdot} = \boldsymbol{C}^* \cdot \dot{\boldsymbol{C}} = (\det \boldsymbol{C})\boldsymbol{C}^{-1} \cdot \dot{\boldsymbol{C}}, \tag{10.20}$$

for arbitrary symmetric $\dot{\boldsymbol{C}}$. Thus,

$$g_{\boldsymbol{C}} = (\det \boldsymbol{C})\boldsymbol{C}^{-1}, \tag{10.21}$$

which reduces to

$$g_{\boldsymbol{C}} = \boldsymbol{C}^{-1} \tag{10.22}$$

on the set of $\boldsymbol{C}'s$ that satisfy the constraint. Writing $\lambda = -\bar{p}$, the associated reactive stress is found, from (10.18), to be of the form

$$\boldsymbol{N} = -\bar{p}\boldsymbol{F}\boldsymbol{C}^{-1}\boldsymbol{F}^t = -\bar{p}\boldsymbol{F}(\boldsymbol{F}^t\boldsymbol{F})^{-1}\boldsymbol{F}^t = -\bar{p}\boldsymbol{I}, \tag{10.23}$$

and is therefore mechanically equivalent to a pressure.

For example, in the case of the Newtonian viscous fluid model (9.60),

$$\boldsymbol{T} = -p\boldsymbol{I} + 2\mu\boldsymbol{D} \tag{10.24}$$

if the constraint is operative, where $p = \tilde{p} + \bar{p}$. The constitutive indeterminacy of $\bar{p}$ implies that p is likewise indeterminate, and we arrive at (9.83).

(2) For the case of inextensibility of embedded fibers having unit tangent $\boldsymbol{M}$ in κ, we use (10.12) to derive

$$g_{\boldsymbol{C}} \cdot \dot{\boldsymbol{C}} = \boldsymbol{M} \cdot \dot{\boldsymbol{C}}\boldsymbol{M} = \boldsymbol{M} \otimes \boldsymbol{M} \cdot \dot{\boldsymbol{C}} \tag{10.25}$$

for all symmetric $\dot{\boldsymbol{C}}$. Thus,

$$g_{\boldsymbol{C}} = \boldsymbol{M} \otimes \boldsymbol{M}. \tag{10.26}$$

Writing $\lambda = \sigma$, we use (10.18) to obtain

$$\boldsymbol{N} = \sigma\boldsymbol{F}(\boldsymbol{M} \otimes \boldsymbol{M})\boldsymbol{F}^t = \sigma\boldsymbol{F}\boldsymbol{M} \otimes \boldsymbol{F}\boldsymbol{M}, \tag{10.27}$$

which reduces to

$$\boldsymbol{N} = \sigma\boldsymbol{m} \otimes \boldsymbol{m} \tag{10.28}$$

when the constraint is in effect (see (2.67)), where $\boldsymbol{m}$ is the unit tangent to a fiber in κ_t. Thus, the reactive stress is mechanically equivalent to a uniaxial stress along the fibers (cf. (6.130)).

(3) In the case of rigidity it follows from (10.13) that

$$\begin{aligned}(g_{AB})_{\boldsymbol{C}} \cdot \dot{\boldsymbol{C}} = \dot{C}_{AB} &= (\hat{\boldsymbol{E}}_A \cdot \boldsymbol{C}\hat{\boldsymbol{E}}_B)^{\cdot} \\ &= (\hat{\boldsymbol{E}}_A \otimes \hat{\boldsymbol{E}}_B) \cdot \dot{\boldsymbol{C}} \\ &= Sym(\hat{\boldsymbol{E}}_A \otimes \hat{\boldsymbol{E}}_B) \cdot \dot{\boldsymbol{C}},\end{aligned} \tag{10.29}$$

on account of the symmetry of $\dot{\boldsymbol{C}}$, where

$$Sym(\hat{\boldsymbol{E}}_A \otimes \hat{\boldsymbol{E}}_B) = \frac{1}{2}(\hat{\boldsymbol{E}}_A \otimes \hat{\boldsymbol{E}}_B + \hat{\boldsymbol{E}}_B \otimes \hat{\boldsymbol{E}}_A). \tag{10.30}$$

Thus,

$$(g_{AB})_{\boldsymbol{C}} = Sym(\hat{\boldsymbol{E}}_A \otimes \hat{\boldsymbol{E}}_B) \tag{10.31}$$

and it follows from (10.19) that

$$\begin{aligned}\boldsymbol{N} = \lambda_{AB}\boldsymbol{Q}[Sym(\hat{\boldsymbol{E}}_A \otimes \hat{\boldsymbol{E}}_B)]\boldsymbol{Q}^t &= \boldsymbol{Q}[\lambda_{(AB)}\hat{\boldsymbol{E}}_A \otimes \hat{\boldsymbol{E}}_B]\boldsymbol{Q}^t, \\ \text{with } \lambda_{(AB)} &= \frac{1}{2}(\lambda_{AB} + \lambda_{BA}),\end{aligned} \tag{10.32}$$

where the usual summation convention is in effect; the constraint has been imposed in the form $\boldsymbol{F} = \boldsymbol{Q}$, a rotation; and where λ_{AB}

are arbitrary scalars. The term in square brackets to the right of the second equality is an arbitrary symmetric referential tensor. Thus, $\boldsymbol{N}$ is an arbitrary symmetric spatial tensor. It follows that the net stress $\boldsymbol{T}$ is constitutively indeterminate. Accordingly, the stress in a rigid body is indeterminate. Nevertheless, the equations of rigid-body dynamics (see Section 6.4.2) suffice to determine the motion in terms of the applied forces and torques, and vice versa.

10.3 Problems

1. Write the constraints of incompressibility, inextensibility and rigidity in the frame of observer $\mathcal{O}^+$.
2. Find an expression for the reactive stress $\boldsymbol{N}^+$, as perceived by $\mathcal{O}^+$, associated with a generic constraint. Show that all observers agree on the values of the various constitutively indeterminate multipliers.
3. Consider a laminated material formed from thin sheets of stiff paper interspersed in a soft matrix material. Take the sheets to be continuously distributed parallel planes spanned by $\hat{\boldsymbol{E}}_1$ and $\hat{\boldsymbol{E}}_2$ in the reference configuration. The constraints are $\hat{\boldsymbol{E}}_1 \cdot \boldsymbol{C}\hat{\boldsymbol{E}}_2 = 0$ and $\hat{\boldsymbol{E}}_1 \cdot \boldsymbol{C}\hat{\boldsymbol{E}}_1 = \hat{\boldsymbol{E}}_2 \cdot \boldsymbol{C}\hat{\boldsymbol{E}}_2 = 1$. Show that there can be no extensional or shear strain in the plane of the sheets, but that transverse normal and shear strains are permitted. What is the form of the reactive stress?
4. The body is laminated as in the previous problem, but now constrained in such a way that the planes experience no change in local surface area as they deform. Find the form of the reactive stress.

Chapter 11

Elastic Solids

We define elastic materials to be the subclass of the idealized models discussed in Chapter 8 for which the constitutive functions are insensitive to the velocity gradient. The elements of the list

$$C = \{\boldsymbol{T}, \boldsymbol{q}, \psi, \eta\} \tag{11.1}$$

of response functions at a material point p are thus presumed to depend on the abbreviated list

$$L = \{\boldsymbol{x}, \boldsymbol{v}, \theta, \boldsymbol{F}, grad\theta\}, \tag{11.2}$$

evaluated at the same point. For example, the Helmholtz energy ψ is given by

$$\psi = \hat{\psi}_\kappa(\boldsymbol{x}, \boldsymbol{v}, \theta, \boldsymbol{F}_\kappa, grad\theta; p), \tag{11.3}$$

in terms of a constitutive function $\hat{\psi}_\kappa$ based on the reference configuration used by observer $\mathcal{O}$ to compute $\boldsymbol{F}_\kappa$.

Considerations of frame invariance, combined with (6.186) and (7.37), imply that

$$\hat{\psi}_\kappa(\boldsymbol{x}, \boldsymbol{v}, \theta, \boldsymbol{F}_\kappa, grad\theta; p) = \hat{\psi}^+_{\kappa^+}(\boldsymbol{x}^+, \boldsymbol{v}^+, \theta^+, \boldsymbol{F}^+_{\kappa^+}, grad^+\theta^+; p), \tag{11.4}$$

where $\hat{\psi}^+_{\kappa^+}$ is the constitutive function pertaining to observer $\mathcal{O}^+$. We may then proceed, as in Sections 9.1 and 10.1, to impose this requirement for two relative motions of the observers (see (9.6)), with $\boldsymbol{Q}_2(t_0) = \boldsymbol{Q}_1(t_0)$ and $\boldsymbol{c}_2(t_0) = \boldsymbol{c}_1(t_0)$, concluding that

$$\hat{\psi}_\kappa(\boldsymbol{x}_2, \boldsymbol{v}_2, \theta, \boldsymbol{F}_2, (grad\theta)_2; p) = \hat{\psi}_\kappa(\boldsymbol{x}_1, \boldsymbol{v}_1, \theta, \boldsymbol{F}_1, (grad\theta)_1; p), \quad (11.5)$$

where $\boldsymbol{x}_2$ and $\boldsymbol{v}_2$ are given in terms of $\boldsymbol{x}_1$ and $\boldsymbol{v}_1$ by (9.12), in which $\boldsymbol{Q} = \boldsymbol{Q}_2^t \boldsymbol{Q}_1$ is a *rotation*, and where

$$\boldsymbol{F}_2 = \boldsymbol{Q}\boldsymbol{F}_1 \quad \text{and} \quad (grad\theta)_2 = \boldsymbol{Q}(grad\theta)_1. \quad (11.6)$$

Choosing $\boldsymbol{Q} = \boldsymbol{I}$, we find, as in Chapter 9, that $\hat{\psi}_\kappa$ does not depend on $\boldsymbol{x}$ or $\boldsymbol{v}$, and hence that

$$\psi = \Psi(\theta, \boldsymbol{F}, grad\theta; p), \quad (11.7)$$

for some function Ψ, where we have lightened the notation by suppressing the subscript κ. This function is such that

$$\Psi(\theta, \boldsymbol{Q}\boldsymbol{F}, \boldsymbol{Q}grad\theta; p) = \Psi(\theta, \boldsymbol{F}, grad\theta; p) \quad \text{for all rotations } \boldsymbol{Q}. \quad (11.8)$$

The same reasoning leads to the conclusion that the constitutive equations for the stress and heat flux are likewise insensitive to $\boldsymbol{x}$ and $\boldsymbol{v}$. For example, the heat flux in the frame of $\mathcal{O}$ is given by

$$\boldsymbol{q} = \hat{\boldsymbol{q}}(\theta, \boldsymbol{F}, grad\theta; p), \quad (11.9)$$

where

$$\hat{\boldsymbol{q}}(\theta, \boldsymbol{Q}\boldsymbol{F}, \boldsymbol{Q}grad\theta; p) = \boldsymbol{Q}\hat{\boldsymbol{q}}(\theta, \boldsymbol{F}, grad\theta; p) \quad \text{for all rotations } \boldsymbol{Q}. \quad (11.10)$$

11.1 Clausius–Duhem inequality

Before exploring the consequences of these restrictions, we first consider the implications of the Clausius-Duhem inequality (6.187) for our constitutive hypotheses. Here we use the chain rule

$$\dot{\psi} = \Psi_\theta \dot{\theta} + \Psi_{\boldsymbol{F}} \cdot \dot{\boldsymbol{F}} + \Psi_{grad\theta} \cdot (grad\theta)^{\cdot}, \quad (11.11)$$

together with

$$\boldsymbol{T} \cdot \boldsymbol{L} = \boldsymbol{T} \cdot \dot{\boldsymbol{F}}\boldsymbol{F}^{-1} = tr(\boldsymbol{T}(\boldsymbol{F}^{-t}\dot{\boldsymbol{F}}^t)) = \boldsymbol{T}\boldsymbol{F}^{-t} \cdot \dot{\boldsymbol{F}} \quad (11.12)$$

to reduce (6.187) — after multiplying by $J(>0)$ and invoking (2.55) and (2.59) — to

$$-\rho_\kappa(\Psi_\theta + \eta)\dot{\theta} + (\boldsymbol{P} - \rho_\kappa \Psi_{\boldsymbol{F}}) \cdot \dot{\boldsymbol{F}} - \rho_\kappa \Psi_{grad\theta} \cdot (grad\theta)^{\cdot}$$
$$-J\theta^{-1}\boldsymbol{q} \cdot grad\theta \geq 0, \qquad (11.13)$$

where $\boldsymbol{P}$ is the Piola stress (see (6.140)). This is again of the form (9.64), now with

$$\boldsymbol{z} = (\theta, \boldsymbol{F}, grad\theta), \quad \boldsymbol{w} = (\dot{\theta}, \dot{\boldsymbol{F}}, (grad\theta)^{\cdot}), \quad \boldsymbol{a}(\boldsymbol{z}) = (-\rho_\kappa(\Psi_\theta + \eta),$$
$$\boldsymbol{P} - \rho_\kappa \Psi_{\boldsymbol{F}}, -\rho_\kappa \Psi_{grad\theta}) \qquad (11.14)$$

and

$$b(\boldsymbol{z}) = -J\theta^{-1}\boldsymbol{q} \cdot grad\theta. \qquad (11.15)$$

From (9.68) we conclude that $\Psi_{grad\theta} = \boldsymbol{0}$, together with

$$\eta = -\Psi_\theta(\theta, \boldsymbol{F}; p), \quad \boldsymbol{P} = W_{\boldsymbol{F}}(\theta, \boldsymbol{F}; \boldsymbol{X}) \quad \text{and} \quad \boldsymbol{q} \cdot grad\theta \leq 0, \qquad (11.16)$$

where we continue to denote the free energy by Ψ, and where

$$W(\theta, \boldsymbol{F}; \boldsymbol{X}) = \rho_\kappa(\boldsymbol{X})\Psi(\theta, \boldsymbol{F}; \boldsymbol{\kappa}^{-1}(\boldsymbol{X})) \qquad (11.17)$$

is the free energy per unit volume of the reference configuration κ. Thus the entropy is independent of the temperature gradient, as in the case of viscous fluids. Further, we have proved that the Piola stress (and hence also the Cauchy stress) is insensitive to the temperature gradient. This stands in contrast to the situation for viscous fluids encountered in Chapter 9, in which such insensitivity was assumed *a priori*.

11.2 Stress response

It follows from (11.8) and (11.17) that

$$W(\theta, \boldsymbol{QF}; \boldsymbol{X}) = W(\theta, \boldsymbol{F}; \boldsymbol{X}) \quad \text{for all rotations } \boldsymbol{Q}. \qquad (11.18)$$

Proceeding as in the passage from (10.8) to (10.9), we conclude that for this to be satisfied it is necessary and sufficient that

$$W(\theta, \boldsymbol{F}; \boldsymbol{X}) = \bar{W}(\theta, \boldsymbol{C}; \boldsymbol{X}) \tag{11.19}$$

for some function $\bar{W}$, where $\boldsymbol{C}$ is the right Cauchy–Green deformation tensor. Accordingly,

$$W_{\boldsymbol{F}} \cdot \dot{\boldsymbol{F}} = \bar{W}_{\boldsymbol{C}} \cdot \dot{\boldsymbol{C}} = \bar{W}_{\boldsymbol{C}} \cdot (\dot{\boldsymbol{F}}^t \boldsymbol{F} + \boldsymbol{F}^t \dot{\boldsymbol{F}}), \tag{11.20}$$

where $\bar{W}_{\boldsymbol{C}} = (\bar{W}_{\boldsymbol{C}})^t$ is the symmetric referential gradient of $\bar{W}$ with respect to $\boldsymbol{C}$. Using the rules

$$\boldsymbol{A}_1 \cdot \boldsymbol{A}_2 \boldsymbol{A}_3 = \boldsymbol{A}_2^t \boldsymbol{A}_1 \cdot \boldsymbol{A}_3 = \boldsymbol{A}_3 \boldsymbol{A}_1^t \cdot \boldsymbol{A}_2^t, \tag{11.21}$$

which follow from the properties of the trace operator, we conclude that $\bar{W}_{\boldsymbol{C}} \cdot \dot{\boldsymbol{F}}^t \boldsymbol{F}$ and $\bar{W}_{\boldsymbol{C}} \cdot \boldsymbol{F}^t \dot{\boldsymbol{F}}$ are both equal to $\boldsymbol{F}(\bar{W}_{\boldsymbol{C}}) \cdot \dot{\boldsymbol{F}}$, and hence that

$$[W_{\boldsymbol{F}} - 2\boldsymbol{F}(\bar{W}_{\boldsymbol{C}})] \cdot \dot{\boldsymbol{F}} = 0, \tag{11.22}$$

in which the expression in brackets is a two-point tensor and $\dot{\boldsymbol{F}}$ is an arbitrary two-point tensor. Because the set of two-point tensors is a linear space, we conclude that the inner product of the bracket with an arbitrary element of this space vanishes, and hence that the bracket vanishes, yielding

$$W_{\boldsymbol{F}} = 2\boldsymbol{F}(\bar{W}_{\boldsymbol{C}}). \tag{11.23}$$

From (6.144) and $(11.16)_2$, the Piola–Kirchhoff stress $\boldsymbol{S}$ is thus given by

$$\boldsymbol{S} = 2\bar{W}_{\boldsymbol{C}}. \tag{11.24}$$

This is automatically symmetric, and (6.145) ensures that the Cauchy stress is also symmetric.

11.2.1 *Material symmetry*

The energy W is subject to the material symmetry condition (cf. (8.16) and (8.17))

$$W(\theta, \boldsymbol{F}; \boldsymbol{X}) = W(\theta, \boldsymbol{F}\boldsymbol{R}; \boldsymbol{X}) \quad \text{for all } \boldsymbol{R} \in \mathcal{G}_{\kappa(p)}, \tag{11.25}$$

where $\mathcal{G}_{\kappa(p)}$ is the symmetry group at $p \in B$ in the configuration κ. From (11.19), $\bar{W}$ is therefore such that

$$\bar{W}(\theta, \boldsymbol{C}; \boldsymbol{X}) = \bar{W}(\theta, \boldsymbol{R}^t \boldsymbol{C} \boldsymbol{R}; \boldsymbol{X}). \tag{11.26}$$

Here again we emphasize that $\boldsymbol{R}$, a referential tensor, must not be confused with the two-point rotation tensor in the polar decomposition of $\boldsymbol{F}$.

For example, if $\mathcal{G}_{\kappa(p)} = U$, the unimodular group, then we may proceed, as in Chapter 9, to conclude that W depends on $\boldsymbol{F}$ via $J = \det \boldsymbol{F}$, i.e., that $W(\theta, \boldsymbol{F}; \boldsymbol{X}) = w(\theta, J; \boldsymbol{X})$ for some function w. This implies, on holding θ fixed (and invoking the result of Problem 2(b) in Chapter 4), that

$$\boldsymbol{P} \cdot \dot{\boldsymbol{F}} = \dot{W} = \dot{w} = w_J \dot{J} = w_J \boldsymbol{F}^* \cdot \dot{\boldsymbol{F}}, \tag{11.27}$$

and hence that $\boldsymbol{P} = w_J \boldsymbol{F}^*$. It then follows from (6.140) that the Cauchy stress is $\boldsymbol{T} = w_J \boldsymbol{I}$. This may be combined with conservation of mass to arrive at the constitutive equation (9.38) for inviscid fluids, with the pressure given by $\tilde{p} = -w_J$.

For elastic solids, recall that there exists a local undistorted configuration relative to which the symmetry group is contained in the group $Orth^+$ of rotation tensors. If this symmetry group coincides with $Orth^+$, then the material is said to be *hemitropic* at the point in question. In this case there is no rotation preceding a deformation at the point in question that can be detected experimentally. Assuming $\kappa(p)$ — a neighborhood of p in configuration κ — to be such a local configuration, we have that

$$W(\theta, \boldsymbol{F}; \boldsymbol{X}) = W(\theta, \boldsymbol{F}\boldsymbol{R}; \boldsymbol{X}) \quad \text{for all } \boldsymbol{R} \in Orth^+, \tag{11.28}$$

and hence, from (11.26), that

$$\bar{W}(\theta, \boldsymbol{C}; \boldsymbol{X}) = \bar{W}(\theta, \boldsymbol{R}^t \boldsymbol{C} \boldsymbol{R}; \boldsymbol{X}) \quad \text{for all } \boldsymbol{R} \in Orth, \tag{11.29}$$

where $Orth$ is the full orthogonal group consisting of all orthogonal tensors. This follows from the fact that $Orth = Orth^+ \cup \{-\hat{\boldsymbol{I}}\}$. Thus, if $\boldsymbol{R} \in Orth^+$ satisfies this restriction for all symmetric, positive definite $\boldsymbol{C}$, then it is also satisfied by $-\boldsymbol{R} \in Orth$, i.e, the sign cancels out. Such functions, with the symmetry group extended to $Orth$, are *isotropic*.

We may thus invoke the result proved in Part 1 of Section 9.1.1 to conclude that

$$\bar{W}(\theta, \boldsymbol{C}; \boldsymbol{X}) = \hat{W}(\theta, I_1(\boldsymbol{C}), I_2(\boldsymbol{C}), I_3(\boldsymbol{C}); \boldsymbol{X}), \tag{11.30}$$

where

$$I_1(\boldsymbol{C}) = tr\boldsymbol{C}, \quad I_2(\boldsymbol{C}) = \frac{1}{2}[I_1^2 - tr(\boldsymbol{C}^2)], \quad I_3(\boldsymbol{C}) = \det \boldsymbol{C} \tag{11.31}$$

are the principal invariants of $\boldsymbol{C}$. We combine this with (11.24), obtaining

$$\left[\frac{1}{2}\boldsymbol{S} - \sum_{k=1}^{3} \hat{W}_k (I_k)_{\boldsymbol{C}}\right] \cdot \dot{\boldsymbol{C}} = 0$$

$$\text{for all symmetric } \dot{\boldsymbol{C}}, \text{where } \hat{W}_k = \partial \hat{W}/\partial I_k, \tag{11.32}$$

and the formulas developed in Section B.3 of Appendix B for the gradients $(I_k)_{\boldsymbol{C}}$ then deliver the Piola–Kirchhoff stress

$$\boldsymbol{S} = 2[(\hat{W}_1 + I_1\hat{W}_2)\hat{\boldsymbol{I}} - \hat{W}_2\boldsymbol{C} + I_3\hat{W}_3\boldsymbol{C}^{-1}]. \tag{11.33}$$

This may be substituted into (6.145) to obtain the Cauchy stress. Thus,

$$J\boldsymbol{T} = 2[(\hat{W}_1 + I_1\hat{W}_2)\boldsymbol{B} - \hat{W}_2\boldsymbol{B}^2 + I_3\hat{W}_3\boldsymbol{I}], \tag{11.34}$$

where $\boldsymbol{B}$ is the left Cauchy–Green deformation tensor.

In connection with these results, we note that the principal invariants of $\boldsymbol{C}$ coincide with those of $\boldsymbol{B}$, i.e., $I_k(\boldsymbol{C}) = I_k(\boldsymbol{B})$; $k = 1, 2, 3$. One way to demonstrate this is to use the spectral decomposition $(3.38)_2$ of the right Cauchy–Green tensor, and that of the left Cauchy–Green tensor, to obtain

$$\begin{aligned} I_1(\boldsymbol{C}) &= \lambda_1^2 + \lambda_2^2 + \lambda_3^2 = I_1(\boldsymbol{B}), \\ I_2(\boldsymbol{C}) &= \lambda_1^2\lambda_2^2 + \lambda_2^2\lambda_3^2 + \lambda_1^2\lambda_3^2 = I_2(\boldsymbol{B}), \quad \text{and} \\ I_3(\boldsymbol{C}) &= \lambda_1^2\lambda_2^2\lambda_3^2 = I_3(\boldsymbol{B}), \end{aligned} \tag{11.35}$$

where λ_k are the principal stretches.

If the material is incompressible, then

$$T = -pI + 2[(W_1^* + I_1 W_2^*)B - W_2^* B^2],$$
$$\text{where } W^*(I_1, I_2; X) = \hat{W}(I_1, I_2, 1; X), \tag{11.36}$$

with $W_k^* = \partial W^*/\partial I_k$; $k = 1, 2$, and where p is a constitutively indeterminate pressure. In particular, the constraint (10.11) of incompressibility is unaltered if C is replaced by $R^t CR$ for all $R \in Orth$. Thus isotropy is compatible with incompressibility, in the sense that an isotropic material could conceivably be incompressible, though of course not every incompressible material is isotropic. That is, (10.11) also remains unaltered if R is an element of the discrete set of rotations constituting the symmetry group of a crystalline solid, for example. In fact, (10.11) is unaltered by any $R \in U$, the unimodular group, and so all kinds of symmetry, including that associated with fluidity, are compatible with incompressibility. The same cannot be said of inextensibility, however. For example, there are infinitely many $R's$ in $Orth$ that do not leave (10.12) invariant. Accordingly a material reinforced by a family of inextensible fibers is not isotropic.

Example: Simple shear

Consider the simple shear deformation discussed in Section 3.5.1. We suppose this deformation is experienced by a uniform isotropic material, i.e., a material that is isotropic at every material point, with the energy $\hat{W}$ (or W^*) not dependent on X explicitly. The relevant left Cauchy–Green tensor, given by (3.54), is uniform. Accordingly, for compressible materials, (11.34), with $J = 1$, delivers a uniform stress with zero divergence, and (6.109) implies that the deformation is equilibrated provided that the body force vanishes. If the material is incompressible, then (11.36) is operative, and again the deformation is equilibrated in the absence of body force, provided that $div(pI) = grad p$ vanishes, i.e., provided that the reactive pressure p is uniform.

Using (3.54) we find that

$$B^2 = [(1 + \gamma^2) + \gamma^2]e_1 \otimes e_1 + (1 + \gamma^2)e_2 \otimes e_2 + e_3 \otimes e_3$$
$$+ \gamma(2 + \gamma^2)(e_1 \otimes e_2 + e_2 \otimes e_1), \tag{11.37}$$

where γ is the uniform shear. Straightforward calculations yield

$$I_1(\boldsymbol{B}) = I_2(\boldsymbol{B}) = 3 + \gamma^2, \tag{11.38}$$

and, of course, $I_3(\boldsymbol{B}) = 1$. Thus the coefficients involving $\hat{W}$ and W^* on the right-hand sides of (11.34) and (11.36), respectively, are functions of γ^2. Resolving the latter equations in the basis $\{\boldsymbol{e}_i \otimes \boldsymbol{e}_j\}$, we collect the associated components T_{ij} in the matrix equation

$$(T_{ij}) = f_0 \begin{pmatrix} 1 & 0 & 0 \\ 0 & 1 & 0 \\ 0 & 0 & 1 \end{pmatrix} + f_1(\gamma^2) \begin{pmatrix} 1+\gamma^2 & \gamma & 0 \\ \gamma & 1 & 0 \\ 0 & 0 & 1 \end{pmatrix}$$

$$+ f_2(\gamma^2) \begin{pmatrix} (1+\gamma^2)+\gamma^2 & \gamma(2+\gamma^2) & 0 \\ \gamma(2+\gamma^2) & 1+\gamma^2 & 0 \\ 0 & 0 & 1 \end{pmatrix}, \tag{11.39}$$

where f_1 and f_2 are constitutive functions and f_0 is also a constitutive function of γ^2 in the case of a compressible material, or the negative of the uniform constraint pressure in the case of an incompressible material.

It follows that, in simple shear, $T_{13} = T_{23} = 0$, whereas

$$T_{12} = \gamma\mu(\gamma^2), \quad \text{where } \mu(\gamma^2) = f_1(\gamma^2) + \gamma(2+\gamma^2) f_2(\gamma^2). \tag{11.40}$$

Assuming the function μ to be differentiable, we have

$$T_{12} = G\gamma + O(\gamma^3) \tag{11.41}$$

where $G = \mu(0)$ is the classical shear modulus for small deformations. We also find that T_{11}, T_{22} and T_{33} are non-zero, and that

$$T_{11} - T_{22} = \gamma T_{12}. \tag{11.42}$$

This result is remarkable in that it does not involve material properties. Accordingly, this *universal relation* holds for all uniform isotropic materials, and therefore, to the extent that it involves only directly measurable quantities, furnishes the experimenter with a critical test of the underlying theory. Although pure simple shear is difficult to attain in the laboratory, other modes of deformation involving simple shear locally, such as torsion (see Section 3.5.4) are experimentally feasible and have been used to corroborate the *normal stress effect* predicted here.

We note that the right-hand side of (11.42) is of order γ^2 for small gamma and is therefore zero according to the the linear theory of elasticity in which nonlinear effects are disregarded. The normal stress effect is thus an inherently nonlinear phenomenon.

11.3 Heat flux

If the material is hemitropic at the material point p, then the constitutive function $\hat{\boldsymbol{q}}$ for the heat flux is such that

$$\hat{\boldsymbol{q}}(\theta, \boldsymbol{F}, grad\theta; p) = \hat{\boldsymbol{q}}(\theta, \boldsymbol{F}\boldsymbol{R}, grad\theta; p) \quad \text{for all } \boldsymbol{R} \in Orth^+. \quad (11.43)$$

To derive a necessary condition for this, let $\boldsymbol{R} = \boldsymbol{R}_F^t \mathbf{1}$, where $\mathbf{1}$ is the shifter and $\boldsymbol{R}_F$ is the two-point rotation tensor in the polar factorization of $\boldsymbol{F}$. From (3.32) (with $\boldsymbol{R}_F$ in place of $\boldsymbol{R}$), we then have

$$\hat{\boldsymbol{q}}(\theta, \boldsymbol{F}, grad\theta; p) = \hat{\boldsymbol{q}}(\theta, \boldsymbol{V}\boldsymbol{I}\mathbf{1}, grad\theta; p) = \hat{\boldsymbol{q}}(\theta, \boldsymbol{V}\mathbf{1}, grad\theta; p), \quad (11.44)$$

where $\boldsymbol{V}$ is the left stretch tensor. Because $\mathbf{1}$ is fixed in the frame of $\mathcal{O}$, and because $\boldsymbol{V}$ is uniquely determined by the left Cauchy–Green tensor $\boldsymbol{B}(= \boldsymbol{F}\boldsymbol{F}^t = \boldsymbol{V}^2)$, it follows that

$$\hat{\boldsymbol{q}}(\theta, \boldsymbol{F}, grad\theta; p) = \boldsymbol{q}^*(\theta, \boldsymbol{B}, grad\theta; \boldsymbol{X}) \quad (11.45)$$

for some function $\boldsymbol{q}^*$. Conversely, if this holds then

$$\begin{aligned}\hat{\boldsymbol{q}}(\theta, \boldsymbol{F}\boldsymbol{R}, grad\theta; p) &= \boldsymbol{q}^*(\theta, \boldsymbol{F}\boldsymbol{R}\boldsymbol{R}^t\boldsymbol{F}^t, grad\theta; \boldsymbol{X}) \\ &= \boldsymbol{q}^*(\theta, \boldsymbol{F}\hat{\boldsymbol{I}}\boldsymbol{F}^t, grad\theta; \boldsymbol{X}) \\ &= \boldsymbol{q}^*(\theta, \boldsymbol{B}, grad\theta; \boldsymbol{X}) = \hat{\boldsymbol{q}}(\theta, \boldsymbol{F}, grad\theta; p) \quad (11.46)\end{aligned}$$

for *all* $\boldsymbol{R} \in Orth^+$, and so (11.45) is both necessary and sufficient for hemitropy.

The restriction (11.10) associated with frame invariance may then be cast as

$$\boldsymbol{q}^*(\theta, \boldsymbol{QBQ}^t, \boldsymbol{Q}grad\theta; \boldsymbol{X}) = \boldsymbol{Qq}^*(\theta, \boldsymbol{B}, grad\theta; \boldsymbol{X})$$

$$\text{for all rotations } \boldsymbol{Q}\text{: } E^3 \to E^3. \tag{11.47}$$

This is very similar to the restriction (9.40) on the constitutive equation for the heat flux in a fluid. However, in that context $\boldsymbol{Q}$ is an arbitrary orthogonal tensor. This is due to having eliminated the effect of the configuration κ by operating within the unimodular symmetry group for fluids (see (9.3) and (9.39)). In contrast, for solids, the configuration κ manifests itself explicitly. Having chosen this to be occupiable, with reference to the discussion leading to (10.8) we then require that $\boldsymbol{Q}$ be a rotation.

However, if we further stipulate that the heat flux be an odd function of the temperature gradient, as in the classical Fourier law (9.61) — this being realistic for most (and perhaps all) materials — i.e., if

$$\boldsymbol{q}^*(\theta, \boldsymbol{B}, -grad\theta; \boldsymbol{X}) = -\boldsymbol{q}^*(\theta, \boldsymbol{B}, grad\theta; \boldsymbol{X}), \tag{11.48}$$

then the restriction to rotations is effectively removed and (11.47) holds for all *orthogonal* $\boldsymbol{Q}$. In this case we may invoke the representation (9.41), with (9.42), to arrive at

$$\boldsymbol{q} = (\phi_0 \boldsymbol{I} + \phi_1 \boldsymbol{B} + \phi_2 \boldsymbol{B}^2) grad\theta, \tag{11.49}$$

where ϕ_0, ϕ_1 and ϕ_2 are functions of θ; of the invariants $I_1(\boldsymbol{B})$, $I_2(\boldsymbol{B})$, $I_3(\boldsymbol{B})$; and of the scalars

$$I_4(\boldsymbol{B}, grad\theta) = grad\theta \cdot \boldsymbol{B}(grad\theta), \quad I_5(\boldsymbol{B}, grad\theta) = grad\theta \cdot \boldsymbol{B}^2(grad\theta)$$

$$\text{and } I_6(grad\theta) = |grad\theta|, \tag{11.50}$$

and also explicitly dependent on $\boldsymbol{X}$ in the case of a non-uniform material.

11.3.1 *General form of the heat flux*

The result (11.49) is a special case of the general relation

$$\boldsymbol{q} = \boldsymbol{K}(\theta, \boldsymbol{F}, grad\theta; \boldsymbol{X}) grad\theta \tag{11.51}$$

derived in [22] for all types of material symmetry, in which $\boldsymbol{K}$ is a tensor-valued function. To establish this we fix θ, $\boldsymbol{F}$ and $\boldsymbol{X}$ (hence p), and define

$$\boldsymbol{h}(\boldsymbol{p}) = \hat{\boldsymbol{q}}(\theta, \boldsymbol{F}, \boldsymbol{p}; p). \tag{11.52}$$

Inequality $(11.16)_3$ takes the form

$$\boldsymbol{p} \cdot \boldsymbol{h}(\boldsymbol{p}) \leq 0 \quad \text{for all } \boldsymbol{p}. \tag{11.53}$$

Thus $\boldsymbol{p}\cdot\boldsymbol{h}(\boldsymbol{p})$ is maximized, and hence stationary, at $\boldsymbol{p} = \boldsymbol{0}$. Assuming $\boldsymbol{h}(\boldsymbol{p})$ to be differentiable, we have

$$0 = d\boldsymbol{p} \cdot \boldsymbol{h}(\boldsymbol{0}) + \{\boldsymbol{p} \cdot [\nabla \boldsymbol{h}(\boldsymbol{p})]d\boldsymbol{p}\}_{|\boldsymbol{p}=\boldsymbol{0}}, \tag{11.54}$$

where $\nabla \boldsymbol{h}(\boldsymbol{p})$ is the tensor-valued gradient of $\boldsymbol{h}$ with respect to $\boldsymbol{p}$. Thus, $d\boldsymbol{p} \cdot \boldsymbol{h}(\boldsymbol{0}) = 0$ for all $d\boldsymbol{p}$, implying that $\boldsymbol{h}(\boldsymbol{0}) = \boldsymbol{0}$, i.e.,

$$\hat{\boldsymbol{q}}(\theta, \boldsymbol{F}, \boldsymbol{0}; p) = \boldsymbol{0}. \tag{11.55}$$

Thus, there is no heat flux in the absence of a temperature gradient.

Following [22], we fix $\boldsymbol{p}$ and define

$$\boldsymbol{g}(x) = \boldsymbol{h}(x\boldsymbol{p}) \quad \text{with } x \in [0, 1]. \tag{11.56}$$

The derivative of this function is

$$\boldsymbol{g}'(x) = [\nabla \boldsymbol{h}(x\boldsymbol{p})](x\boldsymbol{p})' = [\nabla \boldsymbol{h}(x\boldsymbol{p})]\boldsymbol{p}. \tag{11.57}$$

Then, with $\boldsymbol{g}(0) = \boldsymbol{0}$ and $\boldsymbol{g}(1) = \boldsymbol{h}(\boldsymbol{p})$, we have

$$\boldsymbol{h}(\boldsymbol{p}) = \int_0^1 \boldsymbol{g}'(x)dx = [\boldsymbol{K}(\boldsymbol{p})]\boldsymbol{p}, \quad \text{where } \boldsymbol{K}(\boldsymbol{p}) = \int_0^1 \nabla \boldsymbol{h}(x\boldsymbol{p})dx, \tag{11.58}$$

which yields (11.51) on reinstating the passive variables.

In the special case of (11.49), $\boldsymbol{K}$ is given by the symmetric tensor in parentheses. However, in the general case the foregoing derivation does not furnish a symmetric $\boldsymbol{K}$. Nevertheless the dissipation inequality (11.53), i.e.,

$$\boldsymbol{K} \cdot \boldsymbol{p} \otimes \boldsymbol{p} \leq 0 \tag{11.59}$$

involves only the symmetric part of $\boldsymbol{K}$.

11.4 Problems

1. Prove (11.21).
2. Derive (11.34).
3. Prove (11.35).
4. Use (11.34) (or (11.36)) and (11.49) to find expressions for the stress and heat flux in the frame of observer $\mathcal{O}^+$.
5. Are the constraints considered in Problems 3 and 4 of Chapter 10 compatible with isotropy?
6. Verify (11.38).
7. Derive (11.42).
8. Consider the strain-energy function $W^* = \frac{1}{2}G(I_1 - 3)$, where G is a positive constant. This model, called the *neo-Hookean material*, furnishes an accurate model of the behavior of incompressible rubber provided that the principal stretches λ_i lie in the approximate range $\frac{1}{2} < \lambda_i < 2$.

   (a) Show that the Cauchy stress $\boldsymbol{T}$ is given by

   $$\boldsymbol{T} = -p\boldsymbol{I} + G\boldsymbol{B},$$

   where p is a pressure field associated with the incompressibility constraint. Show that for this material, equilibrium without body force requires that

   $$grad p = G div \boldsymbol{B}.$$

   (b) Use the fact that the Piola stress is power-conjugate to the deformation gradient to deduce that G is the ratio of the shear component of the Piola traction to the amount of shear in a simple-shear deformation. Therefore G is the shear modulus.

   (c) Consider the deformation described in Problem 6 of Chapter 4. Find expressions for $\boldsymbol{C}$ and $\boldsymbol{B}$.

   (d) Impose the zero-traction condition $\boldsymbol{Tn} = \boldsymbol{0}$ for any unit vector $\boldsymbol{n}$ such that $\boldsymbol{n} \cdot \boldsymbol{e}_1 = 0$. This simulates the traction-free lateral surface of a prismatic bar with axis $\boldsymbol{e}_1$. Assuming equilibrium without body force, use this condition to show that $p = G/\lambda$.

   (e) Use the previous result to show that

   $$\boldsymbol{T} = T(\lambda)\boldsymbol{e}_1 \otimes \boldsymbol{e}_1, \quad \text{where } T(\lambda) = G(\lambda^2 - \lambda^{-1}).$$

(f) By definition, Young's modulus E is equal to the derivative $T'(\lambda)$, evaluated at $\lambda = 1$. Determine E in terms of G.

(g) Show that the Piola and Piola-Kirchhoff stresses, $\boldsymbol{P}$ and $\boldsymbol{S}$ respectively, are given by

$$\boldsymbol{P} = P(\lambda)\boldsymbol{e}_1 \otimes \hat{\boldsymbol{E}}_1 \quad \text{and} \quad \boldsymbol{S} = S(\lambda)\hat{\boldsymbol{E}}_1 \otimes \hat{\boldsymbol{E}}_1,$$

for certain functions $P(\lambda)$ and $S(\lambda)$. Determine these functions. How are the values of the derivatives $P'(\lambda)_{|\lambda=1}$ and $S'(\lambda)_{|\lambda=1}$ related to Young's modulus?

9. Suppose a circular cylindrical bar, consisting of a neo-Hookean material and with axis $\hat{\boldsymbol{E}}_3$ in κ, is subjected to a static anti-plane shear deformation of the form discussed in Problem 5 of Chapter 3. The bar has inner and outer radii A and B. Assume the bar to be in equilibrium under negligible body force, and that $w(A) = W$, $w(B) = 0$, where W is a specified constant. Obtain the function $w(R)$.
10. Consider an elastic solid that is isotropic at all material points relative to the adopted reference configuration. Suppose this solid is stressed in its *undeformed* state; i.e., $\boldsymbol{T}$ is non-zero when $\boldsymbol{F} = \mathbf{1}$. In this case we say that the undeformed state is *residually stressed.* Suppose the residually stressed undeformed configuration of the body to be in equilibrium without body force. Also, suppose the traction acting on a portion of the boundary of the undeformed body vanishes. Show that the residual stress vanishes identically everywhere in the body.
11. Consider the deformation described by

$$\boldsymbol{x} = \mathbf{1}\boldsymbol{X} + w(\Theta)\boldsymbol{e}_3,$$

where Θ is the azimuthal angle in a cylindrical polar coordinate system $\{R, \Theta, Z\}$ in the reference configuration.

(a) Use $\boldsymbol{x} = r\boldsymbol{e}_r(\theta) + z\boldsymbol{e}_3$ with an appropriate selection of basis vectors to show that we can assume, without loss of generality, that $R = r$, $\Theta = \theta$ and $Z + w = z$.

(b) Show that the deformation gradient is

$$\boldsymbol{F} = \mathbf{1} + R^{-1}w'(\Theta)\boldsymbol{e}_3 \otimes \hat{\boldsymbol{E}}_\Theta(\Theta)$$

and obtain an expression for the left Cauchy–Green tensor $\boldsymbol{B}$. Show that $\det \boldsymbol{F} = 1$.

(c) Suppose this deformation is experienced by a neo-Hookean material occupying an annular cylindrical tube of inner and outer radii A and B. Suppose the material is in equilibrium with zero body force. If the inner and outer cylindrical surfaces are traction free, show that

$$w(\Theta) = C\Theta + D,$$

where C and D are constants. Can you interpret this deformation in physical terms?

12. Suppose a solid circular cylindrical bar composed of a neo-Hookean material is subjected to the torsional deformation described in Section 3.5.4. Assume the bar to be in equilibrium without body force and that the traction on the cylindrical surface of the bar vanishes. Obtain the stress field and use it to compute the net force and torque acting on a cross-section of the bar. Show that the relation between the torque and twist is the same as that predicted by classical linear strength-of-materials theory, and, contrary to that theory, that the axial force is non-zero. Show that the latter is proportional to the square of the twist and is therefore a nonlinear effect. This is a manifestation of the normal-stress effect described in the text.
13. Consider a circular cylinder of radius A with a wedge removed (Figure 11.1). The angular extent of the material in the reference configuration is γ. Suppose the cross section is deformed in such a way that the gap is closed in the deformed configuration. The deformation is described by

$$\boldsymbol{X} = R\hat{\boldsymbol{E}}_r(\Theta) + Z\hat{\boldsymbol{E}}_3, \quad \boldsymbol{x} = r\boldsymbol{e}_r(\theta) + z\boldsymbol{e}_3,$$

where $z = Z$, $r = f(R)$ and $\theta = g(\Theta)$.

(a) Find an expression for $\boldsymbol{F}$ assuming that f and g are known functions.
(b) Assume the deformation to be isochoric and find the functions f and g that satisfy the boundary conditions inferred from the figure. Show that the deformed radius a is proportional to $\sqrt{\gamma}$.
(c) Find $\boldsymbol{R}, \boldsymbol{U}, \boldsymbol{V}$.
(d) Obtain the principal stretches in terms of γ.

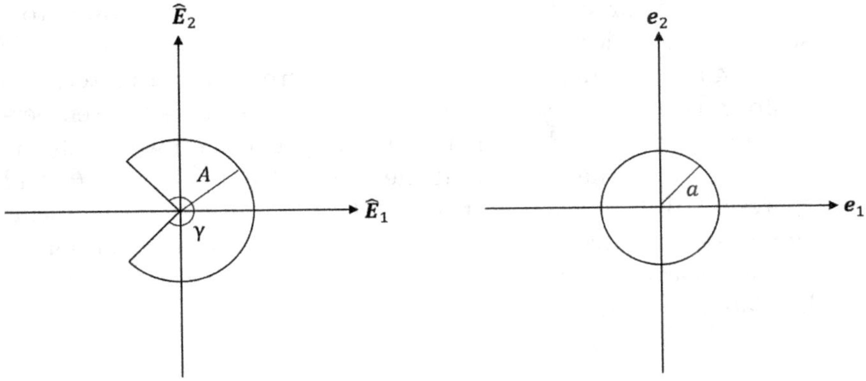

Figure 11.1. Deformation of a circular cylinder with a wedge removed.

(e) Solve the problem of equilibrium without body force for a neo-Hookean material. Find the reactive pressure distribution and show that it is singular at the centerline of the bar. Evaluate the traction, acting at the outer circular boundary in the deformed configuration, required to maintain the deformation. Show that this is an arbitrary uniform pressure. What is the contact traction on the planar surfaces where the two sides of the wedge meet in the deformed configuration? If contact between these surfaces is maintained by glue, what traction must the glue transmit if there is no traction acting at the outer boundary?

14. Derive the energy balance equation

$$\rho_\kappa r = Div\boldsymbol{q}_\kappa - \theta\dot{W}, \quad \text{where } \dot{W} = W_{\theta\theta}\dot{\theta} + W_{\theta\boldsymbol{F}} \cdot \dot{\boldsymbol{F}},$$

where $\boldsymbol{q}_\kappa$ is the referential heat flux (see (6.174)).

15. Show that if the spatial heat flux is given by (11.49), then the referential heat flux is given by

$$\boldsymbol{q}_\kappa = (\psi_0\hat{\boldsymbol{I}} + \psi_1\boldsymbol{C} + \psi_2\boldsymbol{C}^2)Grad\theta,$$

where ψ_0, ψ_1 and ψ_2 are functions of θ; of the invariants $I_1(\boldsymbol{C})$, $I_2(\boldsymbol{C})$, $I_3(\boldsymbol{C})$; and of the scalars $I_4(\boldsymbol{C}, Grad\theta)$, $I_5(\boldsymbol{C}, Grad\theta)$ and $I_6(Grad\theta)$.

16. A solid circular cylindrical bar is subjected to the static torsional deformation described in Section 3.5.4. Suppose the $\phi's$ in (11.49) are independent of temperature and the material is uniform (they may still depend on $I_1 - I_6$). Assume the temperature to be independent of time and dependent only z, with the ends $z = 0, l$ of the bar held at the constant temperatures θ_0 and θ_l, respectively. Assuming this function to be linear, show that the energy balance equation with zero internal heating supply is automatically satisfied, and that the lateral surface of the bar is insulated, i.e., there is no heat flux across this surface.

Chapter 12

Linear Elasticity

The theory of linear elasticity is among the most successful analytical tools in all of engineering. It underlies the vast majority of analyses of engineering structures and materials in applications involving small strains and rotations. Accordingly it has been the subject of a great deal of research and has reached a very high level of development. Our purpose in this final chapter is to outline the essential aspects of this important subject. For the sake of simplicity we confine our attention to the purely mechanical theory in which the free energy is independent of temperature, which we assume to be constant everywhere in the material. It then follows, from (11.51) and the result of Problem 14 of the preceding chapter, that the thermal energy balance is identically satisfied provided that the internal heating supply vanishes.

12.1 Kinematics

The basic premise of linear elasticity theory is that the norm of the displacement gradient $\boldsymbol{H}$ is small everywhere in the body (see (7.22)). Assuming $\boldsymbol{H}$ to be a continuous function of $\boldsymbol{X}$, we then have

$$\varepsilon = \max_{\boldsymbol{X}\in\kappa} |\boldsymbol{H}(\boldsymbol{X},t)| \ll 1, \tag{12.1}$$

where $|\boldsymbol{H}|^2 = tr(\boldsymbol{H}\boldsymbol{H}^t) = tr(\boldsymbol{H}^t\boldsymbol{H})$. In terms of Cartesian components, $|\boldsymbol{H}| = \sqrt{H_{iA}H_{iA}}$. Thus the absolute value of each component is much less than unity: $|H_{iA}| \ll 1$.

It follows from (7.22) that the right Cauchy–Green tensor is

$$\boldsymbol{C} = \hat{\boldsymbol{I}} + \mathbf{1}^t\boldsymbol{H} + \boldsymbol{H}^t\mathbf{1} + \boldsymbol{H}^t\boldsymbol{H}. \tag{12.2}$$

Equivalently,

$$\boldsymbol{C} = \hat{\boldsymbol{I}} + \boldsymbol{G} + \boldsymbol{G}^t + \boldsymbol{G}^t\boldsymbol{G}, \quad \text{where } \boldsymbol{G} = \mathbf{1}^t\boldsymbol{H} \tag{12.3}$$

is a referential tensor, with Cartesian components $G_{AB} = \delta_{iA}H_{iB}$. This yields

$$\boldsymbol{G}d\boldsymbol{X} = \mathbf{1}^t\boldsymbol{H}d\boldsymbol{X} = \mathbf{1}^t d\boldsymbol{u} = d\boldsymbol{w}, \quad \text{where } \boldsymbol{w} = \mathbf{1}^t\boldsymbol{u} \tag{12.4}$$

is the referential displacement, in which $\boldsymbol{u}$ is the spatial displacement and we have used the fact that the shifter is uniform. Thus,

$$\boldsymbol{G} = Grad\boldsymbol{w}, \tag{12.5}$$

and $\boldsymbol{u} = \mathbf{1}\boldsymbol{w}$. In terms of components, $w_A = \delta_{iA}u_i$, $u_i = \delta_{iA}w_A$ and $G_{AB} = w_{A,B}$. Using (7.22) with $\boldsymbol{H} = \mathbf{1}\boldsymbol{G}$, we may write the deformation gradient in the form

$$\boldsymbol{F} = \mathbf{1}(\hat{\boldsymbol{I}} + \boldsymbol{G}). \tag{12.6}$$

It is a simple matter to verify that $|\boldsymbol{H}| = |\boldsymbol{G}|\ (= \sqrt{G_{AB}G_{AB}})$, and hence that (12.1) is equivalent to the requirement that $|\boldsymbol{G}| \ll 1$ everywhere in the body, i.e., that each component is such that $|G_{AB}| \ll 1$. Accordingly $\boldsymbol{C}$ is everywhere close to $\hat{\boldsymbol{I}}$. For this reason, it is more convenient to work with the Lagrange strain $\boldsymbol{E}$ defined by (2.97). This is given in terms of the displacement gradient by

$$\boldsymbol{E} = \frac{1}{2}(\boldsymbol{G} + \boldsymbol{G}^t + \boldsymbol{G}^t\boldsymbol{G}). \tag{12.7}$$

If $|\boldsymbol{G}| \ll 1$, as we assume, then

$$\boldsymbol{E} \simeq \boldsymbol{\epsilon}, \quad \text{where } \boldsymbol{\epsilon} = \frac{1}{2}(\boldsymbol{G} + \boldsymbol{G}^t) \tag{12.8}$$

is the *referential infinitesimal strain tensor*. Here and henceforth we use the notation $\simeq$ to identify statements that are valid to leading order in $|\boldsymbol{G}|$. Thus, $|\boldsymbol{G}| \ll 1$ implies that $|\boldsymbol{E}| \ll 1$, but the converse is not true. That is, a small strain does not necessarily yield a small displacement gradient.

These results pertain to the frame used by observer $\mathcal{O}$, say. It follows easily from $(12.3)_2$, (7.18) and (7.25) that, in the frame of $\mathcal{O}^+$,

$$\boldsymbol{G}^+ = \boldsymbol{K}\boldsymbol{G}\boldsymbol{K}^t, \tag{12.9}$$

where $\boldsymbol{K}$: $\hat{E}^3 \to \hat{E}^{3+}$ is the fixed orthogonal tensor relating the reference configurations adopted by the two observers. This in turn implies that

$$\boldsymbol{E}^+ = \boldsymbol{K}\boldsymbol{E}\boldsymbol{K}^t, \tag{12.10}$$

a result that may be corroborated independently by combining (7.20) and (7.26) with $\boldsymbol{E}^+ = \frac{1}{2}[(\boldsymbol{F}^+)^t\boldsymbol{F}^+ - \hat{\boldsymbol{I}}^+]$. Further, (12.9) yields $|\boldsymbol{G}^+| = |\boldsymbol{G}|$, which combines with

$$\boldsymbol{\epsilon}^+ = \frac{1}{2}[\boldsymbol{G}^+ + (\boldsymbol{G}^+)^t] = \boldsymbol{K}\boldsymbol{\epsilon}\boldsymbol{K}^t \tag{12.11}$$

to establish that (12.8) is equivalent to

$$\boldsymbol{E}^+ \simeq \boldsymbol{\epsilon}^+. \tag{12.12}$$

Thus, if one observer perceives the strain to be small, then the same is true of all observers. This follows from the exact transformation formulas (12.9) and (12.10), and we have proved that it also follows from the approximation (12.8) in the case of small displacement gradients. In contrast, in the extant literature the strain–displacement relation is obtained from (12.2) with the shifter omitted. In particular, for small displacement gradients it yields $2\boldsymbol{E} \simeq \boldsymbol{H} + \boldsymbol{H}^t$ in the frame of $\mathcal{O}$, in which the left-hand side is a referential tensor and the right-hand side involves two-point tensors. Accordingly, this is not a valid tensor equation. Further, (7.25) shows that $\boldsymbol{H}+\boldsymbol{H}^t$ transforms in a manner that is incompatible with (12.10). Therefore, from the perspective of $\mathcal{O}^+$, $\boldsymbol{H}^+ + (\boldsymbol{H}^+)^t$ cannot furnish an approximation to $2\boldsymbol{E}^+$. This observation underscores the necessity of the shifter concept in the formulation of an acceptable theory.

Recall that $\boldsymbol{u}$ is a spatial vector (see (2.77) and (2.78)). Its spatial gradient, $grad\boldsymbol{u}$, is such that $d\boldsymbol{u} = (grad\boldsymbol{u})d\boldsymbol{x}$ at fixed t. Thus, $\boldsymbol{H}d\boldsymbol{X} = d\boldsymbol{u} = (grad\boldsymbol{u})\boldsymbol{F}d\boldsymbol{X}$. Taken together with (12.6), this furnishes

$$\boldsymbol{H} = (grad\boldsymbol{u})\boldsymbol{F} \simeq (grad\boldsymbol{u})\boldsymbol{1}, \tag{12.13}$$

or, in terms of components, $\hat{u}_{i,A} = \tilde{u}_{i,j}F_{jA} = \tilde{u}_{i,j}\delta_{jB}(\delta_{BA} + G_{BA}) \simeq \tilde{u}_{i,j}\delta_{jA}$. This is an example of the general estimate

$$(\cdot)_{,A} \simeq (\cdot)_{,j}\delta_{jA}, \tag{12.14}$$

valid to leading order for small displacement gradients. Using $(12.3)_2$ we also have

$$\boldsymbol{G} \simeq \mathbf{1}^t(grad\boldsymbol{u})\mathbf{1} \quad \text{and} \quad grad\boldsymbol{u} \simeq \mathbf{1}\boldsymbol{G}\mathbf{1}^t, \tag{12.15}$$

and therefore

$$|\boldsymbol{H}| = |\boldsymbol{G}| \simeq |grad\boldsymbol{u}|\,. \tag{12.16}$$

Thus, $|\tilde{u}_{i,j}| \ll 1$. In addition,

$$\boldsymbol{\epsilon} \simeq \mathbf{1}^t\boldsymbol{\varepsilon}\mathbf{1}, \quad \text{where } \boldsymbol{\varepsilon} = \frac{1}{2}[grad\boldsymbol{u} + (grad\boldsymbol{u})^t] \tag{12.17}$$

is the *spatial infinitesimal strain tensor*. The components satisfy the estimates $\epsilon_{AB} \simeq \delta_{iA}\delta_{jB}\varepsilon_{ij}$. Using the exact relation $(12.13)_1$, or either of the estimates $(12.13)_2$ or (12.15), we have that

$$grad^+\boldsymbol{u}^+ = \boldsymbol{Q}(grad\boldsymbol{u})\boldsymbol{Q}^t \quad \text{and} \quad \boldsymbol{\varepsilon}^+ = \boldsymbol{Q}\boldsymbol{\varepsilon}\boldsymbol{Q}^t, \tag{12.18}$$

where $\boldsymbol{Q}$: $E^3 \to E^{3+}$ is the orthogonal tensor in the transformation from the frame of $\mathcal{O}$ to that of $\mathcal{O}^+$. These also follow directly from (7.3) and (7.21).

Recall the polar decomposition $\boldsymbol{F} = \boldsymbol{R}\boldsymbol{U}$, where the $\boldsymbol{U}$ is the symmetric, positive definite referential right stretch tensor and $\boldsymbol{R}$ is a two-point rotation tensor. From (12.3) and (12.8),

$$\boldsymbol{U}^2 - \hat{\boldsymbol{I}} \simeq 2\boldsymbol{\epsilon}, \tag{12.19}$$

and therefore,

$$\boldsymbol{U} - \hat{\boldsymbol{I}} \simeq \boldsymbol{\epsilon}, \quad \boldsymbol{U}^{-1} - \hat{\boldsymbol{I}} \simeq -\boldsymbol{\epsilon} \quad \text{and} \quad \boldsymbol{R} - \mathbf{1} \simeq \mathbf{1}\hat{\boldsymbol{\omega}}, \tag{12.20}$$

where

$$\hat{\boldsymbol{\omega}} = \frac{1}{2}(\boldsymbol{G} - \boldsymbol{G}^t) \tag{12.21}$$

is the *referential infinitesimal rotation* tensor. Equivalently,

$$\boldsymbol{R}-\mathbf{1}\simeq\boldsymbol{\omega}\mathbf{1},\quad \text{where } \boldsymbol{\omega}=\frac{1}{2}[grad\boldsymbol{u}-(grad\boldsymbol{u})^{t}]=\mathbf{1}\hat{\boldsymbol{\omega}}\mathbf{1}^{t} \tag{12.22}$$

is the *spatial infinitesimal rotation* tensor.

The local volume change induced by a deformation is $J-1$, where $J^2=\det\boldsymbol{C}$. On combining (12.3) with the result of Problem 23 in Chapter 1, we have

$$J^{2}=\det(\hat{\boldsymbol{I}}+\boldsymbol{A})=\det(\boldsymbol{A}-\lambda\hat{\boldsymbol{I}})_{|\lambda=-1},\quad \text{where } \boldsymbol{A}=\boldsymbol{G}+\boldsymbol{G}^{t}+\boldsymbol{G}^{t}\boldsymbol{G}. \tag{12.23}$$

Thus,

$$J^{2}-1\simeq tr\boldsymbol{A}\simeq 2tr\boldsymbol{G}\simeq 2tr(grad\boldsymbol{u}), \tag{12.24}$$

where the final estimate on the right follows from (12.15) and $tr[\mathbf{1}^{t}(grad\boldsymbol{u})\mathbf{1}]=tr[\mathbf{1}\mathbf{1}^{t}(grad\boldsymbol{u})]=tr[\boldsymbol{I}(grad\boldsymbol{u})]=tr(grad\boldsymbol{u})$. Thus,

$$J-1\simeq Div\boldsymbol{w}\simeq div\boldsymbol{u}. \tag{12.25}$$

The change of volume, $\Delta V(S,t)$, of an arbitrary part S of the body is thus approximated by

$$\Delta V=\int_{P}(J-1)dV\simeq\int_{P}Div\boldsymbol{w}dV=\int_{\partial P}\boldsymbol{w}\cdot\boldsymbol{N}dA, \tag{12.26}$$

where P is the volume occupied by S in κ, bounded by the surface ∂P with exterior unit normal $\boldsymbol{N}$. Alternatively, with $J^{-1}-1\simeq -div\boldsymbol{u}$ we have

$$\Delta V\simeq\int_{P}div\boldsymbol{u}dV=\int_{P_t}J^{-1}div\boldsymbol{u}dv\simeq\int_{P_t}div\boldsymbol{u}dv=\int_{\partial P_t}\boldsymbol{u}\cdot\boldsymbol{n}da, \tag{12.27}$$

where P_t is the volume occupied by S in κ_t, bounded by the surface ∂P_t with exterior unit normal $\boldsymbol{n}$. Thus the volume change is given, to leading order, by the flux of displacement through the boundary.

Lastly, the change of density induced by a deformation is

$$\Delta\rho=\rho-\rho_{\kappa},\quad \text{where } \rho_{\kappa}=\rho J. \tag{12.28}$$

Thus,

$$\Delta\rho \simeq -\rho_\kappa Div\boldsymbol{w} \simeq -\rho_\kappa div\boldsymbol{u}, \quad \text{yielding } \rho \simeq \rho_\kappa \tag{12.29}$$

at leading order.

12.2 Stress–deformation relation

We seek small-displacement-gradient estimates of the Piola, Piola–Kirchhoff and Cauchy stresses. To this end we introduce the *strain-energy function*

$$U(\boldsymbol{E};\boldsymbol{X}) = \bar{W}(\theta_0, \hat{\boldsymbol{I}} + 2\boldsymbol{E};\boldsymbol{X}), \tag{12.30}$$

where $\bar{W}$ is the free energy per unit reference volume (see (11.19)) and θ_0 is the fixed uniform temperature of the body. Proceeding as in (11.20), we have

$$U_{\boldsymbol{E}} \cdot \dot{\boldsymbol{E}} = \bar{W}_{\boldsymbol{C}} \cdot \dot{\boldsymbol{C}} = 2\bar{W}_{\boldsymbol{C}} \cdot \dot{\boldsymbol{E}} = \boldsymbol{S} \cdot \dot{\boldsymbol{E}}, \tag{12.31}$$

for all symmetric $\dot{\boldsymbol{E}}$, where (11.24) has been invoked and where $U_{\boldsymbol{E}}$ is the symmetric tensor-valued gradient of U with respect to the strain $\boldsymbol{E}$. Thus,

$$\boldsymbol{S} = U_{\boldsymbol{E}}. \tag{12.32}$$

Recalling that a small displacement gradient implies a small strain, we seek an estimate of the strain-energy function for strains close to zero. This estimate is provided by the low-order Taylor expansion [2]

$$U(\boldsymbol{E};\boldsymbol{X}) = U(\boldsymbol{O};\boldsymbol{X}) + \boldsymbol{S}_0(\boldsymbol{X}) \cdot \boldsymbol{E} + \frac{1}{2}\boldsymbol{E} \cdot \mathcal{D}(\boldsymbol{X})[\boldsymbol{E}] + o(|\boldsymbol{E}|^2), \tag{12.33}$$

where

$$\boldsymbol{S}_0(\boldsymbol{X}) = U_{\boldsymbol{E}|\boldsymbol{E}=\boldsymbol{O}} \quad \text{and} \quad \mathcal{D}(\boldsymbol{X}) = U_{\boldsymbol{EE}|\boldsymbol{E}=\boldsymbol{O}}, \tag{12.34}$$

is a referential *fourth-order tensor*. Here $\mathcal{D}(\boldsymbol{X})[\boldsymbol{E}]$ is a linear second-order-tensor-valued function of $\boldsymbol{E}$. Using the basis decomposition

$\boldsymbol{E} = E_{CD}\hat{\boldsymbol{E}}_C \otimes \hat{\boldsymbol{E}}_D$, we have

$$\begin{aligned}\mathcal{D}(\boldsymbol{X})[\boldsymbol{E}] &= \mathcal{D}(\boldsymbol{X})[E_{CD}\hat{\boldsymbol{E}}_C \otimes \hat{\boldsymbol{E}}_C] \\ &= E_{CD}\mathcal{D}(\boldsymbol{X})[\hat{\boldsymbol{E}}_C \otimes \hat{\boldsymbol{E}}_D], \quad \text{by linearity,}\end{aligned} \tag{12.35}$$

and

$$\boldsymbol{E} \cdot \mathcal{D}(\boldsymbol{X})[\boldsymbol{E}] = E_{AB}E_{CD}\mathcal{D}_{ABCD}(\boldsymbol{X}),$$
$$\text{where } \mathcal{D}_{ABCD} = \hat{\boldsymbol{E}}_A \otimes \hat{\boldsymbol{E}}_B \cdot \mathcal{D}[\hat{\boldsymbol{E}}_C \otimes \hat{\boldsymbol{E}}_D] = \hat{\boldsymbol{E}}_A \cdot (\mathcal{D}[\hat{\boldsymbol{E}}_C \otimes \hat{\boldsymbol{E}}_D])\hat{\boldsymbol{E}}_B. \tag{12.36}$$

Observe that, by relabeling summation indices and because ordinary multiplication commutes,

$$E_{AB}E_{CD}\mathcal{D}_{ABCD} = E_{CD}E_{AB}\mathcal{D}_{CDAB} = E_{AB}E_{CD}\mathcal{D}_{CDAB}, \tag{12.37}$$

and so we may assume, without loss of generality, that

$$\mathcal{D}_{ABCD} = \mathcal{D}_{CDAB}. \tag{12.38}$$

Further, on relabeling summation indices and invoking the symmetry $E_{AB} = E_{BA}$, we have

$$E_{BA}E_{CD}\mathcal{D}_{BACD} = E_{AB}E_{CD}\mathcal{D}_{ABCD} = E_{BA}E_{CD}\mathcal{D}_{ABCD}, \tag{12.39}$$

so that we may take

$$\mathcal{D}_{BACD} = \mathcal{D}_{ABCD} \tag{12.40}$$

without loss of generality, and hence also, from (12.38),

$$\mathcal{D}_{ABCD} = \mathcal{D}_{ABDC}. \tag{12.41}$$

Let A_{AB} and B_{CD} respectively be the components of arbitrary referential tensors $\boldsymbol{A}$ and $\boldsymbol{B}$ in the basis $\{\hat{\boldsymbol{E}}_A \otimes \hat{\boldsymbol{E}}_B\}$. Then $(12.36)_2$, (12.38) and the linearity of $\mathcal{D}(\boldsymbol{X})[\cdot]$ yield the *major symmetry* property

$$\boldsymbol{A} \cdot \mathcal{D}[\boldsymbol{B}] = A_{AB}B_{CD}\mathcal{D}_{ABCD} = B_{CD}A_{AB}\mathcal{D}_{CDAB} = \boldsymbol{B} \cdot \mathcal{D}[\boldsymbol{A}] \tag{12.42}$$

for all $\boldsymbol{A}$ and $\boldsymbol{B}$, whereas (12.40) and (12.41), respectively, yield the *minor symmetries*

$$\boldsymbol{A}^t \cdot \mathcal{D}[\boldsymbol{B}] = \boldsymbol{A} \cdot \mathcal{D}[\boldsymbol{B}] \quad \text{and} \quad \boldsymbol{A} \cdot \mathcal{D}[\boldsymbol{B}^t] = \boldsymbol{A} \cdot \mathcal{D}[\boldsymbol{B}]. \tag{12.43}$$

If we retain only terms through quadratic order in (12.33) and invoke major symmetry, together with

$$(\mathcal{D}(\boldsymbol{X})[\boldsymbol{E}])^{\cdot} = \dot{E}_{CD}\mathcal{D}(\boldsymbol{X})[\hat{\boldsymbol{E}}_C \otimes \hat{\boldsymbol{E}}_D] = \mathcal{D}(\boldsymbol{X})[\dot{\boldsymbol{E}}], \tag{12.44}$$

which follows from (12.35), we obtain

$$\begin{aligned} U_{\boldsymbol{E}} \cdot \dot{\boldsymbol{E}} &= \boldsymbol{S}_0(\boldsymbol{X}) \cdot \dot{\boldsymbol{E}} + \frac{1}{2}\{\boldsymbol{E} \cdot \mathcal{D}(\boldsymbol{X})[\dot{\boldsymbol{E}}] + \dot{\boldsymbol{E}} \cdot \mathcal{D}(\boldsymbol{X})[\boldsymbol{E}]\} \\ &= \{\boldsymbol{S}_0(\boldsymbol{X}) + \mathcal{D}(\boldsymbol{X})[\boldsymbol{E}]\} \cdot \dot{\boldsymbol{E}}, \end{aligned} \tag{12.45}$$

for all symmetric $\dot{\boldsymbol{E}}$, and (12.32) and (12.40) then deliver the Piola–Kirchhoff stress

$$\boldsymbol{S} = \boldsymbol{S}_0(\boldsymbol{X}) + \mathcal{D}(\boldsymbol{X})[\boldsymbol{E}]. \tag{12.46}$$

Here, $\boldsymbol{S}_0(\boldsymbol{X})$ is the *residual stress*, i.e., the stress in the absence of strain, whereas $\mathcal{D}(\boldsymbol{X})$, the *stiffness tensor*, also known as the *elastic modulus tensor*, characterizes the linear part of the relationship between $\boldsymbol{S}$ and $\boldsymbol{E}$.

For example, in the case of isotropy, we have (see (11.29))

$$\begin{aligned} U(\boldsymbol{E};\boldsymbol{X}) &= \bar{W}(\theta_0, \hat{\boldsymbol{I}} + 2\boldsymbol{E}; \boldsymbol{X}) = \bar{W}(\theta_0, \boldsymbol{R}^t(\hat{\boldsymbol{I}} + 2\boldsymbol{E})\boldsymbol{R}; \boldsymbol{X}) \\ &= \bar{W}(\theta_0, \hat{\boldsymbol{I}} + 2\boldsymbol{R}^t\boldsymbol{E}\boldsymbol{R}; \boldsymbol{X}) \\ &= U(\boldsymbol{R}^t\boldsymbol{E}\boldsymbol{R}; \boldsymbol{X}) \quad \text{for all } \boldsymbol{R} \in Orth, \end{aligned} \tag{12.47}$$

and thus U depends on $\boldsymbol{E}$ through its principal invariants $I_1(\boldsymbol{E}) = tr\boldsymbol{E} = \hat{\boldsymbol{I}} \cdot \boldsymbol{E}$, $I_2(\boldsymbol{E}) = \frac{1}{2}[I_1^2 - tr(\boldsymbol{E}^2)]$ and $I_3(\boldsymbol{E}) = \det \boldsymbol{E}$; or, equivalently, through $\hat{\boldsymbol{I}} \cdot \boldsymbol{E}$, $\boldsymbol{E} \cdot \boldsymbol{E}$ and $\det \boldsymbol{E}$. Retaining terms through quadratic order, we thus obtain

$$U(\boldsymbol{E};\boldsymbol{X}) = U(\boldsymbol{O};\boldsymbol{X}) + \alpha(\boldsymbol{X})\hat{\boldsymbol{I}} \cdot \boldsymbol{E} + \frac{1}{2}\lambda(\boldsymbol{X})(\hat{\boldsymbol{I}} \cdot \boldsymbol{E})^2 + \mu(\boldsymbol{X})\boldsymbol{E} \cdot \boldsymbol{E}, \tag{12.48}$$

where the functions α, λ and μ are properties of the particular isotropic material at hand. This yields

$$U_{\boldsymbol{E}} \cdot \dot{\boldsymbol{E}} = \{[\alpha + \lambda(tr\boldsymbol{E})]\hat{\boldsymbol{I}} + 2\mu\boldsymbol{E}\} \cdot \dot{\boldsymbol{E}}, \tag{12.49}$$

and (12.46) furnishes

$$\boldsymbol{S}_0(\boldsymbol{X}) = \alpha(\boldsymbol{X})\hat{\boldsymbol{I}} \quad \text{and} \quad \mathcal{D}(\boldsymbol{X})[\boldsymbol{E}] = \lambda(\boldsymbol{X})(tr\boldsymbol{E})\hat{\boldsymbol{I}} + 2\mu(\boldsymbol{X})\boldsymbol{E}. \tag{12.50}$$

Equation (12.46) provides an accurate estimate of the stress when $|\boldsymbol{E}|$ is small. It yields a nonlinear relationship between $\boldsymbol{S}$ and the referential displacement gradient $\boldsymbol{G}$. However, if $|\boldsymbol{G}|$ is also small, as we assume in linear elasticity theory, then from (12.8) this relationship is approximately linear, i.e.,

$$\boldsymbol{S} - \boldsymbol{S}_0(\boldsymbol{X}) \simeq \mathcal{D}(\boldsymbol{X})[\boldsymbol{\epsilon}] = \mathcal{D}(\boldsymbol{X})[\boldsymbol{G}], \tag{12.51}$$

on account of the minor symmetry $(12.43)_2$; thus, $S_{AB} - S_{(0)AB} \simeq \mathcal{D}_{ABCD}w_{C,D}$. Classical linear elasticity theory is concerned with circumstances in which the residual stress vanishes, i.e.,

$$\boldsymbol{S} \simeq \mathcal{D}(\boldsymbol{X})[\boldsymbol{G}]. \tag{12.52}$$

In the case of isotropy this is given simply by

$$\boldsymbol{S} \simeq \lambda(\boldsymbol{X})(tr\boldsymbol{G})\hat{\boldsymbol{I}} + \mu(\boldsymbol{X})(\boldsymbol{G} + \boldsymbol{G}^t). \tag{12.53}$$

To estimate the Piola stress for small displacement gradients we combine (6.144) and (12.6). The latter gives $\boldsymbol{F} \simeq \boldsymbol{1}$, yielding

$$\boldsymbol{P} = \boldsymbol{F}\boldsymbol{S} \simeq \boldsymbol{1}\boldsymbol{S}. \tag{12.54}$$

To the same order of approximation we have $\boldsymbol{F}^* \simeq \boldsymbol{1}$, and either of (6.140) or (6.145) furnishes the estimate

$$\boldsymbol{T} \simeq \boldsymbol{1}\boldsymbol{S}\boldsymbol{1}^t \tag{12.55}$$

of the Cauchy stress. In terms of components, $P_{iA} \simeq \delta_{iB}S_{BA}$ and $T_{ij} \simeq \delta_{iA}\delta_{jB}S_{AB}$.

In the case of vanishing residual stress, (12.15) and (12.52) combine to give

$$\boldsymbol{T} \simeq \mathcal{C}[grad\boldsymbol{u}], \quad \text{where } \mathcal{C}[grad\boldsymbol{u}] = \mathbf{1}\{\mathcal{D}[\mathbf{1}^t(grad\boldsymbol{u})\mathbf{1}]\}\mathbf{1}^t. \quad (12.56)$$

This rather bewildering result is perhaps best expressed in terms of components. Thus,

$$\begin{aligned} T_{ij} &= \boldsymbol{e}_i \cdot \boldsymbol{T}\boldsymbol{e}_j = \mathbf{1}^t\boldsymbol{e}_i \cdot (\mathcal{D}[u_{k,l}\delta_{kC}\delta_{lD}\hat{\boldsymbol{E}}_C \otimes \hat{\boldsymbol{E}}_D])\mathbf{1}^t\boldsymbol{e}_j \\ &= \delta_{iA}\delta_{jB}\hat{\boldsymbol{E}}_A \cdot (\mathcal{D}[\delta_{kC}\delta_{lD}u_{k,l}\hat{\boldsymbol{E}}_C \otimes \hat{\boldsymbol{E}}_D])\hat{\boldsymbol{E}}_B \\ &= \delta_{iA}\delta_{jB}\delta_{kC}\delta_{lD}u_{k,l}\hat{\boldsymbol{E}}_A \cdot (\mathcal{D}[\hat{\boldsymbol{E}}_C \otimes \hat{\boldsymbol{E}}_D])\hat{\boldsymbol{E}}_B, \quad \text{by linearity} \\ &= \mathcal{C}_{ijkl}u_{k,l}, \end{aligned} \quad (12.57)$$

where

$$\mathcal{C}_{ijkl} = \delta_{iA}\delta_{jB}\delta_{kC}\delta_{lD}\mathcal{D}_{ABCD}. \quad (12.58)$$

It is straightforward to show that $\mathcal{C}$ enjoys the same symmetries as $\mathcal{D}$; namely, the major symmetry

$$\mathcal{C}_{ijkl} = \mathcal{C}_{klij}, \quad (12.59)$$

and the two minor symmetries

$$\mathcal{C}_{ijkl} = \mathcal{C}_{jikl} \quad \text{and} \quad \mathcal{C}_{ijkl} = \mathcal{C}_{ijlk}. \quad (12.60)$$

the last of these implying that $T_{ij} = \mathcal{C}_{ijkl}\varepsilon_{kl}$, where ε_{ij} are the components of the spatial infinitesimal strain tensor. In the case of isotropy, we may proceed directly from (12.15) and (12.53) to obtain

$$\boldsymbol{T} \simeq \lambda(div\boldsymbol{u})\boldsymbol{I} + \mu[grad\boldsymbol{u} + (grad\boldsymbol{u})^t]. \quad (12.61)$$

The Piola traction is $\boldsymbol{p} = \boldsymbol{P}\boldsymbol{N}$ and the Cauchy traction is $\boldsymbol{t} = \boldsymbol{T}\boldsymbol{n}$, where $\boldsymbol{N}$ and $\boldsymbol{n}$, respectively, are the unit normals to a material surface in κ and its image in κ_t (see (6.94) and (6.132)). Using the leading-order estimate

$$\boldsymbol{n}da \simeq \mathbf{1}\boldsymbol{N}dA, \quad (12.62)$$

which follows from the Piola–Nanson formula (2.54), we conclude that $da \simeq dA$ and $\boldsymbol{n} \simeq \mathbf{1}\boldsymbol{N}$. Then,

$$\boldsymbol{p} = \boldsymbol{PN} = \boldsymbol{TF}^*\boldsymbol{N} \simeq \boldsymbol{T}\mathbf{1}\boldsymbol{N} \simeq \boldsymbol{Tn} = \boldsymbol{t}, \tag{12.63}$$

i.e., the two tractions coincide at leading order. From (12.54) we see that these estimates are equivalent to

$$\mathbf{1}^t\boldsymbol{p} \simeq \boldsymbol{SN}. \tag{12.64}$$

The counterparts of the foregoing approximations in the frame of $\mathcal{O}^+$ are obtained with the aid of the transformation formulas developed in Chapter 7. Thus,

$$\boldsymbol{S}^+ \simeq \mathcal{D}^+[\boldsymbol{G}^+], \quad \boldsymbol{T}^+ \simeq \mathbf{1}^+\boldsymbol{S}^+(\mathbf{1}^+)^t \simeq \mathcal{C}^+[grad^+\boldsymbol{u}^+]$$
$$\text{and} \quad (\mathbf{1}^+)^t\boldsymbol{p}^+ \simeq \boldsymbol{S}^+\boldsymbol{N}^+, \tag{12.65}$$

where

$$\mathcal{D}^+[\boldsymbol{G}^+] = \boldsymbol{K}(\mathcal{D}[\boldsymbol{K}^t\boldsymbol{G}^+\boldsymbol{K}])\boldsymbol{K}^t$$
$$\text{and} \quad \mathcal{C}^+[grad^+\boldsymbol{u}^+] = \boldsymbol{Q}\{\mathcal{C}[\boldsymbol{Q}^t(grad^+\boldsymbol{u}^+)\boldsymbol{Q}]\}\boldsymbol{Q}^t. \tag{12.66}$$

In the special case of isotropy the latter furnish

$$\mathcal{D}^+[\boldsymbol{G}^+] = \lambda(tr\boldsymbol{G}^+)\hat{\boldsymbol{I}}^+ + \mu[\boldsymbol{G}^+ + (\boldsymbol{G}^+)^t] \quad \text{and}$$
$$\mathcal{C}^+[grad^+\boldsymbol{u}^+] = \lambda(div^+\boldsymbol{u}^+)\boldsymbol{I}^+ + \mu[grad^+\boldsymbol{u}^+ + \left(grad^+\boldsymbol{u}^+\right)^t], \tag{12.67}$$

respectively.

12.3 Equation of motion

Assuming the frame of $\mathcal{O}$ to be inertial, the referential equation of motion is given by (6.136) in which $\boldsymbol{P} = \boldsymbol{FS}$ and

$$\dot{\boldsymbol{v}} = \ddot{\chi}_\kappa = (\boldsymbol{u} + \mathbf{1}\boldsymbol{X})^{\cdot\cdot} = \ddot{\boldsymbol{u}}, \tag{12.68}$$

on recalling, from the discussion in Chapter 7, that $\mathbf{1}$ is fixed. The divergence of the Piola stress appearing in this equation is

$Div\boldsymbol{P} = P_{iA,A}\boldsymbol{e}_i$, where

$$P_{iA,A} = (F_{iB}S_{BA})_{,A} = F_{iB}S_{BA,A} + \delta_{iC}G_{CB,A}S_{BA}$$
$$\simeq \delta_{iB}S_{BA,A} + \delta_{iC}G_{CB,A}S_{BA}. \tag{12.69}$$

In the absence of residual stress (12.52) implies that the non-dimensionalized stresses S_{BA} are of order $O(\varepsilon)$, where ε is the small number representing the maximum of the norm of the displacement gradient in the body (see (12.1)). If we make the further assumption that $G_{CB,A}$, suitably non-dimensionalized by a length scale in the problem at hand, is also of order $O(\varepsilon)$, then $P_{iA,A} \simeq \delta_{iB}S_{BA,A}$ at leading order, i.e.,

$$Div\boldsymbol{P} \simeq \mathbf{1}(Div\boldsymbol{S}). \tag{12.70}$$

The restriction on the derivatives $G_{CB,A}$ leading to this result is tacitly made throughout the literature on linear elasticity theory (see [23], for example). The paper by Carlson [24] may be consulted for an interesting discussion of its implications.

With these estimates, we may approximate the equation of motion in the frame of $\mathcal{O}$ by

$$Div\boldsymbol{S} + \rho_\kappa \boldsymbol{f} = \rho_\kappa \ddot{\boldsymbol{w}}, \tag{12.71}$$

where $\boldsymbol{f} = \mathbf{1}^t\boldsymbol{b}$ is the referential form of the body force, $\ddot{\boldsymbol{w}} = (\mathbf{1}^t\boldsymbol{u})^{\cdot\cdot} = \mathbf{1}^t\ddot{\boldsymbol{u}}$ is the referential acceleration, and

$$\boldsymbol{S} = \mathcal{D}[Grad\boldsymbol{w}]. \tag{12.72}$$

This is a linear second-order partial differential system for the determination of $\boldsymbol{w}(\boldsymbol{X}, t)$. In typical initial-boundary-value problems this system is augmented by initial conditions $\boldsymbol{w}_{|t=0}$ and $\dot{\boldsymbol{w}}_{|t=0}$, together with boundary conditions specifying $\mathbf{1}^t\boldsymbol{p}$ on a part $\partial\kappa_p$ of the boundary $\partial\kappa$ and $\boldsymbol{w}$ on the complementary part $\partial\kappa_w = \partial\kappa \setminus \partial\kappa_p$.

In principle, once a solution to this problem has been obtained, it should be checked to ensure that our assumptions regarding the norm of the displacement gradient, and its gradient, are verified.

It is a simple matter to show that in the frame of $\mathcal{O}^+$, $Div^+\boldsymbol{S}^+$ is given by

$$Div^+\boldsymbol{S}^+ = \boldsymbol{K}(Div\boldsymbol{S}), \tag{12.73}$$

and hence, with the aid of $(7.38)_3$, that

$$Div^+ \boldsymbol{S}^+ + \rho^+_{\kappa^+} \boldsymbol{f}^+ = \rho^+_{\kappa^+} \boldsymbol{K}\ddot{\boldsymbol{w}}, \tag{12.74}$$

with

$$\boldsymbol{f}^+ = (\boldsymbol{1}^+)^t \boldsymbol{b}^+ \quad \text{and} \quad \boldsymbol{S}^+ = \mathcal{D}^+[Grad^+ \boldsymbol{w}^+], \quad \text{where } \boldsymbol{w}^+ = (\boldsymbol{1}^+)^t \boldsymbol{u}^+. \tag{12.75}$$

If the frame of $\mathcal{O}^+$ is also inertial, then it is related to the frame of $\mathcal{O}$ by a Galilean transformation of the form (7.10) in which $\boldsymbol{Q}$ is a fixed orthogonal tensor and $\boldsymbol{V}$ is a fixed vector. In this case it is possible to show that $\boldsymbol{K}\ddot{\boldsymbol{w}} = \ddot{\boldsymbol{w}}^+$, and hence that

$$Div^+ \boldsymbol{S}^+ + \rho^+_{\kappa^+} \boldsymbol{f}^+ = \rho^+_{\kappa^+} \ddot{\boldsymbol{w}}^+. \tag{12.76}$$

Thus linear elasticity theory also holds in the frame of $\mathcal{O}^+$. We leave the derivation of the equations holding in a non-inertial frame to the interested reader.

Finally, while the shifter plays an essential role in establishing the frame invariance of the theory [25], in any particular problem it may be effectively avoided. For example, $\mathcal{O}$ is always free to choose shifted basis vectors $\boldsymbol{E}_A = \boldsymbol{1}\hat{\boldsymbol{E}}_A \in E^3$ (see (2.68)) that coincide with the basis vectors $\boldsymbol{e}_i \in E^3$. In this case the components δ_{iA} are numerically equal to the components of the Kronecker delta, and the components of the various spatial and referential variables discussed in this chapter are then indistinguishable.

12.4 Problems

1. Derive Eqs. (12.9)–(12.11). Show that $|\boldsymbol{G}^+| = |\boldsymbol{G}|$. What is the relationship between $\boldsymbol{w}^+$ and $\boldsymbol{w}$?
2. Show that in the purely mechanical theory, $\frac{d}{dt}\mathcal{E}(S,t) = \mathcal{P}(S,t)$, where $\mathcal{P}$ is the power of the forces acting on $S \subset B$ and

$$\mathcal{E} = \int_P U dV + \mathcal{K}(S,t),$$

where $\mathcal{K}$ is the kinetic energy and $P = \boldsymbol{\kappa}(S)$.

3. Suppose the residually stressed undeformed configuration of an isotropic material to be in equilibrium without body force, and that the associated traction vanishes on a portion of the boundary. Prove that the residual stress vanishes everywhere in the body.
4. Show that

$$\boldsymbol{F}^* - \mathbf{1} \simeq \mathbf{1}[(tr\boldsymbol{G})\hat{\boldsymbol{I}} - \boldsymbol{G}^t]$$

   and hence that $\boldsymbol{F}^* \simeq \mathbf{1}$, as claimed in the text. Use this to confirm that $\boldsymbol{u} \cdot \boldsymbol{n} da \simeq \boldsymbol{w} \cdot \boldsymbol{N} dA$ and hence that the right-hand sides of Eqs. (12.26) and (12.27) agree at leading order.
5. Verify Eqs. (12.59) and (12.60).
6. Establish the fact that in the absence of residual stress, the strain energy U is approximated at leading order by $\frac{1}{2}\boldsymbol{\epsilon}\cdot\mathcal{D}[\boldsymbol{\epsilon}]$, or, equivalently, by $\frac{1}{2}\boldsymbol{\varepsilon}\cdot\mathcal{C}[\boldsymbol{\varepsilon}]$. Thus, demonstrate that $\boldsymbol{S}$ and $\boldsymbol{T}$, respectively, are approximated by the gradients $U_{\boldsymbol{\epsilon}}$ and $U_{\boldsymbol{\varepsilon}}$.
7. Find expressions for the components $\mathcal{C}_{ijkl}$ and $\mathcal{D}_{ABCD}$ for isotropic solids. In the theory of linearly elastic solids it is always assumed that the strain-energy function is positive definite, i.e., that $U \geq 0$, with equality holding if and only if the strain vanishes. Show, for isotropic solids, that this condition is satisfied if and only if $\mu > 0$ and $\kappa > 0$, where $\kappa = \lambda + \frac{2}{3}\mu$.
8. Show that Cauchy's equation of motion in an inertial frame is approximated, at leading order in the small displacement gradient, by

$$div\boldsymbol{T} + \rho_\kappa \boldsymbol{b} = \rho_\kappa \ddot{\boldsymbol{u}}.$$

9. Verify (12.73). Show that $\ddot{\boldsymbol{w}} \to \ddot{\boldsymbol{w}}^+ = \boldsymbol{K}\ddot{\boldsymbol{w}}$ under a Galilean transformation and hence that (12.76) holds if the frame of $\mathcal{O}^+$ is inertial.
10. How is (12.76) modified if the frame of $\mathcal{O}^+$ is not inertial?

Appendix A

Some Facts About Vector and Tensor Algebra

1. Non-zero vectors $\boldsymbol{u}$, $\boldsymbol{v}$ *are linearly dependent if and only if* $\boldsymbol{u} \times \boldsymbol{v} = \mathbf{0}$.

Proof. We have $\boldsymbol{u} \neq \mathbf{0}$ *and* $\boldsymbol{v} \neq \mathbf{0}$. Use the result of Problem 11(a) in Chapter 1. Multiply the equation involving $\boldsymbol{e}$ by $|\boldsymbol{u}|^2$ to get

$$|\boldsymbol{u}|^2 \boldsymbol{v} = (\boldsymbol{v} \cdot \boldsymbol{u})\boldsymbol{u} + \boldsymbol{u} \times (\boldsymbol{v} \times \boldsymbol{u}) \tag{A.1}$$

where $\boldsymbol{u} = |\boldsymbol{u}|\boldsymbol{e}$. Suppose $\boldsymbol{v} \times \boldsymbol{u} = \mathbf{0}$. Then $|\boldsymbol{u}|^2 \boldsymbol{v} = (\boldsymbol{v} \cdot \boldsymbol{u})\boldsymbol{u}$, and dotting with $\boldsymbol{v}$ gives $|\boldsymbol{u}|^2|\boldsymbol{v}|^2 = (\boldsymbol{u} \cdot \boldsymbol{v})^2$, or $\boldsymbol{u} \cdot \boldsymbol{v} = \pm|\boldsymbol{u}||\boldsymbol{v}|$. Substitute back to get $|\boldsymbol{u}|\boldsymbol{v} = \pm|\boldsymbol{v}|\boldsymbol{u}$, so $\boldsymbol{u}$, $\boldsymbol{v}$ are linearly dependent, i.e., there are non-zero $\gamma, \delta \in \mathbb{R}$ such that $\gamma\boldsymbol{u} + \delta\boldsymbol{v} = \mathbf{0}$. Conversely, suppose $\boldsymbol{u}, \boldsymbol{v}$ are linearly dependent. Then $\gamma\boldsymbol{u} + \delta\boldsymbol{v} = \mathbf{0}$ with δ or γ nonzero. Taking cross products, we conclude that $\boldsymbol{u} \times \boldsymbol{v} = \mathbf{0}$. It follows that $\boldsymbol{u}$ and $\boldsymbol{v}$ are linearly dependent if and only if $\boldsymbol{u} \times \boldsymbol{v} = \mathbf{0}$.

2. Three vectors $\boldsymbol{a}$, $\boldsymbol{b}$, *and* $\boldsymbol{c}$ *are linearly independent (in* E^3*) if and only if* $[\boldsymbol{a}, \boldsymbol{b}, \boldsymbol{c}] \neq 0$.

Proof.

1. Let $\boldsymbol{a}$, $\boldsymbol{b}$, and $\boldsymbol{c}$ be linearly dependent. Then there are real numbers α, β, and γ, at least one of which is nonzero, such that $\alpha\boldsymbol{a} + \beta\boldsymbol{b} + \gamma\boldsymbol{c} = \mathbf{0}$. Then the three box products $[\alpha\boldsymbol{a} + \beta\boldsymbol{b} + \gamma\boldsymbol{c}, \boldsymbol{b}, \boldsymbol{c}]$, $[\boldsymbol{a}, \alpha\boldsymbol{a} + \beta\boldsymbol{b} + \gamma\boldsymbol{c}, \boldsymbol{c}]$, and $[\boldsymbol{a}, \boldsymbol{b}, \alpha\boldsymbol{a} + \beta\boldsymbol{b} + \gamma\boldsymbol{c}]$ are all zero. The linearity

of the box operation with respect to each argument individually allows us to reduce these to $(\alpha, \beta, \gamma)[\boldsymbol{a}, \boldsymbol{b}, \boldsymbol{c}]$, respectively. Since at least one of the coefficients is not zero, we have $[\boldsymbol{a}, \boldsymbol{b}, \boldsymbol{c}] = 0$.

Consequently, if $[\boldsymbol{a}, \boldsymbol{b}, \boldsymbol{c}] \neq 0$, then $\boldsymbol{a}$, $\boldsymbol{b}$, and $\boldsymbol{c}$ are linearly independent.

2. Let $[\boldsymbol{a}, \boldsymbol{b}, \boldsymbol{c}] = 0$. Then there are two possibilities:
   (a) $\boldsymbol{b} \times \boldsymbol{c} \neq \boldsymbol{0}$, so $\boldsymbol{b}$ and $\boldsymbol{c}$ are linearly independent by *Statement 1*. Then either: $\boldsymbol{a} = \boldsymbol{0}$, in which case $\alpha\boldsymbol{a} + 0\boldsymbol{b} + 0\boldsymbol{c} = \boldsymbol{0}$ for any α and $\boldsymbol{a}$, $\boldsymbol{b}$, $\boldsymbol{c}$ are linearly dependent, or $\boldsymbol{a} \neq \boldsymbol{0}$ and $\boldsymbol{a}$ is orthogonal to the one-dimensional vector space spanned by $\boldsymbol{b} \times \boldsymbol{c}$. It thus belongs to the two-dimensional orthogonal complement $E^2 = Span\{\boldsymbol{b}, \boldsymbol{c}\}$. Since any set of $n + 1$ vectors in E^n is linearly dependent, the same is true of $\boldsymbol{a}, \boldsymbol{b}, \boldsymbol{c}$.
   (b) $\boldsymbol{b} \times \boldsymbol{c} = \boldsymbol{0}$. Then (i) either $\boldsymbol{b} = \boldsymbol{0}$ or $\boldsymbol{c} = \boldsymbol{0}$; the first yields $0\boldsymbol{a} + \beta\boldsymbol{b} + 0\boldsymbol{c} = \boldsymbol{0}$ for any β and the second possibility is similar, so $\boldsymbol{a}, \boldsymbol{b}, \boldsymbol{c}$ are linearly dependent; or, (ii) $\boldsymbol{b} \neq \boldsymbol{0}$ *and* $\boldsymbol{c} \neq \boldsymbol{0}$. By *Statement 1*, there are non-zero numbers β and γ such that $\beta\boldsymbol{b} + \gamma\boldsymbol{c} = \boldsymbol{0}$, so $0\boldsymbol{a} + \beta\boldsymbol{b} + \gamma\boldsymbol{c} = \boldsymbol{0}$ and $\boldsymbol{a}, \boldsymbol{b}, \boldsymbol{c}$ are linearly dependent.

Summarizing, $[\boldsymbol{a}, \boldsymbol{b}, \boldsymbol{c}] = 0$ if and only if $\boldsymbol{a}$, $\boldsymbol{b}$, and $\boldsymbol{c}$ are linearly dependent, and so $\boldsymbol{a}$, $\boldsymbol{b}$, and $\boldsymbol{c}$ are linearly independent if and only if $[\boldsymbol{a}, \boldsymbol{b}, \boldsymbol{c}] \neq 0$.

3. *A tensor* $\boldsymbol{A}$ *is invertible if and only if* $\det \boldsymbol{A} \neq 0$.

Proof. Recall that invertibility of $\boldsymbol{A}$ means that the function $\boldsymbol{f}(\boldsymbol{a}) = \boldsymbol{A}\boldsymbol{a}$ is one-to-one. In other words, invertibility means that if $\boldsymbol{A}\boldsymbol{a}_1 = \boldsymbol{b}_1$ and $\boldsymbol{A}\boldsymbol{a}_2 = \boldsymbol{b}_2$, then $\boldsymbol{b}_1 = \boldsymbol{b}_2 \Leftrightarrow \boldsymbol{a}_1 = \boldsymbol{a}_2$. Since $\boldsymbol{A}\boldsymbol{0} = \boldsymbol{0}$, these yield the implication: $\boldsymbol{A}\boldsymbol{a} = \boldsymbol{0} \Leftrightarrow \boldsymbol{a} = \boldsymbol{0}$. Thus, $\boldsymbol{A}$ is *not* invertible if and only if $\boldsymbol{A}\boldsymbol{a} = \boldsymbol{0}$ has a *non-zero* solution $\boldsymbol{a} \neq \boldsymbol{0}$, and so it is enough to show that $\boldsymbol{A}\boldsymbol{a} = \boldsymbol{0}$ for some $\boldsymbol{a} \neq \boldsymbol{0}$ if and only if $\det \boldsymbol{A} = 0$.

1. Suppose $\boldsymbol{A}\boldsymbol{a} = \boldsymbol{0}$ and $\boldsymbol{a} \neq \boldsymbol{0}$. Let $\boldsymbol{a}, \boldsymbol{b}, \boldsymbol{c}$ be linearly independent. Then $[\boldsymbol{a}, \boldsymbol{b}, \boldsymbol{c}] \neq 0$ by *Statement 2* above, and $\det \boldsymbol{A} = \frac{[\boldsymbol{0}, \boldsymbol{A}\boldsymbol{b}, \boldsymbol{A}\boldsymbol{c}]}{[\boldsymbol{a}, \boldsymbol{b}, \boldsymbol{c}]} = 0$.
2. Suppose $\det \boldsymbol{A} = 0$ and let $\boldsymbol{a}, \boldsymbol{b}, \boldsymbol{c}$ be linearly independent. Then $[\boldsymbol{A}\boldsymbol{a}, \boldsymbol{A}\boldsymbol{b}, \boldsymbol{A}\boldsymbol{c}] = 0$ so $\boldsymbol{A}\boldsymbol{a}$, $\boldsymbol{A}\boldsymbol{b}$, and $\boldsymbol{A}\boldsymbol{c}$ are linearly dependent by *Statement 2*. Thus there are α, β, γ, at least one of which

is nonzero, such that $\mathbf{0} = \alpha \boldsymbol{A}\boldsymbol{a} + \beta \boldsymbol{A}\boldsymbol{b} + \gamma \boldsymbol{A}\boldsymbol{c} = \boldsymbol{A}\boldsymbol{d}$, where $\boldsymbol{d} = \alpha \boldsymbol{a} + \beta \boldsymbol{b} + \gamma \boldsymbol{c}$ is non-zero by the assumed linear independence. So, there is $\boldsymbol{d} \neq \mathbf{0}$ such that $\boldsymbol{A}\boldsymbol{d} = \mathbf{0}$.

4. The functions $I_{1,2,3}(\boldsymbol{A})$ are well defined.

Recall that the box-product definitions of the $I_k(\boldsymbol{A})$ involves vectors $\boldsymbol{a}, \boldsymbol{b}, \boldsymbol{c}$. Here, "well-defined" means that the I_k thus computed are independent of $\boldsymbol{a}, \boldsymbol{b}, \boldsymbol{c}$. Were it otherwise then the I_k would not be functions of $\boldsymbol{A}$ alone, i.e., they would not be intrinsic properties of the considered tensor. We prove this for $I_3 = \det \boldsymbol{A}$ and $I_1 = tr \boldsymbol{A}$. The proof for I_2 may be based on that for I_1 together with the result of Problem 22 of Chapter 1. We use $\boldsymbol{A}\boldsymbol{u} = A_{ij}\boldsymbol{e}_i(\boldsymbol{u} \cdot \boldsymbol{e}_j) = A_{ij}u_j\boldsymbol{e}_i$ and $e_{123} = 1$.

1. We have

$$I_3(\boldsymbol{A})[\boldsymbol{a}, \boldsymbol{b}, \boldsymbol{c}] = [\boldsymbol{A}\boldsymbol{a}, \boldsymbol{A}\boldsymbol{b}, \boldsymbol{A}\boldsymbol{c}] = [A_{ij}a_j\boldsymbol{e}_i, A_{kl}b_l\boldsymbol{e}_k, A_{mn}c_n\boldsymbol{e}_m] \tag{A.2}$$

and by linearity we get

$$\begin{aligned}
[A_{ij}a_j\boldsymbol{e}_i, A_{kl}b_l\boldsymbol{e}_k, A_{mn}c_n\boldsymbol{e}_m] &= A_{ij}A_{kl}A_{mn}c_nb_la_j[\boldsymbol{e}_i, \boldsymbol{e}_k, \boldsymbol{e}_m] \\
&= (e_{ikm}A_{ij}A_{kl}A_{mn})\, c_nb_la_j \\
&= Ae_{jln}c_nb_la_j \\
&= A[\boldsymbol{a}, \boldsymbol{b}, \boldsymbol{c}]
\end{aligned} \tag{A.3}$$

where $A = \det(A_{ij})$ (the matrix determinant). Let $\boldsymbol{a}, \boldsymbol{b}, \boldsymbol{c}$ be linearly independent, so that $[\boldsymbol{a}, \boldsymbol{b}, \boldsymbol{c}] \neq 0$. Then $I_3(\boldsymbol{A}) = \det \boldsymbol{A} = A$. We have shown that the determinant of a tensor that maps a vector space to itself is equal to the determinant of its matrix relative to a right-handed orthonormal basis. If we use another linearly independent set $\boldsymbol{u}, \boldsymbol{v}, \boldsymbol{w}$ in place of $\boldsymbol{a}, \boldsymbol{b}, \boldsymbol{c}$ and repeat the argument we arrive at precisely the same result. Thus $\det \boldsymbol{A}$ is independent of the vectors used in its definition. This has the important practical consequence that one may choose *any* convenient basis to compute $\det \boldsymbol{A}$ and be assured that the result obtained is valid in general.

2. We have

$$I_1(\boldsymbol{A})[\boldsymbol{a},\boldsymbol{b},\boldsymbol{c}] = [\boldsymbol{Aa},\boldsymbol{b},\boldsymbol{c}] + [\boldsymbol{a},\boldsymbol{Ab},\boldsymbol{c}] + [\boldsymbol{a},\boldsymbol{b},\boldsymbol{Ac}] = a_i b_j c_k A_{ijk} \tag{A.4}$$

where

$$A_{ijk} = [\boldsymbol{Ae}_i, \boldsymbol{e}_j, \boldsymbol{e}_k] + [\boldsymbol{e}_i, \boldsymbol{Ae}_j, \boldsymbol{e}_k] + [\boldsymbol{e}_i, \boldsymbol{e}_j, \boldsymbol{Ae}_k] \tag{A.5}$$

and we have used the linearity of the box product. Using the cyclic symmetry of the box product (it is unchanged by a cyclic permutation of the arguments) we find that $A_{ijk} = A_{kij} = A_{jki}$. The skew property of the box product (interchanging any two arguments reverses the sign) yields $A_{ijk} = -A_{jik} = A_{jki}$. We also verify that $A_{ijk} = 0$ if any two subscripts are the same. Thus we conclude that $A_{ijk} = e_{ijk}A_{123}$. Then $a_i b_j c_k A_{ijk} = [\boldsymbol{a},\boldsymbol{b},\boldsymbol{c}]A_{123}$ and choosing linearly independent $\{\boldsymbol{a},\boldsymbol{b},\boldsymbol{c}\}$ yields

$$\begin{aligned} I_1(\boldsymbol{A}) = A_{123} &= [\boldsymbol{Ae}_1, \boldsymbol{e}_2, \boldsymbol{e}_3] + [\boldsymbol{e}_1, \boldsymbol{Ae}_2, \boldsymbol{e}_3] + [\boldsymbol{e}_1, \boldsymbol{e}_2, \boldsymbol{Ae}_3] \\ &= A_{i1}[\boldsymbol{e}_i, \boldsymbol{e}_2, \boldsymbol{e}_3] + A_{j2}[\boldsymbol{e}_1, \boldsymbol{e}_j, \boldsymbol{e}_3] + A_{k3}[\boldsymbol{e}_1, \boldsymbol{e}_2, \boldsymbol{e}_k] \\ &= A_{i1}e_{i23} + A_{j2}e_{1j3} + A_{k3}e_{12k}. \end{aligned} \tag{A.6}$$

The permutation symbols take non-zero values only when all subscripts are distinct and therefore $I_1(\boldsymbol{A}) = A_{11} + A_{22} + A_{33} = A_{ii}$, the trace of the matrix $\boldsymbol{A}$ relative to the basis. For this reason we usually write $I_1(\boldsymbol{A}) = tr\boldsymbol{A}$. Repeating the argument with another linearly independent set yields the same result and so the trace of a tensor is intrinsic.

Note, however, that (A.4) does not apply to two-point tensors because the factors in the box products are then elements of different vector spaces, i.e., the scalar triple products are not defined. Thus the trace of a two-point tensor is not a meaningful concept.

Some properties of I_1:

1. (Linearity) Replacing $\boldsymbol{A}$ by $\alpha\boldsymbol{A} + \beta\boldsymbol{B}$ in the definition and using the linearity of the box product furnishes $I_1(\alpha\boldsymbol{A} + \beta\boldsymbol{B}) = \alpha I_1(\boldsymbol{A}) + \beta I_1(\boldsymbol{B})$ and so the trace is a linear scalar-valued function (this also follows directly from the foregoing matrix representation). It must therefore be of the form discussed in Problem 13 of Chapter 1. The specific function is obtained on

noting that $A_{ii} = \delta_{ij}A_{ij} = \boldsymbol{I} \cdot \boldsymbol{A}$; thus, $I_1(\boldsymbol{A}) = \boldsymbol{I} \cdot \boldsymbol{A}$. Note that $I_3(\boldsymbol{A})$ is *not* a linear function.

2. (Symmetry) $I_1(\boldsymbol{A}^t) = \boldsymbol{I} \cdot \boldsymbol{A}^t = \delta_{ij}A_{ji} = A_{ii} = \delta_{ij}A_{ij} = \boldsymbol{I} \cdot \boldsymbol{A} = I_1(\boldsymbol{A})$, even when $\boldsymbol{A}^t \neq \boldsymbol{A}$; $I_1(\boldsymbol{BA}) = B_{ij}A_{ji} = A_{ji}B_{ij} = I_1(\boldsymbol{AB})$, even when $\boldsymbol{AB} \neq \boldsymbol{BA}$.
3. $I_1(\boldsymbol{a} \otimes \boldsymbol{b}) = a_i b_i = \boldsymbol{a} \cdot \boldsymbol{b}$. Combining this with the linearity of the trace gives $tr\boldsymbol{A} = tr(A_{ij}\boldsymbol{e}_i \otimes \boldsymbol{e}_j) = A_{ij}tr(\boldsymbol{e}_i \otimes \boldsymbol{e}_j) = A_{ij}\boldsymbol{e}_i \cdot \boldsymbol{e}_j = A_{ij}\delta_{ij}$, as before. In other words, we could equally well *define* the trace to be a linear function such that $I_1(\boldsymbol{a} \otimes \boldsymbol{b}) = a_i b_i = \boldsymbol{a} \cdot \boldsymbol{b}$. This is equivalent to the box-product definition and is in fact the definition adopted in most texts.

5. *The cofactor.*

The cofactor $\boldsymbol{A}^*$ of a tensor $\boldsymbol{A}$ is the tensor defined by

$$\boldsymbol{A}^*(\boldsymbol{a} \times \boldsymbol{b}) = \boldsymbol{Aa} \times \boldsymbol{Ab}, \tag{A.7}$$

for all vectors $\boldsymbol{a}$ and $\boldsymbol{b}$ for which the right-hand side is well-defined. This is meaningful for tensors that map a vector space to itself, and also for two-point tensors that map one space to another. The cofactor is a tensor of the same type as the original tensor.

For example, if $\boldsymbol{A}$ maps E^3 to itself, then

$$\begin{aligned} \boldsymbol{A}^*(\boldsymbol{e}_i \times \boldsymbol{e}_j) &= \boldsymbol{Ae}_i \times \boldsymbol{Ae}_j \\ &= A_{ki}A_{lj}\boldsymbol{e}_k \times \boldsymbol{e}_l \\ &= A_{ki}A_{lj}e_{mkl}\boldsymbol{e}_m. \end{aligned} \tag{A.8}$$

On the left-hand side we have

$$\boldsymbol{A}^*(\boldsymbol{e}_i \times \boldsymbol{e}_j) = e_{nij}\boldsymbol{A}^*\boldsymbol{e}_n = e_{nij}A^*_{mn}\boldsymbol{e}_m. \tag{A.9}$$

Thus,

$$e_{nij}A^*_{mn} = A_{ki}A_{lj}e_{mkl}. \tag{A.10}$$

To isolate the components A^*_{ij} we multiply by e_{pij} and use $e_{pij}e_{nij} = 2\delta_{pn}$ (see (1.37)) to obtain

$$A^*_{mp} = \frac{1}{2}e_{mkl}e_{pij}A_{ki}A_{lj}. \tag{A.11}$$

Appendix B

Derivatives of Scalars with Respect to Tensors

B.1 Gradients of scalar-valued functions

The gradient of a scalar-valued function of tensors is defined in exactly the same way as for functions of vectors or positions. Let $g(\boldsymbol{A})$ be such a function and suppose it is differentiable at $\boldsymbol{A}_1$. This means that for each $\boldsymbol{A}_2$ in an open set (in the space of tensors) containing $\boldsymbol{A}_1$, there is a *linear* function $f(\boldsymbol{B})$ such that

$$g(\boldsymbol{A}_2) = g(\boldsymbol{A}_1) + f(\boldsymbol{A}_2 - \boldsymbol{A}_1) + o(|\boldsymbol{A}_2 - \boldsymbol{A}_1|). \tag{B.1}$$

From Problem 13 in Chapter 1, we know that $f(\boldsymbol{B})$ is expressible as the inner product of a unique tensor with $\boldsymbol{B}$; we call this tensor $g_{\boldsymbol{A}}(\boldsymbol{A}_1)$. Thus,

$$g(\boldsymbol{A}_2) = g(\boldsymbol{A}_1) + g_{\boldsymbol{A}}(\boldsymbol{A}_1) \cdot (\boldsymbol{A}_2 - \boldsymbol{A}_1) + o(|\boldsymbol{A}_2 - \boldsymbol{A}_1|). \tag{B.2}$$

Applying this to two-point tensors for illustrative purposes, let $\boldsymbol{A} = A_{iB}\boldsymbol{e}_i \otimes \hat{\boldsymbol{E}}_B$. Then, $\boldsymbol{A}_{1,2} = A_{iB}^{(1,2)}\boldsymbol{e}_i \otimes \hat{\boldsymbol{E}}_B$. Letting $\overline{g}(A_{jC}) = g(A_{iB}\boldsymbol{e}_i \otimes \hat{\boldsymbol{E}}_B)$, we re-write (B.2) in the form

$$\overline{g}(A_{jC}^{(2)}) = \overline{g}(A_{jC}^{(1)}) + (A_{iB}^{(2)} - A_{iB}^{(1)})\boldsymbol{e}_i \otimes \hat{\boldsymbol{E}}_B \cdot g_{\boldsymbol{A}}(\boldsymbol{A}_1) + o(|\boldsymbol{A}_2 - \boldsymbol{A}_1|). \tag{B.3}$$

This must hold for all $\boldsymbol{A}_{1,2}$. Imposing it for $\boldsymbol{A}_2 - \boldsymbol{A}_1 = A\boldsymbol{e}_1 \otimes \hat{\boldsymbol{E}}_2$, for example (i.e., $A_{iB}^{(2)} - A_{iB}^{(1)} = A\delta_{i1}\delta_{B2}$), yields

$$\overline{g}(A_{jC}^{(1)} + A\delta_{j1}\delta_{C2}) - \overline{g}(A_{jC}^{(1)}) = A\boldsymbol{e}_1 \otimes \hat{\boldsymbol{E}}_2 \cdot g_{\boldsymbol{A}}(\boldsymbol{A}_1) + o(A). \tag{B.4}$$

Dividing by A and passing to the limit, we get

$$\boldsymbol{e}_1 \cdot [g_{\boldsymbol{A}}(\boldsymbol{A}_1)]\hat{\boldsymbol{E}}_2 = \left(\frac{\partial \overline{g}}{\partial A_{12}}\right)_{|\boldsymbol{A}_1}, \tag{B.5}$$

wherein we hold fixed all components other than A_{12}. In general, we then have

$$g_{\boldsymbol{A}}(\boldsymbol{A}_1) = \frac{\partial \overline{g}}{\partial A_{iB}} \boldsymbol{e}_i \otimes \hat{\boldsymbol{E}}_B, \tag{B.6}$$

provided that all the derivatives are *independent.* This would not be true if there were any *a priori* relation among the components, as is the case, for example, with symmetric or skew tensors that map a vector space to itself.

B.2 Chain rule

Consider a curve in the set of tensors described by a differentiable function $\boldsymbol{A}(t)$ where t is a parameter in some open interval (a, b). Let $\tilde{g}(t) = g(\boldsymbol{A}(t))$. Suppose that $\tilde{g}$ is differentiable with respect to t and that is differentiable with respect to $\boldsymbol{A}$. Further, let $\boldsymbol{A}_{1,2} = \boldsymbol{A}(t_{1,2})$. Then, from (B.2),

$$\tilde{g}(t_2) = \tilde{g}(t_1) + g_{\boldsymbol{A}}(\boldsymbol{A}_1) \cdot (\boldsymbol{A}_2 - \boldsymbol{A}_1) + o(|\boldsymbol{A}_2 - \boldsymbol{A}_1|) \tag{B.7}$$

We also have

$$\boldsymbol{A}_2 - \boldsymbol{A}_1 = (t_2 - t_1)\dot{\boldsymbol{A}}(t_1) + o(t_2 - t_1), \tag{B.8}$$

and therefore,

$$|\boldsymbol{A}_2 - \boldsymbol{A}_1| = O(t_2 - t_1). \tag{B.9}$$

Thus,

$$\tilde{g}(t_2) - \tilde{g}(t_1) = (t_2 - t_1) g_{\boldsymbol{A}}(\boldsymbol{A}_1) \cdot \dot{\boldsymbol{A}}(t_1) + o(t_2 - t_1). \tag{B.10}$$

Dividing by $t_2 - t_1$ and passing to the limit, we obtain the chain rule:

$$\dot{g} = g_{\boldsymbol{A}}(\boldsymbol{A}_1) \cdot \dot{\boldsymbol{A}}. \tag{B.11}$$

B.3 Gradients of the invariants of a symmetric tensor

We need formulas for the gradients of the invariants $I_k(\boldsymbol{A})$ with respect to a symmetric tensor $\boldsymbol{A}$. The symmetry of $\boldsymbol{A}$ implies that its off-diagonal components are equal and therefore not independent. This means that a formula like (B.6) is not applicable. We therefore resort to an alternative method based on the chain rule.

Let $\boldsymbol{A}(t)$ describe a curve in Sym, and consider

$$I_1(\boldsymbol{A}) = tr\boldsymbol{A} = \boldsymbol{I} \cdot \boldsymbol{A}. \tag{B.12}$$

Then,

$$(I_1)_{\boldsymbol{A}} \cdot \dot{\boldsymbol{A}} = \dot{I}_1 = \boldsymbol{I} \cdot \dot{\boldsymbol{A}}. \tag{B.13}$$

The symmetry of $\boldsymbol{A}(t)$ implies that $\dot{\boldsymbol{A}}$ is symmetric too (the proof is an immediate consequence of the definition of the derivative). If we decompose the tensor $(I_1)_{\boldsymbol{A}}$ into the sum of symmetric and skew parts, and then form the inner product with $\dot{\boldsymbol{A}}$ we find that

$$(I_1)_{\boldsymbol{A}} \cdot \dot{\boldsymbol{A}} = Sym(I_1)_{\boldsymbol{A}} \cdot \dot{\boldsymbol{A}}, \tag{B.14}$$

where

$$2Sym\boldsymbol{A} = \boldsymbol{A} + \boldsymbol{A}^t. \tag{B.15}$$

Then (B.13) yields

$$[Sym(I_1)_{\boldsymbol{A}} - \boldsymbol{I}] \cdot \dot{\boldsymbol{A}} = 0 \tag{B.16}$$

for all symmetric $\dot{\boldsymbol{A}}$. The term in brackets is a symmetric tensor, and the condition says that it is orthogonal to every element of the set of symmetric tensors. This set is a linear space (see Problem 14 in Chapter 1). Therefore, the term in brackets must be the zero tensor, yielding

$$Sym(I_1)_{\boldsymbol{A}} = \boldsymbol{I}. \tag{B.17}$$

Note that the derivation yields no information about the skew part of $(I_1)_{\boldsymbol{A}}$, which may be arbitrary. However, because the skew part, if any, is not present in (B.14), we lose no information if we simply set this part to zero and take $(I_1)_{\boldsymbol{A}}$ to be symmetric. We do this henceforth when taking derivatives of scalars with respect to symmetric tensors. Thus, $(I_1)_{\boldsymbol{A}} = \boldsymbol{I}$.

Next, consider

$$I_2(\boldsymbol{A}) = tr\boldsymbol{A}^* = \frac{1}{2}\left[(tr\boldsymbol{A})^2 - tr(\boldsymbol{A}^2)\right] = \frac{1}{2}\left(I_1^2 - \boldsymbol{I}\cdot\boldsymbol{A}^2\right). \quad \text{(B.18)}$$

Then,

$$(I_2)_{\boldsymbol{A}} \cdot \dot{\boldsymbol{A}} = \dot{I}_2 = I_1\dot{I}_1 - \frac{1}{2}\boldsymbol{I}\cdot\left(\dot{\boldsymbol{A}}\boldsymbol{A} + \boldsymbol{A}\dot{\boldsymbol{A}}\right). \quad \text{(B.19)}$$

Using the trace definition of the inner product we can show that

$$\boldsymbol{I}\cdot(\dot{\boldsymbol{A}}\boldsymbol{A}) = \boldsymbol{I}\cdot(\boldsymbol{A}\dot{\boldsymbol{A}}) = \boldsymbol{A}\cdot\dot{\boldsymbol{A}}. \quad \text{(B.20)}$$

Thus,

$$[(I_2)_{\boldsymbol{A}} - (I_1\boldsymbol{I} - \boldsymbol{A})]\cdot\dot{\boldsymbol{A}} = 0 \quad \text{(B.21)}$$

for all symmetric $\dot{\boldsymbol{A}}$, yielding

$$(I_2)_{\boldsymbol{A}} = I_1\boldsymbol{I} - \boldsymbol{A}. \quad \text{(B.22)}$$

Finally, recall that (see Problem 2 in Chapter 4)

$$\dot{J} = \boldsymbol{F}^*\cdot\dot{\boldsymbol{F}}, \quad \text{(B.23)}$$

where $\boldsymbol{F}^*$ is the cofactor of $\boldsymbol{F}$ and $J = \det\boldsymbol{F}$. By the same reasoning, with

$$I_3(\boldsymbol{A}) = \det\boldsymbol{A} \quad \text{(B.24)}$$

we get $\dot{I}_3 = \boldsymbol{A}^*\cdot\dot{\boldsymbol{A}}$, and therefore

$$[(I_3)_{\boldsymbol{A}} - \boldsymbol{A}^*]\cdot\dot{\boldsymbol{A}} = 0, \quad \text{(B.25)}$$

yielding

$$(I_3)_{\boldsymbol{A}} = \boldsymbol{A}^*, \quad \text{(B.26)}$$

where

$$\boldsymbol{A}^* = I_3\boldsymbol{A}^{-1}, \quad \text{(B.27)}$$

provided that $\boldsymbol{A}$ is invertible.

Appendix C

Rotation Tensors and the Rodrigues Representation

If $\boldsymbol{Q}$ is orthogonal then $\boldsymbol{Q}^t\boldsymbol{Q} = \boldsymbol{I} = \boldsymbol{Q}\boldsymbol{Q}^t$ and therefore $\det \boldsymbol{Q} = \pm 1$. Also,

$$\boldsymbol{Q}^t (\boldsymbol{Q} - \boldsymbol{I}) = \boldsymbol{I} - \boldsymbol{Q}^t = -(\boldsymbol{Q} - \boldsymbol{I})^t.$$

Here, we confine attention to rotations that map a vector space to itself. Accordingly two-point rotations such as those discussed in Chapters 2 and 3 are excluded.

Taking determinants, we have

$$(\det \boldsymbol{Q}) \det (\boldsymbol{Q} - \boldsymbol{I}) = (-1)^3 \det (\boldsymbol{Q} - \boldsymbol{I}) = -\det (\boldsymbol{Q} - \boldsymbol{I}).$$

If $\boldsymbol{Q}$ is a rotation, then $\det \boldsymbol{Q} = 1$ and this reduces to $\det(\boldsymbol{Q} - \boldsymbol{I}) = 0$. It follows that $\boldsymbol{Q} - \boldsymbol{I}$ is *not* invertible, and hence that there is a non-zero vector $\boldsymbol{p}$ such that $(\boldsymbol{Q} - \boldsymbol{I})\boldsymbol{p} = \boldsymbol{0}$, i.e.,

$$\boldsymbol{Q}\boldsymbol{p} = \boldsymbol{p}.$$

The vector $\boldsymbol{p}$ is called the *axis* of the rotation $\boldsymbol{Q}$. Note that this equation is unchanged if we multiply it by $|\boldsymbol{p}|^{-1}$, so we can take $|\boldsymbol{p}| = 1$ without loss of generality.

Let $\{\boldsymbol{p}, \boldsymbol{q}, \boldsymbol{r}\}$ be a right-handed orthonormal set. Then

$$\boldsymbol{Q}\boldsymbol{I} = \boldsymbol{Q}(\boldsymbol{p}\otimes\boldsymbol{p}+\boldsymbol{q}\otimes\boldsymbol{q}+\boldsymbol{r}\otimes\boldsymbol{r}) = \boldsymbol{Q}\boldsymbol{p}\otimes\boldsymbol{p}+\boldsymbol{Q}\boldsymbol{q}\otimes\boldsymbol{q}+\boldsymbol{Q}\boldsymbol{r}\otimes\boldsymbol{r}. \quad \text{(C.1)}$$

Note that

$$0 = \boldsymbol{p} \cdot \boldsymbol{q} = \boldsymbol{Q}\boldsymbol{p} \cdot \boldsymbol{Q}\boldsymbol{q} = \boldsymbol{Q}\boldsymbol{q} \cdot \boldsymbol{p},$$

and, similarly,

$$0 = \boldsymbol{p} \cdot \boldsymbol{r} = \boldsymbol{Q}\boldsymbol{r} \cdot \boldsymbol{p}.$$

Thus,

$$\boldsymbol{Q}\boldsymbol{q}, \boldsymbol{Q}\boldsymbol{r} \in Span\{\boldsymbol{q}, \boldsymbol{r}\}. \tag{C.2}$$

Further, $\boldsymbol{Q}\boldsymbol{q} \cdot \boldsymbol{Q}\boldsymbol{r} = \boldsymbol{q} \cdot \boldsymbol{r} = 0$ and so $\{\boldsymbol{Q}\boldsymbol{p}, \boldsymbol{Q}\boldsymbol{q}, \boldsymbol{Q}\boldsymbol{r}\} = \{\boldsymbol{p}, \boldsymbol{Q}\boldsymbol{q}, \boldsymbol{Q}\boldsymbol{r}\}$ is an orthonormal set. Lastly,

$$[\boldsymbol{p}, \boldsymbol{Q}\boldsymbol{q}, \boldsymbol{Q}\boldsymbol{r}] = [\boldsymbol{Q}\boldsymbol{p}, \boldsymbol{Q}\boldsymbol{q}, \boldsymbol{Q}\boldsymbol{r}] = (\det \boldsymbol{Q})[\boldsymbol{p}, \boldsymbol{q}, \boldsymbol{r}] = [\boldsymbol{p}, \boldsymbol{q}, \boldsymbol{r}] = 1,$$

so $\{\boldsymbol{p}, \boldsymbol{Q}\boldsymbol{q}, \boldsymbol{Q}\boldsymbol{r}\}$ is right-handed, i.e.,

$$\boldsymbol{Q}\boldsymbol{r} = \boldsymbol{p} \times \boldsymbol{Q}\boldsymbol{q}. \tag{C.3}$$

Now, (C.2) implies that $\boldsymbol{Q}\boldsymbol{q} = a\boldsymbol{q} + b\boldsymbol{r}$ for some $a, b \in \mathbb{R}$, and $|\boldsymbol{Q}\boldsymbol{q}| = |\boldsymbol{q}| = 1$ implies that $a^2 + b^2 = 1$. Thus, there is $\theta \in \mathbb{R}$ such that $a = \cos\theta$ and $b = \sin\theta$, i.e.,

$$\boldsymbol{Q}\boldsymbol{q} = \cos\theta \boldsymbol{q} + \sin\theta \boldsymbol{r},$$

and (C.3) yields

$$\boldsymbol{Q}\boldsymbol{r} = -\sin\theta \boldsymbol{q} + \cos\theta \boldsymbol{r}.$$

Put these results into (C.1) to get the *Rodrigues representation formula*:

$$\boldsymbol{Q} = \boldsymbol{p} \otimes \boldsymbol{p} + \cos\theta\,(\boldsymbol{q} \otimes \boldsymbol{q} + \boldsymbol{r} \otimes \boldsymbol{r}) + \sin\theta\,(\boldsymbol{r} \otimes \boldsymbol{q} - \boldsymbol{q} \otimes \boldsymbol{r}). \tag{C.4}$$

Here, θ is the angle of rotation about the axis $\boldsymbol{p}$.

The first two terms comprise the symmetric part of the rotation, and the third is the skew part. We remark that such a decomposition is well defined only for tensors that map a vector space to itself. In particular, two-point rotations do not admit such a decomposition.

A natural question to ask is: Is the axis of rotation unique? To answer this we need to characterize all non-zero $\boldsymbol{v}$ such that $\boldsymbol{Q}\boldsymbol{v} = \boldsymbol{v}$. We will show that this requires either $\boldsymbol{Q} = \boldsymbol{I}$, in which case $\boldsymbol{v}$ is arbitrary (note that $\boldsymbol{I}$ is trivially a rotation); or, if $\boldsymbol{Q} \neq \boldsymbol{I}$ then $\boldsymbol{v}$ is parallel to $\boldsymbol{p}$, i.e., the axis is unique. Using (C.4) we write $\boldsymbol{v} = \boldsymbol{Q}\boldsymbol{v}$ in the form

$$(\boldsymbol{p}\cdot\boldsymbol{v})\boldsymbol{p} + (\boldsymbol{q}\cdot\boldsymbol{v})\boldsymbol{q} + (\boldsymbol{r}\cdot\boldsymbol{v})\boldsymbol{r} = (\boldsymbol{p}\cdot\boldsymbol{v})\boldsymbol{p} + \cos\theta\,[(\boldsymbol{q}\cdot\boldsymbol{v})\,\boldsymbol{q} + (\boldsymbol{r}\cdot\boldsymbol{v})\,\boldsymbol{r}] \\ + \sin\theta\,[(\boldsymbol{q}\cdot\boldsymbol{v})\,\boldsymbol{r} - (\boldsymbol{r}\cdot\boldsymbol{v})\,\boldsymbol{q}], \quad \text{(C.5)}$$

or

$$(\boldsymbol{q}\cdot\boldsymbol{v})\boldsymbol{q} + (\boldsymbol{r}\cdot\boldsymbol{v})\boldsymbol{r} = \cos\theta\,[(\boldsymbol{q}\cdot\boldsymbol{v})\,\boldsymbol{q} + (\boldsymbol{r}\cdot\boldsymbol{v})\,\boldsymbol{r}] \\ + \sin\theta\,[(\boldsymbol{q}\cdot\boldsymbol{v})\,\boldsymbol{r} - (\boldsymbol{r}\cdot\boldsymbol{v})\,\boldsymbol{q}]. \quad \text{(C.6)}$$

We can collect the coefficients of $\boldsymbol{q}$ and $\boldsymbol{r}$ to form the matrix equation

$$\begin{pmatrix} \cos\theta - 1 & -\sin\theta \\ \sin\theta & \cos\theta - 1 \end{pmatrix} \begin{Bmatrix} \boldsymbol{q}\cdot\boldsymbol{v} \\ \boldsymbol{r}\cdot\boldsymbol{v} \end{Bmatrix} = \begin{Bmatrix} 0 \\ 0 \end{Bmatrix}. \quad \text{(C.7)}$$

We then have $(\boldsymbol{q}\cdot\boldsymbol{v}, \boldsymbol{r}\cdot\boldsymbol{v}) \neq (0,0)$ if and only if the determinant of the matrix is zero. This determinant is $2(1-\cos\theta)$, and so $(\boldsymbol{q}\cdot\boldsymbol{v}, \boldsymbol{r}\cdot\boldsymbol{v}) \neq (0,0)$ if and only if $\cos\theta = 1$ and hence $\sin\theta = 0$. In this case (C.4) reduces to

$$\boldsymbol{Q} = \boldsymbol{p}\otimes\boldsymbol{p} + \boldsymbol{q}\otimes\boldsymbol{q} + \boldsymbol{r}\otimes\boldsymbol{r} = \boldsymbol{I}. \quad \text{(C.8)}$$

If $\cos\theta \neq 1$, i.e., if the determinant is not zero and hence $\boldsymbol{Q} \neq \boldsymbol{I}$, then we require $(\boldsymbol{q}\cdot\boldsymbol{v}, \boldsymbol{r}\cdot\boldsymbol{v}) = (0,0)$ and $\boldsymbol{v} = (\boldsymbol{v}\cdot\boldsymbol{p})\,\boldsymbol{p}$. Because $\boldsymbol{v} \neq \boldsymbol{0}$, then $\boldsymbol{v}\cdot\boldsymbol{p} \neq 0$, and we can multiply $\boldsymbol{Q}\boldsymbol{v} = \boldsymbol{v}$ by $(\boldsymbol{p}\cdot\boldsymbol{v})^{-1}$ to obtain $\boldsymbol{Q}\boldsymbol{p} = \boldsymbol{p}$. Thus, every rotation $\boldsymbol{Q} \neq \boldsymbol{I}$ has a unique axis.

More Facts:

(1) Let

$$\boldsymbol{W} = \boldsymbol{r}\otimes\boldsymbol{q} - \boldsymbol{q}\otimes\boldsymbol{r}. \quad \text{(C.9)}$$

Then $\boldsymbol{W}$ is skew and (C.4) gives $\sin\theta\boldsymbol{W} = Skw\boldsymbol{Q}$. Let $\boldsymbol{w}$ be the axial vector of $\boldsymbol{W}$, i.e., the unique vector such that $\boldsymbol{W}\boldsymbol{v} = \boldsymbol{w}\times\boldsymbol{v}$ for all $\boldsymbol{v}$. Note that $\boldsymbol{W}\boldsymbol{p} = \boldsymbol{0}$; thus, $\boldsymbol{w}\times\boldsymbol{p} = \boldsymbol{0}$ and hence $\boldsymbol{w} = \lambda\boldsymbol{p}$ for some $\lambda \in \mathbb{R}$. Further, $\boldsymbol{r} = \boldsymbol{W}\boldsymbol{q} = \boldsymbol{w}\times\boldsymbol{q} = \lambda\boldsymbol{p}\times\boldsymbol{q} = \lambda\boldsymbol{r}$, so $\lambda = 1$ and $\boldsymbol{w} = \boldsymbol{p}$, i.e., the axis of $\boldsymbol{Q}$ coincides with the axial vector of $\boldsymbol{W}$.

(2) Consider $\boldsymbol{W}^2 = \boldsymbol{W}\boldsymbol{r} \otimes \boldsymbol{q} - \boldsymbol{W}\boldsymbol{q} \otimes \boldsymbol{r} = -\boldsymbol{q} \otimes \boldsymbol{q} - \boldsymbol{r} \otimes \boldsymbol{r}$. Then $\boldsymbol{p} \otimes \boldsymbol{p} = \boldsymbol{I} - \boldsymbol{q} \otimes \boldsymbol{q} - \boldsymbol{r} \otimes \boldsymbol{r} = \boldsymbol{I} + \boldsymbol{W}^2$. Put these results into (C.4) to get

$$\boldsymbol{Q} = \boldsymbol{I} + \sin\theta \boldsymbol{W} + (1 - \cos\theta)\boldsymbol{W}^2. \tag{C.10}$$

(3) Recall the exponential function $\exp x = \sum_{n=0}^{\infty} \frac{1}{n!} x^n$ with $x^0 = 1$. We can borrow this definition to define the tensor exponential

$$\exp \boldsymbol{A} = \sum_{n=0}^{\infty} \frac{1}{n!} \boldsymbol{A}^n, \quad \text{where} \quad \boldsymbol{A}^n = \boldsymbol{A}\boldsymbol{A}^{n-1} \quad \text{and} \quad \boldsymbol{A}^0 = \boldsymbol{I}. \tag{C.11}$$

Thus,

$$\exp \boldsymbol{A} = \boldsymbol{I} + \boldsymbol{A} + \frac{1}{2!}\boldsymbol{A}^2 + \frac{1}{3!}\boldsymbol{A}^3 + \cdots \tag{C.12}$$

Consider

$$\begin{aligned}
\exp(\theta \boldsymbol{W}) &= \sum_{n=0}^{\infty} \frac{1}{n!} \theta^n \boldsymbol{W}^n \\
&= \sum_{n=0,\text{even}}^{\infty} \frac{1}{n!} \theta^n \boldsymbol{W}^n + \sum_{n=1,\text{odd}}^{\infty} \frac{1}{n!} \theta^n \boldsymbol{W}^n \\
&= \boldsymbol{I} + \sum_{n=2,\text{even}}^{\infty} \frac{1}{n!} \theta^n \boldsymbol{W}^n + \sum_{n=1,\text{odd}}^{\infty} \frac{1}{n!} \theta^n \boldsymbol{W}^n \\
&= \boldsymbol{I} + \sum_{m=1}^{\infty} \frac{1}{(2m)!} \theta^{2m} \boldsymbol{W}^{2m} \\
&\quad + \sum_{m=0}^{\infty} \frac{1}{(2m+1)!} \theta^{2m+1} \boldsymbol{W}^{2m+1}.
\end{aligned} \tag{C.13}$$

- **Claim 1:** We have

$$\boldsymbol{W}^{2m+1} = (-1)^m \boldsymbol{W}; \quad m = 1, 2, 3, \ldots. \tag{C.14}$$

Proof (by induction). We have $\boldsymbol{W}^2 = -\boldsymbol{q} \otimes \boldsymbol{q} - \boldsymbol{r} \otimes \boldsymbol{r}$. Then,

$$\boldsymbol{W}^3 = -\boldsymbol{W}\boldsymbol{q} \otimes \boldsymbol{q} - \boldsymbol{W}\boldsymbol{r} \otimes \boldsymbol{r} = -\boldsymbol{r} \otimes \boldsymbol{q} + \boldsymbol{q} \otimes \boldsymbol{r} = -\boldsymbol{W}, \tag{C.15}$$

so the claim is true for $m = 1$. Assume it is true for m. Then,

$$\begin{aligned} \boldsymbol{W}^{2(m+1)+1} &= \boldsymbol{W}^{(2m+1)+2} = \boldsymbol{W}^{(2m+1)}\boldsymbol{W}^2 = (-1)^m\boldsymbol{W}^3 \\ &= -(-1)^m\boldsymbol{W}, \end{aligned} \tag{C.16}$$

i.e., $\boldsymbol{W}^{2(m+1)+1} = (-1)^{m+1}\boldsymbol{W}$. So, the claim is true for $m+1$, hence for all $m = 1, 2, 3, \ldots$.

- **Claim 2:** We have

$$\boldsymbol{W}^{2m} = (-1)^{m+1}\boldsymbol{W}^2; \quad m = 1, 2, 3, \ldots \tag{C.17}$$

Proof (by induction). Note that this is an identity for $m = 1$. Suppose that it is true for m. Then

$$\begin{aligned} \boldsymbol{W}^{2(m+1)} &= \boldsymbol{W}^{2m}\boldsymbol{W}^2 = (-1)^{m+1}\boldsymbol{W}^4 = (-1)^{m+1}\boldsymbol{W}^3\boldsymbol{W} \\ &= -(-1)^{m+1}\boldsymbol{W}^2, \end{aligned} \tag{C.18}$$

from (C.15), i.e., $\boldsymbol{W}^{2(m+1)} = (-1)^{(m+1)+1}\boldsymbol{W}^2$, and the claim is true for $m+1$, hence for all $m = 1, 2, 3, \ldots$. Using these results we get

$$\begin{aligned} \exp(\theta\boldsymbol{W}) = \boldsymbol{I} &+ \left[\sum_{m=1}^{\infty} \frac{(-1)^{m+1}}{(2m)!}\theta^{2m}\right]\boldsymbol{W}^2 \\ &+ \left[\sum_{m=0}^{\infty} \frac{(-1)^{m}}{(2m+1)!}\theta^{2m+1}\right]\boldsymbol{W}. \end{aligned} \tag{C.19}$$

Now, by using the expansions

$$\sin\theta = \theta - \frac{1}{3!}\theta^3 + \frac{1}{5!}\theta^5 + \cdots = \sum_{m=0}^{\infty} \frac{(-1)^m}{(2m+1)!}\theta^{2m+1}, \tag{C.20}$$

$$\cos\theta = 1 - \frac{1}{2!}\theta^2 + \frac{1}{4!}\theta^4 + \cdots = 1 - \sum_{m=1}^{\infty} \frac{(-1)^{m+1}}{(2m)!}\theta^{2m} \tag{C.21}$$

we find

$$\exp(\theta\boldsymbol{W}) = \boldsymbol{I} + \sin\theta\boldsymbol{W} + (1 - \cos\theta)\boldsymbol{W}^2 = \boldsymbol{Q}. \tag{C.22}$$

We have thus shown that every rotation $\boldsymbol{Q}$ is expressible as

$$\boldsymbol{Q}(\theta) = \exp(\theta \boldsymbol{W}) = \boldsymbol{I} + \theta \boldsymbol{W} + \frac{1}{2!}\theta^2 \boldsymbol{W}^2 + \cdots, \tag{C.23}$$

where θ is the rotation angle and $\boldsymbol{W}$ is a fixed (i.e., independent of θ) skew tensor.

(4) Note that $\boldsymbol{Q}(0) = \boldsymbol{I}$ and

$$\boldsymbol{Q}'(\theta) = \boldsymbol{W} + \theta \boldsymbol{W}^2 + \cdots = \boldsymbol{W}(\boldsymbol{I} + \theta \boldsymbol{W} + \cdots) = \boldsymbol{W}\boldsymbol{Q}(\theta). \tag{C.24}$$

Thus, the solution to the initial-value problem $\boldsymbol{Q}'(\theta) = \boldsymbol{W}\boldsymbol{Q}(\theta)$ and $\boldsymbol{Q}(0) = \boldsymbol{I}$, where $\boldsymbol{W}$ is a fixed skew tensor, is $\boldsymbol{Q}(\theta) = \exp(\theta \boldsymbol{W})$. The theory of ordinary differential equations ensures that this is the unique solution.

Appendix D

Spectral Representation of a Symmetric Tensor

We proved, in Chapter 3, that if all principal values $\lambda_i \in \mathbb{R}$ of a symmetric tensor $\boldsymbol{A}$ are distinct, then

$$\boldsymbol{A} = \sum \lambda_i \boldsymbol{u}_i \otimes \boldsymbol{u}_i, \tag{D.1}$$

where the orthonormal set $\{\boldsymbol{u}_i\}$ consists of the unique (apart from a multiplicative factor ± 1) principal vectors of $\boldsymbol{A}$. What if the principal values are not distinct? We need to consider two cases:

1. $\lambda_1 \neq \lambda_2$ and $\lambda_3 = \lambda_1$ or λ_2 (any other case of two and only two coincident principal values can be reduced to this case simply by relabeling the principal values).
   We have

   $$\boldsymbol{A}\boldsymbol{u}_1 = \lambda_1 \boldsymbol{u}_1, \quad \boldsymbol{A}\boldsymbol{u}_2 = \lambda_2 \boldsymbol{u}_2 \quad \text{and} \quad \boldsymbol{u}_1 \cdot \boldsymbol{u}_2 = 0. \tag{D.2}$$

   Let $\boldsymbol{u}_3 = \boldsymbol{u}_1 \times \boldsymbol{u}_2$, so that $\{\boldsymbol{u}_i\}$ is an orthonormal set and hence a basis for E^3. Then,

   $$\boldsymbol{A}\boldsymbol{u}_3 = (\boldsymbol{u}_i \cdot \boldsymbol{A}\boldsymbol{u}_3)\boldsymbol{u}_i = (\boldsymbol{u}_3 \cdot \boldsymbol{A}^t \boldsymbol{u}_i)\boldsymbol{u}_i = (\boldsymbol{u}_3 \cdot \boldsymbol{A}\boldsymbol{u}_i)\boldsymbol{u}_i = \alpha \boldsymbol{u}_3, \tag{D.3}$$

   for some $\alpha \in \mathbb{R}$, yielding

   $$\boldsymbol{A} = \boldsymbol{A}\boldsymbol{I} \tag{D.4}$$

   $$= \boldsymbol{A}\boldsymbol{u}_i \otimes \boldsymbol{u}_i \tag{D.5}$$

   $$= \lambda_1 \boldsymbol{u}_1 \otimes \boldsymbol{u}_1 + \lambda_2 \boldsymbol{u}_2 \otimes \boldsymbol{u}_2 + \alpha \boldsymbol{u}_3 \otimes \boldsymbol{u}_3 \tag{D.6}$$

   $$= (\lambda_1 - \alpha)\, \boldsymbol{u}_1 \otimes \boldsymbol{u}_1 + (\lambda_2 - \alpha)\, \boldsymbol{u}_2 \otimes \boldsymbol{u}_2 + \alpha \boldsymbol{I}. \tag{D.7}$$

Because $\boldsymbol{A}\boldsymbol{u}_3 = \alpha \boldsymbol{u}_3$ it follows that α is a principal value of $\boldsymbol{A}$, so $\alpha \in \{\lambda_1, \lambda_2\}$ because there are only two distinct principal values. Thus, (D.6) is a special case of (D.1), with $\lambda_3 = \alpha$.

We have two sub-cases:

- $\alpha = \lambda_1$. Then,

$$\begin{aligned} \boldsymbol{A} &= (\lambda_2 - \lambda_1)\, \boldsymbol{u}_2 \otimes \boldsymbol{u}_2 + \lambda_1 \boldsymbol{I} \\ &= \lambda_2 \boldsymbol{u}_2 \otimes \boldsymbol{u}_2 + \lambda_1 (\boldsymbol{I} - \boldsymbol{u}_2 \otimes \boldsymbol{u}_2). \end{aligned} \tag{D.8}$$

  This means that $\boldsymbol{A}\boldsymbol{v} = \lambda_1 \boldsymbol{v}$ for all $\boldsymbol{v}$ such that $\boldsymbol{v} \cdot \boldsymbol{u}_2 = 0$, i.e., every vector in the plane perpendicular to $\boldsymbol{u}_2$ is a principal vector corresponding to principal value λ_1.

- $\alpha = \lambda_2$. Then,

$$\begin{aligned} \boldsymbol{A} &= (\lambda_1 - \lambda_2)\, \boldsymbol{u}_1 \otimes \boldsymbol{u}_1 + \lambda_2 \boldsymbol{I} \\ &= \lambda_1 \boldsymbol{u}_1 \otimes \boldsymbol{u}_1 + \lambda_2 (\boldsymbol{I} - \boldsymbol{u}_1 \otimes \boldsymbol{u}_1). \end{aligned} \tag{D.9}$$

  This means that $\boldsymbol{A}\boldsymbol{v} = \lambda_2 \boldsymbol{v}$ for all $\boldsymbol{v}$ such that $\boldsymbol{v} \cdot \boldsymbol{u}_1 = 0$, i.e., every vector in the plane perpendicular to $\boldsymbol{u}_1$ is a principal vector corresponding to principal value λ_2.

2. All λ_i are equal, i.e., $\lambda_i = \lambda_1$. Then there is a unit vector, call it $\boldsymbol{u}_1$, such that $\boldsymbol{A}\boldsymbol{u}_1 = \lambda_1 \boldsymbol{u}_1$. Pick any vector $\boldsymbol{u}_2$ orthogonal to $\boldsymbol{u}_1$ and use it to obtain $\boldsymbol{u}_3 = \boldsymbol{u}_1 \times \boldsymbol{u}_2$. Then $\{\boldsymbol{u}_i\}$ is an orthonormal basis for E^3, and

$$\boldsymbol{A}\boldsymbol{u}_2 = (\boldsymbol{u}_i \cdot \boldsymbol{A}\boldsymbol{u}_2)\boldsymbol{u}_i = (\boldsymbol{u}_2 \cdot \boldsymbol{A}^t \boldsymbol{u}_i)\boldsymbol{u}_i = (\boldsymbol{u}_2 \cdot \boldsymbol{A}\boldsymbol{u}_i)\boldsymbol{u}_i = \alpha \boldsymbol{u}_2 + \beta \boldsymbol{u}_3, \tag{D.10}$$

   for some $\alpha, \beta \in \mathbb{R}$. Similarly,

$$\boldsymbol{A}\boldsymbol{u}_3 = (\boldsymbol{u}_i \cdot \boldsymbol{A}\boldsymbol{u}_3)\boldsymbol{u}_i = (\boldsymbol{u}_3 \cdot \boldsymbol{A}\boldsymbol{u}_i)\boldsymbol{u}_i = \gamma \boldsymbol{u}_2 + \delta \boldsymbol{u}_3, \tag{D.11}$$

   with $\gamma, \delta \in \mathbb{R}$ and

$$\gamma = \boldsymbol{u}_2 \cdot \boldsymbol{A}\boldsymbol{u}_3 = \boldsymbol{u}_3 \cdot \boldsymbol{A}\boldsymbol{u}_2 = \beta. \tag{D.12}$$

   Then

$$\begin{aligned} \boldsymbol{A} = \boldsymbol{A}\boldsymbol{u}_i \otimes \boldsymbol{u}_i &= \lambda_1 \boldsymbol{u}_1 \otimes \boldsymbol{u}_1 + \alpha \boldsymbol{u}_2 \otimes \boldsymbol{u}_2 \\ &+ \beta (\boldsymbol{u}_2 \otimes \boldsymbol{u}_3 + \boldsymbol{u}_3 \otimes \boldsymbol{u}_2) + \delta \boldsymbol{u}_3 \otimes \boldsymbol{u}_3. \end{aligned} \tag{D.13}$$

   Now, λ_1 is a triple root of the cubic characteristic equation, i.e.,

$$0 = -(\lambda - \lambda_1)^3 = -\lambda^3 + 3\lambda_1 \lambda^2 - 3\lambda_1^2 \lambda + \lambda_1^3 \tag{D.14}$$

$$= -\lambda^3 + I_1 \lambda^2 - I_2 \lambda + I_3, \tag{D.15}$$

in which the second line is the general form of the characteristic polynomial. Because the third equality is an identity for all real λ, using (D.13), we obtain

$$3\lambda_1 = I_1 = tr\boldsymbol{A} = \lambda_1 + \alpha + \delta, \tag{D.16}$$

$$3\lambda_1^2 = I_2 = \frac{1}{2}\left[I_1^2 - tr(\boldsymbol{A}^2)\right] \tag{D.17}$$

$$= \frac{1}{2}\left[(\lambda_1 + \alpha + \delta)^2 - \left(\lambda_1^2 + \alpha^2 + \delta^2 + 2\beta^2\right)\right] \tag{D.18}$$

$$= \lambda_1(\alpha + \delta) + \alpha\delta - \beta^2, \tag{D.19}$$

$$\lambda_1^3 = I_3 = \det \boldsymbol{A} = [\boldsymbol{A}\boldsymbol{u}_1, \boldsymbol{A}\boldsymbol{u}_2, \boldsymbol{A}\boldsymbol{u}_3] \tag{D.20}$$

$$= \lambda_1 \boldsymbol{u}_1 \cdot \left(\alpha\delta - \beta^2\right)\boldsymbol{u}_1 = \lambda_1\left(\alpha\delta - \beta^2\right). \tag{D.21}$$

The first and third results yield

$$\alpha + \delta = 2\lambda_1 \quad \text{and} \quad \alpha\delta - \beta^2 = \lambda_1^2, \tag{D.22}$$

provided that $\lambda_1 \neq 0$, and the second result is then redundant as it is implied by (D.22). Combining equations (D.22), we obtain

$$(\alpha - \delta)^2 = -4\beta^2, \tag{D.23}$$

so that $\beta^2 \leq 0$. Because $\beta \in \mathbb{R}$ this requires that $\beta = 0$. Therefore, $\delta = \alpha$ and equations (D.22) yield $\delta = \alpha = \lambda_1$. The representation (D.13) reduces to

$$\boldsymbol{A} = \lambda_1 \boldsymbol{u}_i \otimes \boldsymbol{u}_i = \lambda_1 \boldsymbol{I}, \tag{D.24}$$

which is again a special case of (D.1) with $\lambda_i = \lambda_1$. Note that $\boldsymbol{A}\boldsymbol{v} = \lambda_1 \boldsymbol{v}$ in this case, for arbitrary $\boldsymbol{v}$. Every vector is then a principal vector of $\boldsymbol{A}$. In the case, $\lambda_1 = 0$, we get $\delta = -\alpha$ and $\alpha\delta - \beta^2 = 0$. These imply that $\alpha^2 + \beta^2 = 0$ and hence that $\alpha = \beta = \delta = 0$. Equation (D.13) reduces to $\boldsymbol{A} = \boldsymbol{O}$, which is again a special case of (D.1) with all $\lambda_i = 0$.

Appendix E

Solutions to Selected Problems

Chapter 1

1.

(a) Let $\boldsymbol{v} = (\boldsymbol{a} \cdot \boldsymbol{a})\boldsymbol{b} - (\boldsymbol{a} \cdot \boldsymbol{b})\boldsymbol{a}$. Then,

$$\begin{aligned} 0 \leq |\boldsymbol{v}|^2 = \boldsymbol{v} \cdot \boldsymbol{v} &= |\boldsymbol{a}|^4 |\boldsymbol{b}|^2 - 2|\boldsymbol{a}|^2 (\boldsymbol{a} \cdot \boldsymbol{b})^2 + |\boldsymbol{a}|^2 (\boldsymbol{a} \cdot \boldsymbol{b})^2 \\ &= |\boldsymbol{a}|^2 [|\boldsymbol{a}|^2 |\boldsymbol{b}|^2 - (\boldsymbol{a} \cdot \boldsymbol{b})^2], \end{aligned}$$

so that $|\boldsymbol{a}|^2 |\boldsymbol{b}|^2 \geq (\boldsymbol{a} \cdot \boldsymbol{b})^2$, and therefore

$$|\boldsymbol{a}|\,|\boldsymbol{b}| \geq |\boldsymbol{a} \cdot \boldsymbol{b}|.$$

(b)

$$\begin{aligned} |\boldsymbol{a} + \boldsymbol{b}|^2 &= (\boldsymbol{a} + \boldsymbol{b}) \cdot (\boldsymbol{a} + \boldsymbol{b}) = |\boldsymbol{a}|^2 + |\boldsymbol{b}|^2 + 2\boldsymbol{a} \cdot \boldsymbol{b} \\ &= |\boldsymbol{a}|^2 + |\boldsymbol{b}|^2 + 2|\boldsymbol{a}|\,|\boldsymbol{b}| - 2[|\boldsymbol{a}|\,|\boldsymbol{b}| - \boldsymbol{a} \cdot \boldsymbol{b}] \\ &\leq (|\boldsymbol{a}| + |\boldsymbol{b}|)^2, \end{aligned} \qquad (*)$$

from part (a). Therefore,

$$|\boldsymbol{a} + \boldsymbol{b}| \leq |\boldsymbol{a}| + |\boldsymbol{b}|.$$

(c) From (*), we get Pythagoras' formula for $\boldsymbol{a} \cdot \boldsymbol{b} = 0$: $|\boldsymbol{a} + \boldsymbol{b}|^2 = |\boldsymbol{a}|^2 + |\boldsymbol{b}|^2$.

2.

$f(\boldsymbol{v}) = |\boldsymbol{v}|$. Then, $f(\alpha\boldsymbol{v}) = |\alpha\boldsymbol{v}| = \sqrt{\alpha^2\boldsymbol{v}\cdot\boldsymbol{v}} = |\alpha|\,|\boldsymbol{v}| \neq \alpha\,|\boldsymbol{v}|$. Thus, $f(\alpha\boldsymbol{v}) \neq \alpha f(\boldsymbol{v})$, i.e., $f(\boldsymbol{v})$ is not linear.

3.

$\boldsymbol{m} = \frac{\sqrt{2}}{2}(\boldsymbol{e}_1 + \boldsymbol{e}_2)$ and we seek $\boldsymbol{A}$ such that $\boldsymbol{f}(\boldsymbol{v}) = \boldsymbol{A}\boldsymbol{v} = \boldsymbol{m}$ for all $\boldsymbol{v}$. But $\boldsymbol{f}(\alpha\boldsymbol{v}) = \boldsymbol{m} \neq \alpha\boldsymbol{f}(\boldsymbol{v})$, so $\boldsymbol{f}(\boldsymbol{v})$ is not linear and there is no such $\boldsymbol{A}$. That is, there is no tensor that maps all input vectors to a fixed output vector.

4.

$\boldsymbol{A}\boldsymbol{v} = (\boldsymbol{v}\cdot\boldsymbol{n})\boldsymbol{m} = (\boldsymbol{m}\otimes\boldsymbol{n})\boldsymbol{v}$. Thus, $\boldsymbol{A} = \boldsymbol{m}\otimes\boldsymbol{n}$ and $A_{ij} = m_i n_j$, where $m_1 = m_2 = -n_1 = n_3 = \sqrt{2}/2$ and $m_3 = n_2 = 0$. Then,

$$\boldsymbol{A} = A_{ij}\boldsymbol{e}_i\otimes\boldsymbol{e}_j = \frac{1}{2}(-\boldsymbol{e}_1\otimes\boldsymbol{e}_1 + \boldsymbol{e}_1\otimes\boldsymbol{e}_3 - \boldsymbol{e}_2\otimes\boldsymbol{e}_1 + \boldsymbol{e}_2\otimes\boldsymbol{e}_3).$$

5.

Suppose $A_{ji} = A_{ij}$. Then, $e_{ijk}A_{jk} = e_{ijk}A_{kj} = -e_{ikj}A_{kj} = -e_{ijk}A_{jk}$, so $e_{ijk}A_{jk} = 0$. Conversely, if $e_{ijk}A_{jk} = 0$, then for $i = 1, 2, 3$, we have

$$0 = e_{1jk}A_{jk} = e_{123}(A_{23} - A_{32}) = A_{23} - A_{32},$$
$$0 = e_{2jk}A_{jk} = e_{231}(A_{31} - A_{13}) = A_{31} - A_{13},$$
$$0 = e_{3jk}A_{jk} = e_{312}(A_{12} - A_{21}) = A_{12} - A_{21},$$

so that $A_{ij} = A_{ji}$.

6.

(a) $A_{ij}a_ia_j = -A_{ji}a_ia_j = -A_{ji}a_ja_i$, so that $A_{ij}a_ia_j = -A_{ij}a_ia_j$ and therefore $A_{ij}a_ia_j = 0$.

(b) $0 = T_{ij}S_{ij}$ for all S_{ij}. Pick $S_{11} \neq 0$, all other $S_{ij} = 0$. Then, $0 = T_{11}S_{11}$ and therefore $T_{11} = 0$. Do the same for S_{12}, etc. to conclude that all $T_{ij} = 0$.

(c) $0 = \boldsymbol{T}\cdot\boldsymbol{S} = T_{ij}S_{ij} = T_{ij}S_{ji}$. Pick $S_{12} = S_{21} \neq 0$, with all others equal to zero. Then, $0 = (T_{12}+T_{21})S_{12}$ and therefore $T_{21} = -T_{12}$, etc. Now, pick $S_{11} \neq 0$ and all others zero. Then, $0 = T_{11}S_{11}$ and therefore $T_{11} = 0$, etc. Thus, $T_{ji} = -T_{ij}$.

(d) $0 = T_{ij}S_{ij}$. Pick $S_{21} = -S_{12} \neq 0$ and all others zero. Then $0 = (T_{21} - T_{12})S_{21}$ and therefore $T_{21} = T_{12}$, etc. Thus, $T_{ji} = T_{ij}$.

(e) (i) $0 = A_{ij}a_ia_j = A_{11}a_1^2 + (A_{21} + A_{12})a_1a_2 + \ldots$. Pick $a_1 \neq 0$ and all other $a_i = 0$. Then, $A_{11} = 0$, etc. Pick $a_1a_2 \neq 0$ and $a_3 = 0$. Then, $A_{21} + A_{12} = 0$, etc.

(ii) $\boldsymbol{O} \cdot \boldsymbol{S} = 0_{ij}S_{ij} = \sum\sum(0)S_{ij} = 0$.

(iii) $\boldsymbol{T} \cdot \boldsymbol{S} = T_{ij}S_{ij} = T_{ij}S_{ji} = -T_{ji}S_{ji} = -T_{ij}S_{ij}$, so $T_{ij}S_{ij} = 0$.

(iv) $\boldsymbol{T} \cdot \boldsymbol{S} = T_{ij}S_{ij} = -T_{ij}S_{ji} = -T_{ji}S_{ij} = -T_{ij}S_{ij}$, so $T_{ij}S_{ij} = 0$.

7.

No. For $\boldsymbol{A} = \boldsymbol{I}$, we get $\boldsymbol{u} \times \boldsymbol{v} = 2\boldsymbol{u} \times \boldsymbol{v}$, which of course is not true.

8.

$\boldsymbol{e}_i = a_{ij}\boldsymbol{e}'_j = a_{ij}a'_{jk}\boldsymbol{e}_k$, so $a_{ij}a'_{jk} = \delta_{ik}$. Also, $\boldsymbol{e}_i \cdot \boldsymbol{e}'_k = a_{ij}\boldsymbol{e}'_j \cdot \boldsymbol{e}'_k = a_{ij}\delta_{jk} = a_{ik}$ whereas $\boldsymbol{e}_k \cdot \boldsymbol{e}'_i = a'_{ij}\boldsymbol{e}_j \cdot \boldsymbol{e}_k = a'_{ij}\delta_{jk} = a'_{ik}$. Then, $a'_{ki} = a_{ik}$ and thus $a_{ij}a_{kj} = \delta_{ik}$.

9.

$A'_{ij} = a_{ki}a_{lj}A_{kl}$, so $A'_{ii} = a_{ki}a_{li}A_{kl} = \delta_{kl}A_{kl} = A_{kk}$. Then, for $A_{ij} = u_iv_j$ we get $u'_iv'_i = u_kv_k$.

10.

From the figure, we have that $\boldsymbol{v} = \boldsymbol{v}_P + (\boldsymbol{v} \cdot \boldsymbol{n})\boldsymbol{n}$, so that $\boldsymbol{v}_P = \boldsymbol{v} - (\boldsymbol{v} \cdot \boldsymbol{n})\boldsymbol{n} = (\boldsymbol{I} - \boldsymbol{n} \otimes \boldsymbol{n})\boldsymbol{v} = \mathbb{P}_{(\boldsymbol{n})}\boldsymbol{v}$, where $\mathbb{P}_{(\boldsymbol{n})} = \boldsymbol{I} - \boldsymbol{n} \otimes \boldsymbol{n}$. You can easily verify that $\mathbb{P}^t_{(\boldsymbol{n})} = \mathbb{P}_{(\boldsymbol{n})}$ and

$$\mathbb{P}^2_{(\boldsymbol{n})} = \boldsymbol{I} - 2\boldsymbol{n} \otimes \boldsymbol{n} + (\boldsymbol{n} \cdot \boldsymbol{n})\boldsymbol{n} \otimes \boldsymbol{n} = \boldsymbol{I} - \boldsymbol{n} \otimes \boldsymbol{n} = \mathbb{P}_{(\boldsymbol{n})}.$$

The same figure indicates that

$$\boldsymbol{v}_R = \boldsymbol{v}_P - (\boldsymbol{v} \cdot \boldsymbol{n})\boldsymbol{n} = \mathbb{P}_{(\boldsymbol{n})}\boldsymbol{v} - (\boldsymbol{n} \otimes \boldsymbol{n})\boldsymbol{v} = \mathbb{R}_{(\boldsymbol{n})}\boldsymbol{v},$$

where $\mathbb{R}_{(\boldsymbol{n})} = \boldsymbol{I} - 2\boldsymbol{n} \otimes \boldsymbol{n}$. Note that $-\mathbb{R}_{(\boldsymbol{n})}\boldsymbol{v}$ is a 180° rotation of $\boldsymbol{v}$ about the unit normal $\boldsymbol{n}$.

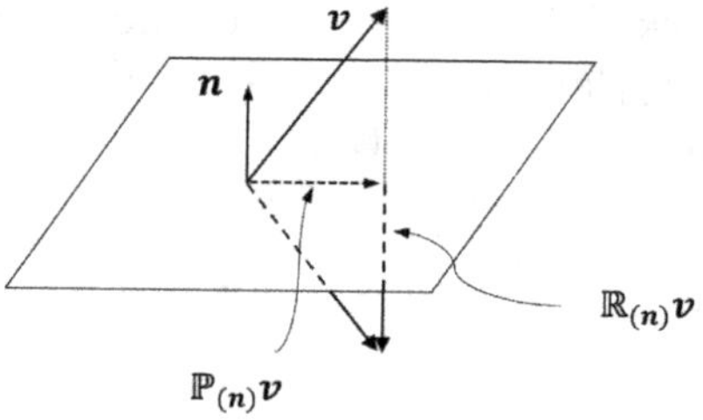

11.

(a) Let $\boldsymbol{d} = \boldsymbol{b} \times \boldsymbol{c} = b_i c_j \boldsymbol{e}_i \times \boldsymbol{e}_j = e_{ijk} b_i c_j \boldsymbol{e}_k$. Then, $\boldsymbol{a} \times (\boldsymbol{b} \times \boldsymbol{c}) = \boldsymbol{a} \times \boldsymbol{d} = a_l d_k e_{lkm} \boldsymbol{e}_m = e_{ijk} e_{mlk} a_l b_i c_j \boldsymbol{e}_m = (\delta_{im}\delta_{jl} - \delta_{il}\delta_{jm}) a_l b_i c_j \boldsymbol{e}_m = (a_j c_j) b_i \boldsymbol{e}_i - (a_i b_i) c_j \boldsymbol{e}_j$, or $\boldsymbol{a} \times (\boldsymbol{b} \times \boldsymbol{c}) = (\boldsymbol{a} \cdot \boldsymbol{c})\boldsymbol{b} - (\boldsymbol{a} \cdot \boldsymbol{b})\boldsymbol{c}$. Let $\boldsymbol{b} = \boldsymbol{v}$ and $\boldsymbol{a} = \boldsymbol{c} = \boldsymbol{e}$ with $|\boldsymbol{e}| = 1$. Then, $\boldsymbol{e} \times (\boldsymbol{v} \times \boldsymbol{e}) = \boldsymbol{v} - (\boldsymbol{e} \cdot \boldsymbol{v})\boldsymbol{e}$.

(b) $\boldsymbol{e} \times (\boldsymbol{v} \times \boldsymbol{e}) = \boldsymbol{v} - (\boldsymbol{e} \cdot \boldsymbol{v})\boldsymbol{e} = (\boldsymbol{I} - \boldsymbol{e} \otimes \boldsymbol{e})\boldsymbol{v} = \mathbb{P}_{(\boldsymbol{e})}\boldsymbol{v}$, so $\boldsymbol{A} = \mathbb{P}_{(\boldsymbol{e})}$.

(c) $\boldsymbol{f}(\boldsymbol{v}) = \boldsymbol{v} + \boldsymbol{a} \times \boldsymbol{v} = \boldsymbol{A}\boldsymbol{v}$ (linearity is trivial), where $\boldsymbol{A} = A_{ij} \boldsymbol{e}_i \otimes \boldsymbol{e}_j$ with $A_{ij} = \boldsymbol{e}_i \cdot \boldsymbol{A}\boldsymbol{e}_j = \boldsymbol{e}_i \cdot \boldsymbol{f}(\boldsymbol{e}_j) = \boldsymbol{e}_i \cdot (\boldsymbol{e}_j + \boldsymbol{a} \times \boldsymbol{e}_j) = \delta_{ij} + \boldsymbol{a} \times \boldsymbol{e}_j \cdot \boldsymbol{e}_i = \delta_{ij} + \boldsymbol{a} \cdot \boldsymbol{e}_j \times \boldsymbol{e}_i = \delta_{ij} + e_{jik} \boldsymbol{a} \cdot \boldsymbol{e}_k$. Thus, $A_{ij} = \delta_{ij} + e_{jik} a_k$.

12.

$\boldsymbol{f}(\boldsymbol{v}) = \boldsymbol{A}\boldsymbol{v}$. Let $\boldsymbol{v} = v_A \hat{\boldsymbol{E}}_A$. Then, $\boldsymbol{f}(\boldsymbol{v}) = v_A \boldsymbol{f}(\hat{\boldsymbol{E}}_A)$, by linearity. Since $\boldsymbol{f}(\hat{\boldsymbol{E}}_A) \in E^3$, we can write $\boldsymbol{f}(\hat{\boldsymbol{E}}_A) = A_{iA} \boldsymbol{e}_i$, where $A_{iA} = \boldsymbol{e}_i \cdot \boldsymbol{f}(\hat{\boldsymbol{E}}_A)$. Then, $\boldsymbol{A}\boldsymbol{v} = \boldsymbol{f}(\boldsymbol{v}) = A_{iA} v_A \boldsymbol{e}_i = A_{iA}(\hat{\boldsymbol{E}}_A \cdot \boldsymbol{v})\boldsymbol{e}_i = (A_{iA} \boldsymbol{e}_i \otimes \hat{\boldsymbol{E}}_A)\boldsymbol{v}$ and thus $\boldsymbol{A} = A_{iA} \boldsymbol{e}_i \otimes \hat{\boldsymbol{E}}_A$.

13.

$f(\boldsymbol{V}) = f(V_{ij} \boldsymbol{e}_i \otimes \boldsymbol{e}_j) = V_{ij} A_{ij}$, by linearity, where $A_{ij} = f(\boldsymbol{e}_i \otimes \boldsymbol{e}_j)$. Thus, $f(\boldsymbol{V}) = \boldsymbol{A} \cdot \boldsymbol{V}$.

14.

$\boldsymbol{A} = \boldsymbol{S} + \boldsymbol{W}$, where $\boldsymbol{S} = \frac{1}{2}(\boldsymbol{A} + \boldsymbol{A}^t)$ and $\boldsymbol{W} = \frac{1}{2}(\boldsymbol{A} - \boldsymbol{A}^t)$. Clearly, $\boldsymbol{S}^t = \frac{1}{2}(\boldsymbol{A}^t + \boldsymbol{A}) = \boldsymbol{S}$ and $\boldsymbol{W}^t = \frac{1}{2}(\boldsymbol{A}^t - \boldsymbol{A}) = -\boldsymbol{W}$, so $\boldsymbol{S} \in Sym$ and $\boldsymbol{W} \in Skw$. Let $\boldsymbol{A} = \boldsymbol{S}' + \boldsymbol{W}'$ with $\boldsymbol{S}' \in Sym$ and $\boldsymbol{W}' \in Skw$. Then, $\boldsymbol{A}^t = \boldsymbol{S}' - \boldsymbol{W}'$ and $2\boldsymbol{S} = \boldsymbol{A} + \boldsymbol{A}^t = 2\boldsymbol{S}'$ and $2\boldsymbol{W} = \boldsymbol{A} - \boldsymbol{A}^t = 2\boldsymbol{W}'$, so $\boldsymbol{S}' = \boldsymbol{S}$ and $\boldsymbol{W}' = \boldsymbol{W}$. This establishes uniqueness.

Further, $\boldsymbol{S} \cdot \boldsymbol{W} = 0$ by Problems 6(c-e), so $Sym \perp Skw$. Also, if $\boldsymbol{S}_1, \boldsymbol{S}_2 \in Sym$, then $\alpha_1 \boldsymbol{S}_1 + \alpha_2 \boldsymbol{S}_2 \in Sym$, and if $\boldsymbol{W}_1, \boldsymbol{W}_2 \in Skw$, then $\alpha_1 \boldsymbol{W}_1 + \alpha_2 \boldsymbol{W}_2 \in Skw$. Finally, $\boldsymbol{O} \in Sym$ and $\boldsymbol{O} \in Skw$, so Sym and Skw are linear spaces, and, denoting E^9 by Lin (the linear space of all tensors), it follows that $Lin = Sym \oplus Skw$.

15.

If $\boldsymbol{T} = \boldsymbol{A} + \alpha \boldsymbol{I}$ with $tr\boldsymbol{A} = 0$, then $3\alpha = \alpha tr\boldsymbol{I} = tr\boldsymbol{T}$. Let $\boldsymbol{T} = \boldsymbol{B} + \beta \boldsymbol{I}$ with $tr\boldsymbol{B} = 0$. Then, $tr\boldsymbol{T} = \beta tr\boldsymbol{I} = 3\beta$, so $\beta = \alpha$ and $\boldsymbol{B} = \boldsymbol{T} - \beta\boldsymbol{I} = \boldsymbol{T} - \alpha\boldsymbol{I} = \boldsymbol{A}$. Thus we have uniqueness. Now, if $\boldsymbol{C} \in Sph$ then $\boldsymbol{C} = \lambda \boldsymbol{I}$ for some $\lambda \in \mathbb{R}$, and if $\boldsymbol{C}_1, \boldsymbol{C}_2 \in Sph$ then $\boldsymbol{C}_1 = \lambda_1 \boldsymbol{I}$, $\boldsymbol{C}_2 = \lambda_2 \boldsymbol{I}$ and $\alpha_1 \boldsymbol{C}_1 + \alpha_2 \boldsymbol{C}_2 = (\alpha_1 \lambda_1 + \alpha_2 \lambda_2)\boldsymbol{I} \in Sph$. Since $\boldsymbol{O} \in Sph$ it then follows that Sph is a linear space.

Similarly, if $\boldsymbol{D}_1, \boldsymbol{D}_2 \in Dev$ then $tr(\alpha_1 \boldsymbol{D}_1 + \alpha_2 \boldsymbol{D}_2) = \alpha_1 tr\boldsymbol{D}_1 + \alpha_2 tr\boldsymbol{D}_2 = 0$, so $\alpha_1 \boldsymbol{D}_1 + \alpha_2 \boldsymbol{D}_2 \in Dev$. Since $\boldsymbol{O} \in Dev$ it then follows that Dev is also a linear space.

Lastly, if $\boldsymbol{C} \in Sph$ and $\boldsymbol{D} \in Dev$, then $\boldsymbol{C} \cdot \boldsymbol{D} = \lambda \boldsymbol{I} \cdot \boldsymbol{D} = \lambda tr\boldsymbol{D} = 0$, so $Sph \perp Dev$ and $Lin = Sph \oplus Dev$.

16.

We have $\boldsymbol{f}(\boldsymbol{v}) = \boldsymbol{\omega} \times \boldsymbol{v}$, which is linear. It follows from this that there is a tensor $\boldsymbol{\Omega}$ such that $\boldsymbol{\Omega v} = \boldsymbol{\omega} \times \boldsymbol{v}$ for all $\boldsymbol{v}$. Then, $\boldsymbol{\Omega e}_j = \boldsymbol{\omega} \times \boldsymbol{e}_j = \omega_i \boldsymbol{e}_i \times \boldsymbol{e}_j = \omega_i e_{ijk} \boldsymbol{e}_k$ and $\Omega_{ij} = \boldsymbol{e}_i \cdot \boldsymbol{\Omega e}_j = \omega_l e_{ljk} \delta_{ki} = e_{jik} \omega_k$ (*). Note that $\Omega_{ji} = e_{ijk}\omega_k = -e_{jik}\omega_k = -\Omega_{ij}$, so $\boldsymbol{\Omega} \in Skw$. Multiply (*) by e_{ijl} and sum, using the fact that $e_{jil} e_{jik} = 2\delta_{kl}$, to obtain $\omega_k = \frac{1}{2}\Omega_{ij} e_{jik}$.

17.

(a) $\boldsymbol{u} \cdot (\boldsymbol{a} \otimes \boldsymbol{b})^t \boldsymbol{v} = (\boldsymbol{a} \otimes \boldsymbol{b})\boldsymbol{u} \cdot \boldsymbol{v} = (\boldsymbol{b} \cdot \boldsymbol{u})(\boldsymbol{a} \cdot \boldsymbol{v}) = \boldsymbol{u} \cdot (\boldsymbol{b} \otimes \boldsymbol{a})\boldsymbol{v}$ for all $\boldsymbol{u}, \boldsymbol{v}$. Thus, $(\boldsymbol{a} \otimes \boldsymbol{b})^t \boldsymbol{v} = (\boldsymbol{b} \otimes \boldsymbol{a})\boldsymbol{v}$ for all $\boldsymbol{v}$ and therefore $(\boldsymbol{a} \otimes \boldsymbol{b})^t = \boldsymbol{b} \otimes \boldsymbol{a}$.

(b) $[\boldsymbol{A}(\boldsymbol{a} \otimes \boldsymbol{b})]\boldsymbol{v} = \boldsymbol{A}[(\boldsymbol{a} \otimes \boldsymbol{b})\boldsymbol{v}] = \boldsymbol{A a}(\boldsymbol{b} \cdot \boldsymbol{v}) = (\boldsymbol{A a} \otimes \boldsymbol{b})\boldsymbol{v}$, and therefore $\boldsymbol{A}(\boldsymbol{a} \otimes \boldsymbol{b}) = \boldsymbol{A a} \otimes \boldsymbol{b}$.

(c) $[(\boldsymbol{a} \otimes \boldsymbol{b})\boldsymbol{A}]\boldsymbol{v} = (\boldsymbol{a} \otimes \boldsymbol{b})(\boldsymbol{A v}) = \boldsymbol{a}(\boldsymbol{b} \cdot \boldsymbol{A v}) = \boldsymbol{a}(\boldsymbol{A}^t \boldsymbol{b} \cdot \boldsymbol{v}) = (\boldsymbol{a} \otimes \boldsymbol{A}^t \boldsymbol{b})\boldsymbol{v}$, so $(\boldsymbol{a} \otimes \boldsymbol{b})\boldsymbol{A} = \boldsymbol{a} \otimes \boldsymbol{A}^t \boldsymbol{b}$.

(d) $\boldsymbol{A} = \boldsymbol{A I} = \boldsymbol{A}(\boldsymbol{e}_i \otimes \boldsymbol{e}_i) = \boldsymbol{A e}_i \otimes \boldsymbol{e}_i$, by (b). Also, $\boldsymbol{A} = \boldsymbol{I A} = (\boldsymbol{e}_i \otimes \boldsymbol{e}_i)\boldsymbol{A} = \boldsymbol{e}_i \otimes \boldsymbol{A}^t \boldsymbol{e}_i$, by (c).

18.

$\boldsymbol{A}^{-1}\boldsymbol{A} = \boldsymbol{I}$. From the fact that $\boldsymbol{I} \in Sym$ we have $\boldsymbol{I} = \boldsymbol{I}^t = (\boldsymbol{A}^{-1}\boldsymbol{A})^t = \boldsymbol{A}^t(\boldsymbol{A}^{-1})^t$ (see Problem 20). Thus, $(\boldsymbol{A}^{-1})^t = (\boldsymbol{A}^t)^{-1}$.

19.

$(\boldsymbol{BA})^{-1}(\boldsymbol{BA}) = \boldsymbol{I}$. But $\boldsymbol{A}^{-1}\boldsymbol{B}^{-1}(\boldsymbol{BA}) = \boldsymbol{A}^{-1}(\boldsymbol{B}^{-1}\boldsymbol{B})\boldsymbol{A} = \boldsymbol{A}^{-1}\boldsymbol{A} = \boldsymbol{I}$, so $(\boldsymbol{BA})^{-1} = \boldsymbol{A}^{-1}\boldsymbol{B}^{-1}$.

20.

$\boldsymbol{u}\cdot(\boldsymbol{BA})^t\boldsymbol{v} = (\boldsymbol{BA})\boldsymbol{u}\cdot\boldsymbol{v}$, and $\boldsymbol{u}\cdot(\boldsymbol{A}^t\boldsymbol{B}^t)\boldsymbol{v} = \boldsymbol{Au}\cdot\boldsymbol{B}^t\boldsymbol{v} = \boldsymbol{B}(\boldsymbol{Au})\cdot\boldsymbol{v} = (\boldsymbol{BA})\boldsymbol{u}\cdot\boldsymbol{v}$, so $\boldsymbol{u}\cdot(\boldsymbol{BA})^t\boldsymbol{v} = \boldsymbol{u}\cdot\boldsymbol{A}^t\boldsymbol{B}^t\boldsymbol{v}$ for all $\boldsymbol{u},\boldsymbol{v}$. Thus, $(\boldsymbol{BA})^t\boldsymbol{v} = (\boldsymbol{A}^t\boldsymbol{B}^t)\boldsymbol{v}$ for all $\boldsymbol{v}$, and therefore $(\boldsymbol{BA})^t = \boldsymbol{A}^t\boldsymbol{B}^t$.

21.

(a) $[\boldsymbol{a},\boldsymbol{b},\boldsymbol{c}]\det(\alpha\boldsymbol{A}) = [\alpha\boldsymbol{Aa},\alpha\boldsymbol{Ab},\alpha\boldsymbol{Ac}] = \alpha\boldsymbol{Aa}\cdot\alpha\boldsymbol{Ab}\times\alpha\boldsymbol{Ac} = \alpha^3\boldsymbol{Aa}\cdot\boldsymbol{Ab}\times\boldsymbol{Ac} = \alpha^3[\boldsymbol{Aa},\boldsymbol{Ab},\boldsymbol{Ac}] = \alpha^3[\boldsymbol{a},\boldsymbol{b},\boldsymbol{c}]\det\boldsymbol{A}$, so $\det(\alpha\boldsymbol{A}) = \alpha^3\det\boldsymbol{A}$.

(b) $[\boldsymbol{a},\boldsymbol{b},\boldsymbol{c}]\det(\boldsymbol{AB}) = [\boldsymbol{A}(\boldsymbol{Ba}),\boldsymbol{A}(\boldsymbol{Bb}),\boldsymbol{A}(\boldsymbol{Bc})] = (\det\boldsymbol{A})[\boldsymbol{Ba},\boldsymbol{Bb},\boldsymbol{Bc}] = (\det\boldsymbol{A})(\det\boldsymbol{B})[\boldsymbol{a},\boldsymbol{b},\boldsymbol{c}]$, so $\det(\boldsymbol{AB}) = (\det\boldsymbol{A})(\det\boldsymbol{B})$.

(c) $[\boldsymbol{a},\boldsymbol{b},\boldsymbol{c}](\det\boldsymbol{I}) = [\boldsymbol{Ia},\boldsymbol{Ib},\boldsymbol{Ic}] = [\boldsymbol{a},\boldsymbol{b},\boldsymbol{c}]$, so $\det\boldsymbol{I} = 1$.

(d) $[\boldsymbol{a},\boldsymbol{b},\boldsymbol{c}] = [\boldsymbol{a},\boldsymbol{b},\boldsymbol{c}](\det\boldsymbol{I}) = [\boldsymbol{a},\boldsymbol{b},\boldsymbol{c}]\det(\boldsymbol{AA}^{-1}) = [\boldsymbol{a},\boldsymbol{b},\boldsymbol{c}](\det\boldsymbol{A})(\det\boldsymbol{A}^{-1})$, by (b). Thus, $\det\boldsymbol{A}^{-1} = (\det\boldsymbol{A})^{-1}$.

(e) $\det(\boldsymbol{a}\otimes\boldsymbol{b})[\boldsymbol{u},\boldsymbol{v},\boldsymbol{w}] = [\boldsymbol{a}(\boldsymbol{b}\cdot\boldsymbol{u}),\boldsymbol{a}(\boldsymbol{b}\cdot\boldsymbol{v}),\boldsymbol{a}(\boldsymbol{b}\cdot\boldsymbol{w})] = (\boldsymbol{b}\cdot\boldsymbol{u})(\boldsymbol{b}\cdot\boldsymbol{v})(\boldsymbol{b}\cdot\boldsymbol{w})[\boldsymbol{a},\boldsymbol{a},\boldsymbol{a}] = 0$, so $\det(\boldsymbol{a}\otimes\boldsymbol{b}) = 0$.

The following solutions are valid for a tensor that maps a vector space to itself.

(f) $[\boldsymbol{u},\boldsymbol{v},\boldsymbol{w}]\det(\boldsymbol{I}+\boldsymbol{a}\otimes\boldsymbol{b}) = [\boldsymbol{u}+(\boldsymbol{b}\cdot\boldsymbol{u})\boldsymbol{a},\boldsymbol{v}+(\boldsymbol{b}\cdot\boldsymbol{v})\boldsymbol{a},\boldsymbol{w}+(\boldsymbol{b}\cdot\boldsymbol{w})\boldsymbol{a}] = [\boldsymbol{u},\boldsymbol{v},\boldsymbol{w}] + (\boldsymbol{b}\cdot\boldsymbol{u})[\boldsymbol{a},\boldsymbol{v},\boldsymbol{w}] + (\boldsymbol{b}\cdot\boldsymbol{v})[\boldsymbol{u},\boldsymbol{a},\boldsymbol{w}] + (\boldsymbol{b}\cdot\boldsymbol{w})[\boldsymbol{u},\boldsymbol{v},\boldsymbol{a}]$ (*). Recalling the definition of I_1, or the trace, we also have $[\boldsymbol{u},\boldsymbol{v},\boldsymbol{w}]tr(\boldsymbol{a}\otimes\boldsymbol{b}) = [(\boldsymbol{a}\otimes\boldsymbol{b})\boldsymbol{u},\boldsymbol{v},\boldsymbol{w}] + [\boldsymbol{u},(\boldsymbol{a}\otimes\boldsymbol{b})\boldsymbol{v},\boldsymbol{w}] + [\boldsymbol{u},\boldsymbol{v},(\boldsymbol{a}\otimes\boldsymbol{b})\boldsymbol{w}] = (\boldsymbol{b}\cdot\boldsymbol{u})[\boldsymbol{a},\boldsymbol{v},\boldsymbol{w}]+(\boldsymbol{b}\cdot\boldsymbol{v})[\boldsymbol{u},\boldsymbol{a},\boldsymbol{w}]+(\boldsymbol{b}\cdot\boldsymbol{w})[\boldsymbol{u},\boldsymbol{v},\boldsymbol{a}]$. Compare with (*) to get $[\boldsymbol{u},\boldsymbol{v},\boldsymbol{w}]\det(\boldsymbol{I}+\boldsymbol{a}\otimes\boldsymbol{b}) = (1+tr(\boldsymbol{a}\otimes\boldsymbol{b}))[\boldsymbol{u},\boldsymbol{v},\boldsymbol{w}]$, so that $\det(\boldsymbol{I}+\boldsymbol{a}\otimes\boldsymbol{b}) = 1+\boldsymbol{a}\cdot\boldsymbol{b}$.

(g) $\boldsymbol{A}+\boldsymbol{a}\otimes\boldsymbol{b} = \boldsymbol{A}(\boldsymbol{I}+\boldsymbol{A}^{-1}\boldsymbol{a}\otimes\boldsymbol{b})$. Thus, $\det(\boldsymbol{A}+\boldsymbol{a}\otimes\boldsymbol{b}) = (\det\boldsymbol{A})\det(\boldsymbol{I}+\boldsymbol{A}^{-1}\boldsymbol{a}\otimes\boldsymbol{b}) = (\det\boldsymbol{A})(1+\boldsymbol{A}^{-1}\boldsymbol{a}\cdot\boldsymbol{b}) = (\det\boldsymbol{A})(1+\boldsymbol{a}\cdot\boldsymbol{A}^{-t}\boldsymbol{b})$.

We can extend this result to two-point tensors as follows. Let $\boldsymbol{a},\boldsymbol{b},\boldsymbol{c}\in\hat{E}^3$. The steps in (b) remain valid if $\boldsymbol{A}$ is an invertible two-point tensor, mapping $\hat{E}^3$ to E^3, and if $\boldsymbol{B}$ maps $\hat{E}^3$ to itself. In particular, all the scalar triple products are well defined. Now, let $\boldsymbol{a}\in E^3$ and $\boldsymbol{b}\in\hat{E}^3$, and let $\boldsymbol{B} = \hat{\boldsymbol{I}}+\boldsymbol{A}^{-1}\boldsymbol{a}\otimes\boldsymbol{b}$, where $\hat{\boldsymbol{I}}$ is the identity for $\hat{E}^3$. This is meaningful because $\boldsymbol{A}^{-1}\boldsymbol{a} \in \hat{E}^3$. Further, $\boldsymbol{A}\boldsymbol{A}^{-1} = \boldsymbol{I}$, the identity for E^3. Thus, $\boldsymbol{A} + \boldsymbol{a} \otimes \boldsymbol{b} = \boldsymbol{A}(\hat{\boldsymbol{I}} + \boldsymbol{A}^{-1}\boldsymbol{a} \otimes \boldsymbol{b})$ and the result follows on noting that $\det(\hat{\boldsymbol{I}} + \boldsymbol{A}^{-1}\boldsymbol{a} \otimes \boldsymbol{b}) = 1 + \boldsymbol{A}^{-1}\boldsymbol{a} \cdot \boldsymbol{b} = 1 + \boldsymbol{a} \cdot \boldsymbol{A}^{-t}\boldsymbol{b}$.

22.

The solutions given for (a)–(d) are valid if $\boldsymbol{A}$ maps E^3 to itself.

(a) $[\boldsymbol{a}, \boldsymbol{b}, \boldsymbol{c}]I_2(\boldsymbol{A}) = [\boldsymbol{a}, \boldsymbol{A}\boldsymbol{b}, \boldsymbol{A}\boldsymbol{c}] + [\boldsymbol{A}\boldsymbol{a}, \boldsymbol{b}, \boldsymbol{A}\boldsymbol{c}] + [\boldsymbol{A}\boldsymbol{a}, \boldsymbol{A}\boldsymbol{b}, \boldsymbol{c}] = \boldsymbol{a} \cdot \boldsymbol{A}^*(\boldsymbol{b} \times \boldsymbol{c}) + \boldsymbol{b} \cdot \boldsymbol{A}^*(\boldsymbol{c} \times \boldsymbol{a}) + \boldsymbol{c} \cdot \boldsymbol{A}^*(\boldsymbol{a} \times \boldsymbol{b}) = (\boldsymbol{A}^*)^t\boldsymbol{a} \cdot \boldsymbol{b} \times \boldsymbol{c} + (\boldsymbol{A}^*)^t\boldsymbol{b} \cdot \boldsymbol{c} \times \boldsymbol{a} + (\boldsymbol{A}^*)^t\boldsymbol{c} \cdot \boldsymbol{a} \times \boldsymbol{b} = [(\boldsymbol{A}^*)^t\boldsymbol{a}, \boldsymbol{b}, \boldsymbol{c}] + [\boldsymbol{a}, (\boldsymbol{A}^*)^t\boldsymbol{b}, \boldsymbol{c}] + [\boldsymbol{a}, \boldsymbol{b}, (\boldsymbol{A}^*)^t\boldsymbol{c}] = [\boldsymbol{a}, \boldsymbol{b}, \boldsymbol{c}]I_1((\boldsymbol{A}^*)^t)$, so $I_2(\boldsymbol{A}) = I_1((\boldsymbol{A}^*)^t) = I_1(\boldsymbol{A}^*)$ since $tr(\boldsymbol{B}^t) = tr(\boldsymbol{B})$.

(b) $A^*_{kp} = \frac{1}{2}e_{kij}e_{pmn}A_{im}A_{jn}$ (see Appendix A), so $I_2(\boldsymbol{A}) = tr\boldsymbol{A}^* = A^*_{kk} = \frac{1}{2}e_{kij}e_{kmn}A_{im}A_{jn} = \frac{1}{2}(\delta_{im}\delta_{jn} - \delta_{in}\delta_{jm})A_{im}A_{jn} = \frac{1}{2}(A_{ii}A_{jj} - A_{ij}A_{ji}) = \frac{1}{2}[(tr\boldsymbol{A})^2 - tr(\boldsymbol{A}^2)]$.

(c) $\boldsymbol{A}^t\boldsymbol{A}^* = A_{ji}A^*_{jm}\boldsymbol{e}_i \otimes \boldsymbol{e}_m = A_{ji}(\frac{1}{2}e_{jpq}e_{mrs}A_{pr}A_{qs})\boldsymbol{e}_i \otimes \boldsymbol{e}_m = \frac{1}{2}e_{mrs}(e_{irs}\det\boldsymbol{A})\boldsymbol{e}_i \otimes \boldsymbol{e}_m = \frac{1}{2}(\det\boldsymbol{A})(2\delta_{im}\boldsymbol{e}_i \otimes \boldsymbol{e}_m) = (\det\boldsymbol{A})\boldsymbol{I}$.

(d) Take the trace of the result in result in (c) to get $3\det\boldsymbol{A} = tr(\boldsymbol{A}^t\boldsymbol{A}^*) = A_{ji}A^*_{ji} = A_{ji}(\frac{1}{2}e_{jmn}e_{irs}A_{mr}A_{ns})$, so $\det\boldsymbol{A} = \frac{1}{6}e_{jmn}e_{irs}A_{ji}A_{mr}A_{ns}$.

We can extend the result of (c) to two-point tensors $\boldsymbol{A}$: $\hat{E}^3 \to E^3$ as follows. This will be needed in subsequent chapters. Thus, let $\boldsymbol{a}, \boldsymbol{b}, \boldsymbol{c} \in \hat{E}^3$ be arbitrary. Then, $(\det\boldsymbol{A})(\boldsymbol{a} \times \boldsymbol{b}) \cdot \boldsymbol{c} = \boldsymbol{A}\boldsymbol{a} \times \boldsymbol{A}\boldsymbol{b} \cdot \boldsymbol{A}\boldsymbol{c} = \boldsymbol{A}^t\boldsymbol{A}^*(\boldsymbol{a} \times \boldsymbol{b}) \cdot \boldsymbol{c}$, which makes sense provided that $\boldsymbol{A}^t\boldsymbol{A}^*$ maps $\hat{E}^3$ to itself. Because $\boldsymbol{c}$ is arbitrary this yields $\boldsymbol{A}^t\boldsymbol{A}^*(\boldsymbol{a} \times \boldsymbol{b}) = (\det\boldsymbol{A})\boldsymbol{a} \times \boldsymbol{b} = (\det\boldsymbol{A})\hat{\boldsymbol{I}}(\boldsymbol{a} \times \boldsymbol{b})$, where $\hat{\boldsymbol{I}}$ is the identity for $\hat{E}^3$. Then, because $\boldsymbol{a} \times \boldsymbol{b}$ is arbitrary, we have $\boldsymbol{A}^t\boldsymbol{A}^* = (\det\boldsymbol{A})\hat{\boldsymbol{I}}$, so that, indeed, $\boldsymbol{A}^t\boldsymbol{A}^*$ maps $\hat{E}^3$ to itself, and the result follows, i.e., $\boldsymbol{A}^* = (\det\boldsymbol{A})\boldsymbol{A}^{-t}$, if $\boldsymbol{A}$ is invertible.

23.

Let $\boldsymbol{B} = \boldsymbol{A} - \lambda\boldsymbol{I}$. Then,

$$
\begin{aligned}
&[\boldsymbol{a}, \boldsymbol{b}, \boldsymbol{c}]\det\boldsymbol{B} \\
&= [\boldsymbol{B}\boldsymbol{a}, \boldsymbol{B}\boldsymbol{b}, \boldsymbol{B}\boldsymbol{c}] \\
&= [(\boldsymbol{A} - \lambda\boldsymbol{I})\boldsymbol{a}, \boldsymbol{B}\boldsymbol{b}, \boldsymbol{B}\boldsymbol{c}]
\end{aligned}
$$

$$
\begin{aligned}
&= [\boldsymbol{Aa}, \boldsymbol{Bb}, \boldsymbol{Bc}] - \lambda[\boldsymbol{a}, \boldsymbol{Bb}, \boldsymbol{Bc}] \\
&= [\boldsymbol{Aa}, (\boldsymbol{A} - \lambda\boldsymbol{I})\boldsymbol{b}, \boldsymbol{Bc}] - \lambda[\boldsymbol{a}, \boldsymbol{Bb}, \boldsymbol{Bc}] \\
&= [\boldsymbol{Aa}, \boldsymbol{Ab}, \boldsymbol{Bc}] - \lambda[\boldsymbol{Aa}, \boldsymbol{b}, \boldsymbol{Bc}] - \lambda[\boldsymbol{a}, \boldsymbol{Bb}, \boldsymbol{Bc}] \\
&= [\boldsymbol{Aa}, \boldsymbol{Ab}, (\boldsymbol{A} - \lambda\boldsymbol{I})\boldsymbol{c}] - \lambda[\boldsymbol{Aa}, \boldsymbol{b}, (\boldsymbol{A} - \lambda\boldsymbol{I})\boldsymbol{c}] - \lambda[\boldsymbol{a}, (\boldsymbol{A} - \lambda\boldsymbol{I})\boldsymbol{b}, \boldsymbol{Bc}] \\
&= [\boldsymbol{Aa}, \boldsymbol{Ab}, \boldsymbol{Ac}] - \lambda[\boldsymbol{Aa}, \boldsymbol{Ab}, \boldsymbol{c}] - \lambda[\boldsymbol{Aa}, \boldsymbol{b}, \boldsymbol{Ac}] - \lambda[\boldsymbol{a}, \boldsymbol{Ab}, \boldsymbol{Bc}] \\
&\quad + \lambda^2[\boldsymbol{Aa}, \boldsymbol{b}, \boldsymbol{c}] + \lambda^2[\boldsymbol{a}, \boldsymbol{b}, \boldsymbol{Bc}] \\
&= [\boldsymbol{Aa}, \boldsymbol{Ab}, \boldsymbol{Ac}] - \lambda\{[\boldsymbol{Aa}, \boldsymbol{Ab}, \boldsymbol{c}] + [\boldsymbol{Aa}, \boldsymbol{b}, \boldsymbol{Ac}] + [\boldsymbol{a}, \boldsymbol{Ab}, \boldsymbol{Ac}]\} \\
&\quad + \lambda^2\{[\boldsymbol{Aa}, \boldsymbol{b}, \boldsymbol{c}] + [\boldsymbol{a}, \boldsymbol{b}, \boldsymbol{Ac}] + [\boldsymbol{a}, \boldsymbol{Ab}, \boldsymbol{c}]\} - \lambda^3[\boldsymbol{a}, \boldsymbol{b}, \boldsymbol{c}] \\
&= [\boldsymbol{a}, \boldsymbol{b}, \boldsymbol{c}]\{I_3(\boldsymbol{A}) - \lambda I_2(\boldsymbol{A}) + \lambda^2 I_1(\boldsymbol{A}) - \lambda^3\}.
\end{aligned}
$$

Thus, $\det(\boldsymbol{A} - \lambda\boldsymbol{I}) = -\lambda^3 + I_1(\boldsymbol{A})\lambda^2 - I_2(\boldsymbol{A})\lambda + I_3(\boldsymbol{A})$. Examination of the derivation shows that the result is valid for tensors that map a vector space to itself. Otherwise, the box products are not defined. Accordingly this formula is not valid for two-point tensors.

24.

(d) $v_i = \phi_{,i}$ so $curl\boldsymbol{v} = e_{kij}v_{j,i}\boldsymbol{e}_k = e_{kij}\phi_{,ji}\boldsymbol{e}_k = \boldsymbol{0}$, since $e_{kij} = -e_{kji}$ and $\phi_{,ji} = \phi_{,ij}$.

(e) $div(curl\boldsymbol{v}) = (e_{kij}v_{j,i})_{,k} = e_{kij}v_{j,ik} = 0$, because $v_{j,ik} = v_{j,ki}$. (g) $(u_i v_j)_{,j} = u_{i,j}v_j + v_{j,j}u_i$.

25.

$df = f_{,r}dr + f_{,\theta}d\theta + f_{,\phi}d\phi$. From $\boldsymbol{x} = r\boldsymbol{e}_\rho(\theta, \phi)$, obtain $d\boldsymbol{x} = dr\boldsymbol{e}_\rho + r\boldsymbol{e}_{\rho,\theta}d\theta + r\boldsymbol{e}_{\rho,\phi}d\phi$. We have $\boldsymbol{e}_\rho = \cos\phi\boldsymbol{e}_r(\theta) + \sin\phi\boldsymbol{e}_3$, so $\boldsymbol{e}_{\rho,\theta} = \cos\phi\boldsymbol{e}_\theta(\theta)$ and $\boldsymbol{e}_{\rho,\phi} = -\sin\phi\boldsymbol{e}_r(\theta) + \cos\phi\boldsymbol{e}_3$. Then, using $\boldsymbol{e}_\phi = \boldsymbol{e}_\rho \times \boldsymbol{e}_\theta = \cos\phi\boldsymbol{e}_3 - \sin\phi\boldsymbol{e}_r$, we get $\boldsymbol{e}_{\rho,\phi} = \boldsymbol{e}_\phi$ and $d\boldsymbol{x} = dr\boldsymbol{e}_\rho + r\cos\phi\boldsymbol{e}_\theta d\theta + r\boldsymbol{e}_\phi d\phi$. Thus, $dr = \boldsymbol{e}_\rho \cdot d\boldsymbol{x}, r\cos\phi d\theta = \boldsymbol{e}_\theta \cdot d\boldsymbol{x}$ and $rd\phi = \boldsymbol{e}_\phi \cdot d\boldsymbol{x}$. Therefore, $df = f_{,r}\boldsymbol{e}_\rho \cdot d\boldsymbol{x} + (r\cos\phi)^{-1}f_{,\theta}\boldsymbol{e}_\theta \cdot d\boldsymbol{x} + r^{-1}f_{,\phi}\boldsymbol{e}_\phi \cdot d\boldsymbol{x} = \nabla f \cdot d\boldsymbol{x}$, where

$$\nabla f = f_{,r}\boldsymbol{e}_\rho + (r\cos\phi)^{-1}f_{,\theta}\boldsymbol{e}_\theta + r^{-1}f_{,\phi}\boldsymbol{e}_\phi.$$

26.

$df(r) = f'(r)dr$ and $r^2 = \boldsymbol{x} \cdot \boldsymbol{x}$, so $dr = r^{-1}\boldsymbol{x} \cdot d\boldsymbol{x}$. Then, $df = r^{-1}f'(r)\boldsymbol{x} \cdot d\boldsymbol{x}$, so $\nabla f = r^{-1}f'(r)\boldsymbol{x}$. This is the same as the result of the previous problem if we set $f_{,\theta} = f_{,\phi} = 0$. Let $g(r) = r^{-1}f'(r)$. Then, $\nabla f = g(r)\boldsymbol{x}$. Let $\boldsymbol{v} = \nabla f = g(r)\boldsymbol{x}$. Then, $d\boldsymbol{v} = gd\boldsymbol{x} + g'(r)\boldsymbol{x}dr = gd\boldsymbol{x} + r^{-1}g'(r)\boldsymbol{x}(\boldsymbol{x}\cdot d\boldsymbol{x}) = [g\boldsymbol{I} + r^{-1}g'(r)\boldsymbol{x}\otimes\boldsymbol{x}]d\boldsymbol{x}$, so $\nabla\boldsymbol{v} = g\boldsymbol{I} + r^{-1}g'(r)\boldsymbol{x}\otimes\boldsymbol{x}$. Then, $div(\nabla f) = div\boldsymbol{v} = tr(\nabla\boldsymbol{v}) = 3g + rg'(r) = r[g'(r) + (3/r)g(r)]$. If $div(\nabla f) = 0$ we then have $0 = g'(r) + (3/r)g(r) = r^{-3}(r^3 g)' = 0$, and therefore $g = C/r^3$, where C is a constant. This gives $f'(r) = rg(r) = C/r^2$, which integrates to $f(r) = A + B/r$, where A, B are constants.

From the boundary conditions we have $f_a = f(a) = A + B/a$ and $f_b = f(b) = A + B/b$. Solve for A, B to get

$$f(r) = \frac{1}{a-b}[(af_a - bf_b) + \frac{ab}{r}(f_b - f_a)].$$

27.

From (1.165) and the properties of the permutation symbol, we have

$$\begin{aligned}\boldsymbol{u}\cdot curl\boldsymbol{v} &= e_{jki}u_j v_{i,k} = e_{jki}[(u_j v_i)_{,k} - u_{j,k}v_i] \\ &= (e_{kij}v_i u_j)_{,k} + v_i e_{ikj}u_{j,k} \\ &= div(\boldsymbol{v}\times\boldsymbol{u}) + \boldsymbol{v}\cdot curl\boldsymbol{u}.\end{aligned}$$

We seek the coefficients in the representation

$$curl\boldsymbol{u} = (\boldsymbol{e}_r\cdot curl\boldsymbol{u})\boldsymbol{e}_r + (\boldsymbol{e}_\theta\cdot curl\boldsymbol{u})\boldsymbol{e}_\theta + (\boldsymbol{k}\cdot curl\boldsymbol{u})\boldsymbol{k},$$

of the curl of the vector field (1.143), where

$$\begin{aligned}\boldsymbol{e}_r\cdot curl\boldsymbol{u} &= \boldsymbol{u}\cdot curl\boldsymbol{e}_r - div(u_\theta\boldsymbol{k} - u_z\boldsymbol{e}_\theta), \\ \boldsymbol{e}_\theta\cdot curl\boldsymbol{u} &= \boldsymbol{u}\cdot curl\boldsymbol{e}_\theta - div(u_z\boldsymbol{e}_r - u_r\boldsymbol{k}) \quad \text{and} \\ \boldsymbol{k}\cdot curl\boldsymbol{u} &= \boldsymbol{u}\cdot curl\boldsymbol{k} - div(u_r\boldsymbol{e}_\theta - u_\theta\boldsymbol{e}_r),\end{aligned}$$

and, of course, $curl\boldsymbol{k} = \boldsymbol{0}$. To obtain $curl\boldsymbol{e}_r$ and $curl\boldsymbol{e}_\theta$ we use (1.116) with (1.165). Thus,

$$curl\boldsymbol{e}_r = [(\sin\theta)_{,1} - (\cos\theta)_{,2}]\boldsymbol{k} \quad \text{and} \quad curl\boldsymbol{e}_\theta = [(\cos\theta)_{,1} + (\sin\theta)_{,2}]\boldsymbol{k},$$

where $\cos\theta = x_1/r$ and $\sin\theta = x_2/r$, with $r = \sqrt{x_1^2 + x_2^2}$. Thus, $(\sin\theta)_{,1} = (\cos\theta)_{,2} = -x_1 x_2/r^3$, $(\cos\theta)_{,1} = \frac{1}{r}(1 - x_1^2/r^2)$ and

$(\sin\theta)_{,2} = \frac{1}{r}(1 - x_2^2/r^2)$. Then,

$$curl\boldsymbol{e}_r = \boldsymbol{0} \quad \text{and} \quad curl\boldsymbol{e}_\theta = \frac{1}{r}\boldsymbol{k}.$$

Using (1.158) to calculate the divergences, we finally arrive at

$$curl\boldsymbol{u} = \frac{1}{r}u_{z,\theta}\boldsymbol{e}_r - u_{z,r}\boldsymbol{e}_\theta + \left[u_{\theta,r} + \frac{1}{r}(u_\theta - u_{r,\theta})\right]\boldsymbol{k}.$$

Chapter 2

1.

Proceeding as in Appendix A, we use $\boldsymbol{F}^*(\boldsymbol{a}\times\boldsymbol{b}) = \boldsymbol{F}\boldsymbol{a}\times\boldsymbol{F}\boldsymbol{b}$ for all $\boldsymbol{a},\boldsymbol{b}\in\hat{E}^3$ to get

$$\begin{aligned} e_{ABC}F^*_{kC}\boldsymbol{e}_k &= e_{ABC}\boldsymbol{F}^*\hat{\boldsymbol{E}}_C \\ &= \boldsymbol{F}^*(\hat{\boldsymbol{E}}_A\times\hat{\boldsymbol{E}}_B) = F_{iA}F_{jB}\boldsymbol{e}_i\times\boldsymbol{e}_j = e_{kij}F_{iA}F_{jB}\boldsymbol{e}_k. \end{aligned}$$

Thus,

$$e_{ABC}F^*_{kC} = e_{kij}F_{iA}F_{jB}$$

and the result follows from

$$2F^*_{kD} = 2\delta_{DC}F^*_{kC} = e_{ABD}e_{ABC}F^*_{kC} = e_{kij}e_{DAB}F_{iA}F_{jB}.$$

2.

(a)

$$\begin{aligned} F^*_{iA} &= \frac{1}{2}e_{ijk}e_{ABC}F_{jB}F_{kC} = \frac{1}{2}e_{ijk}e_{ABC}\chi_{j,B}\chi_{k,C} \\ &= \left(\frac{1}{2}e_{ijk}e_{ABC}\chi_j\chi_{k,C}\right)_{,B} - \frac{1}{2}e_{ijk}e_{ABC}\chi_j\chi_{k,CB} \end{aligned}$$

in which the second term is zero because e_{ABC} is skew in BC, whereas $\chi_{k,CB}$ is symmetric in BC; thus the double sum over B, C gives zero.
Thus, $F^*_{iA} = G_{iAB,B}$, where

$$G_{iAB} = \frac{1}{2}e_{ijk}e_{ABC}\chi_j\chi_{k,C}.$$

(b) $G_{iBA} = \frac{1}{2}e_{ijk}e_{BAC}\chi_j\chi_{k,C} = -\frac{1}{2}e_{ijk}e_{ABC}\chi_j\chi_{k,C} = -G_{iAB}$, so

$$G_{iAB,BA} = G_{iAB,AB} = -G_{iBA,AB} = -G_{iAB,BA},$$

and therefore

$$F^*_{iA,A} = G_{iAB,BA} = 0.$$

(c) From the above extension of Problem 22(c) in Chapter 1 to two-point tensors, we have $\boldsymbol{F}^t\boldsymbol{F}^* = (\det \boldsymbol{F})\hat{\boldsymbol{I}}$. Take the trace to get $3\det \boldsymbol{F} = tr(F_{iA}F^*_{iB}\hat{\boldsymbol{E}}_A \otimes \hat{\boldsymbol{E}}_B) = F_{iA}F^*_{iB}\delta_{AB}$, and combine with the result of Problem 1 above to get

$$\det \boldsymbol{F} = \frac{1}{3}F_{iA}F^*_{iA} = \frac{1}{6}e_{ijk}e_{ABC}F_{iA}F_{jB}F_{kC}.$$

Thus,

$$\det \boldsymbol{F} = \frac{1}{3}F_{iA}G_{iAB,B} = \left(\frac{1}{3}F_{iA}G_{iAB}\right)_{,B} - \frac{1}{3}F_{iA,B}G_{iAB},$$

where $F_{iA,B}G_{iAB} = \chi_{i,AB}G_{iAB} = 0$ because $\chi_{i,AB} = \chi_{i,BA}$ and $G_{iAB} = -G_{iBA}$. Finally,

$$\det \boldsymbol{F} = Div\boldsymbol{H} = H_{B,B}, \quad \text{where} \quad H_B = \frac{1}{3}F_{iA}G_{iAB}.$$

(d) Using $\boldsymbol{F}^t = F_{iA}\hat{\boldsymbol{E}}_A \otimes \boldsymbol{e}_i$, we have

$$\begin{aligned} e_{ijk}\det \boldsymbol{F}^t &= (\det \boldsymbol{F}^t)[\boldsymbol{e}_i, \boldsymbol{e}_j, \boldsymbol{e}_k] = [\boldsymbol{F}^t\boldsymbol{e}_i, \boldsymbol{F}^t\boldsymbol{e}_j, \boldsymbol{F}^t\boldsymbol{e}_k] \\ &= F_{iA}F_{jB}F_{kC}[\hat{\boldsymbol{E}}_A, \hat{\boldsymbol{E}}_B, \hat{\boldsymbol{E}}_C] = e_{ABC}F_{iA}F_{jB}F_{kC}. \end{aligned}$$

Thus,

$$\det \boldsymbol{F}^t = \frac{1}{6}e_{ijk}e_{ABC}F_{iA}F_{jB}F_{kC} = \det \boldsymbol{F},$$

a result that also holds for tensors that map a vector space to itself.

3.

Since the bases are orthonormal, we have $\hat{\boldsymbol{E}}_B = A_{BA}\hat{\boldsymbol{E}}'_A$ and $\boldsymbol{e}_j = a_{ji}\boldsymbol{e}'_i$, where $A_{BA} = \hat{\boldsymbol{E}}_B \cdot \hat{\boldsymbol{E}}'_A$ and $a_{ji} = \boldsymbol{e}_j \cdot \boldsymbol{e}'_i$. Then, $\boldsymbol{F} = F_{jB}\boldsymbol{e}_j \otimes \hat{\boldsymbol{E}}_B = a_{ji}F_{jB}A_{BA}\boldsymbol{e}'_i \otimes \hat{\boldsymbol{E}}'_A$, yielding $F'_{iA} = a_{ji}A_{BA}F_{jB}$.

Chapter 3

1.

We have $\boldsymbol{x} = \lambda_1 X_1 \boldsymbol{e}_1 + \lambda_2 X_2 \boldsymbol{e}_2 + \lambda_3 X_3 \boldsymbol{e}_3$, and therefore $\boldsymbol{F} d\boldsymbol{X} = d\boldsymbol{x} = \lambda_1 dX_1 \boldsymbol{e}_1 + \ldots = \lambda_1 \boldsymbol{e}_1 (\hat{\boldsymbol{E}}_1 \cdot d\boldsymbol{X}) + \ldots$, yielding

$$\boldsymbol{F} = \lambda_1 \boldsymbol{e}_1 \otimes \hat{\boldsymbol{E}}_1 + \lambda_2 \boldsymbol{e}_2 \otimes \hat{\boldsymbol{E}}_2 + \lambda_3 \boldsymbol{e}_3 \otimes \hat{\boldsymbol{E}}_3.$$

(a) $0 < J = \det \boldsymbol{F} = [\boldsymbol{F}\hat{\boldsymbol{E}}_1, \boldsymbol{F}\hat{\boldsymbol{E}}_2, \boldsymbol{F}\hat{\boldsymbol{E}}_3] = \lambda_1\lambda_2\lambda_3[\boldsymbol{e}_1, \boldsymbol{e}_2, \boldsymbol{e}_3] = \lambda_1\lambda_2\lambda_3$. Thus, $\boldsymbol{F}$ is invertible and

$$\boldsymbol{F}^{-1} = \lambda_1^{-1} \hat{\boldsymbol{E}}_1 \otimes \boldsymbol{e}_1 + \lambda_2^{-1} \hat{\boldsymbol{E}}_2 \otimes \boldsymbol{e}_2 + \lambda_3^{-1} \hat{\boldsymbol{E}}_3 \otimes \boldsymbol{e}_3.$$

(b) Let $\lambda = \lambda_1$ and $\mu = \lambda_2 = \lambda_3$. Then $J = \lambda\mu^2$ and

$$\boldsymbol{F}^* = \mu^2 \boldsymbol{e}_1 \otimes \hat{\boldsymbol{E}}_1 + \lambda\mu(\boldsymbol{e}_2 \otimes \hat{\boldsymbol{E}}_2 + \boldsymbol{e}_3 \otimes \hat{\boldsymbol{E}}_3).$$

We can write $\boldsymbol{N} = \cos\phi \hat{\boldsymbol{E}}_1 - \sin\phi \hat{\boldsymbol{E}}_r(\Theta)$ for some angle Θ, where $\hat{\boldsymbol{E}}_r(\Theta) = \cos\Theta \hat{\boldsymbol{E}}_2 + \sin\Theta \hat{\boldsymbol{E}}_3$. (It may be helpful to verify this by drawing a figure). If the deformation is isochoric ($J = 1$), then, if α is the area ratio, it follows from the Piola–Nanson formula that

$$\begin{aligned} \alpha \boldsymbol{n} = \boldsymbol{F}^* \boldsymbol{N} &= \mu^2 \cos\phi \boldsymbol{e}_1 - \lambda\mu \sin\phi \boldsymbol{e}_r(\theta) \\ &= \lambda^{-1} \cos\phi \boldsymbol{e}_1 - \lambda^{1/2} \sin\phi \boldsymbol{e}_r(\theta), \end{aligned}$$

where $\boldsymbol{e}_r(\theta) = \cos\theta \boldsymbol{e}_2 + \sin\theta \boldsymbol{e}_3$ and $\theta = \Theta$. Thus,

$$\alpha = \sqrt{\lambda^{-2}\cos^2\phi + \lambda \sin^2\phi}.$$

(c) $\boldsymbol{n} = \alpha^{-1}[\lambda^{-1}\cos\phi \boldsymbol{e}_1 - \lambda^{1/2}\sin\phi \boldsymbol{e}_r(\theta)]$.

2.

If $\boldsymbol{A}$ and $\boldsymbol{B}$ are orthogonal, then they are invertible, and $(\boldsymbol{AB})^{-1} = \boldsymbol{B}^{-1}\boldsymbol{A}^{-1} = \boldsymbol{B}^t\boldsymbol{A}^t = (\boldsymbol{AB})^t$. Thus, $\boldsymbol{AB}$ is orthogonal. Further, $\det(\boldsymbol{AB}) = (\det \boldsymbol{A})(\det \boldsymbol{B}) = 1$ if $\boldsymbol{A}$ and $\boldsymbol{B}$ are rotations. Thus, $\boldsymbol{AB}$ is a rotation.

3.

$\boldsymbol{C} = \hat{\boldsymbol{G}}^t\hat{\boldsymbol{G}} = \hat{\boldsymbol{I}} + \gamma(\hat{\boldsymbol{E}}_3 \otimes \hat{\boldsymbol{E}}_\Theta + \hat{\boldsymbol{E}}_\Theta \otimes \hat{\boldsymbol{E}}_3) + \gamma^2 \hat{\boldsymbol{E}}_3 \otimes \hat{\boldsymbol{E}}_3$, where $\gamma = R\tau$, and $\boldsymbol{B} = \boldsymbol{G}\boldsymbol{G}^t = \boldsymbol{I} + \gamma(\boldsymbol{e}_3 \otimes \boldsymbol{e}_\theta + \boldsymbol{e}_\theta \otimes \boldsymbol{e}_3) + \gamma^2 \boldsymbol{e}_3 \otimes \boldsymbol{e}_3$.

4.

(a) We have $\boldsymbol{x} = \lambda(R)\mathbf{1}\boldsymbol{X}$, where $R = |\boldsymbol{X}|$. Assume $\lambda > 0$. This deformation maps the sphere $R = const.$ to the sphere $r = const.$, where $r = |\boldsymbol{x}|$.
We have

$$\boldsymbol{F}d\boldsymbol{X} = d\boldsymbol{x} = \lambda'(R)\mathbf{1}\boldsymbol{X}dR + \lambda(R)\mathbf{1}d\boldsymbol{X}.$$

Use $R^2 = \boldsymbol{X} \cdot \boldsymbol{X}$ to get $RdR = \boldsymbol{X} \cdot d\boldsymbol{X}$. Then,

$$\boldsymbol{F}d\boldsymbol{X} = R^{-1}\lambda'(R)\mathbf{1}\boldsymbol{X}(\boldsymbol{X} \cdot d\boldsymbol{X}) + \lambda(R)\mathbf{1}d\boldsymbol{X},$$

yielding

$$\boldsymbol{F} = \mathbf{1}[R^{-1}\lambda'(R)\boldsymbol{X} \otimes \boldsymbol{X} + \lambda(R)\hat{\boldsymbol{I}}].$$

(b) Let $\{\boldsymbol{u}, \boldsymbol{v}, \boldsymbol{w}\}$, with $\boldsymbol{u} = R^{-1}\boldsymbol{X}$, be a right-handed orthonormal basis for $\hat{E}^3$. Then,

$$\boldsymbol{F} = \mathbf{1}[(R\lambda' + \lambda)\boldsymbol{u} \otimes \boldsymbol{u} + \lambda(\boldsymbol{v} \otimes \boldsymbol{v} + \boldsymbol{w} \otimes \boldsymbol{w})].$$

Let $r = |\boldsymbol{x}|$. Then, $r(R) = R\lambda(R)$ and $r'(R) = R\lambda' + \lambda$. Thus,

$$\boldsymbol{F} = \mathbf{1}[r'(R)\boldsymbol{u} \otimes \boldsymbol{u} + \lambda(\boldsymbol{v} \otimes \boldsymbol{v} + \boldsymbol{w} \otimes \boldsymbol{w})],$$

and

$$\begin{aligned} 0 < J = \det \boldsymbol{F} = [\boldsymbol{F}\boldsymbol{u}, \boldsymbol{F}\boldsymbol{v}, \boldsymbol{F}\boldsymbol{w}] &= \lambda^2 r'(R)[\mathbf{1}\boldsymbol{u}, \mathbf{1}\boldsymbol{v}, \mathbf{1}\boldsymbol{w}] \\ &= \lambda^2 r'(R)(\det \mathbf{1})[\boldsymbol{u}, \boldsymbol{v}, \boldsymbol{w}] = \lambda^2 r'(R), \end{aligned}$$

and so we require $r'(R) > 0$, i.e., $r(R)$ is an increasing function.

(c) If $J = 1$, then $r^2 r' = R^2$, which integrates to

$$r^3 - a^3 = R^3 - A^3,$$

implying, as expected, that the volume of the spherical annulus $[A, R]$ equals that of the spherical annulus $[a, r]$.

5.

(a) We have

$$\boldsymbol{F}d\boldsymbol{X} = d\boldsymbol{x} = \mathbf{1}d\boldsymbol{X} + \gamma\boldsymbol{e}_3 dR = (\mathbf{1} + \gamma\boldsymbol{e}_3 \otimes \hat{\boldsymbol{E}}_R)d\boldsymbol{X},$$

$$\text{where} \quad \gamma = \partial w/\partial R; \quad \text{thus,} \quad \boldsymbol{F} = \mathbf{1} + \gamma\boldsymbol{e}_3 \otimes \hat{\boldsymbol{E}}_R.$$

Then,

$$\begin{aligned}\boldsymbol{C} = \boldsymbol{F}^t\boldsymbol{F} &= (\mathbf{1}^t + \gamma\hat{\boldsymbol{E}}_R \otimes \boldsymbol{e}_3)(\mathbf{1} + \gamma\boldsymbol{e}_3 \otimes \hat{\boldsymbol{E}}_R) \\ &= \hat{\boldsymbol{I}} + \gamma(\hat{\boldsymbol{E}}_R \otimes \mathbf{1}^t\boldsymbol{e}_3 + \mathbf{1}^t\boldsymbol{e}_3 \otimes \boldsymbol{E}_R) + \gamma^2\hat{\boldsymbol{E}}_R \otimes \hat{\boldsymbol{E}}_R.\end{aligned}$$

If we choose $\{\boldsymbol{E}_A\} = \{\boldsymbol{e}_i\}$, where $\boldsymbol{E}_A = \mathbf{1}\hat{\boldsymbol{E}}_A$, then $\mathbf{1}^t\boldsymbol{e}_3 = \mathbf{1}^t\boldsymbol{E}_3 = \hat{\boldsymbol{E}}_3$ and

$$\boldsymbol{C} = \hat{\boldsymbol{I}} + \gamma(\hat{\boldsymbol{E}}_R \otimes \hat{\boldsymbol{E}}_3 + \hat{\boldsymbol{E}}_3 \otimes \hat{\boldsymbol{E}}_R) + \gamma^2\hat{\boldsymbol{E}}_R \otimes \hat{\boldsymbol{E}}_R.$$

Similarly, we obtain

$$\boldsymbol{B} = \boldsymbol{F}\boldsymbol{F}^t = \boldsymbol{I} + \gamma(\boldsymbol{e}_r \otimes \boldsymbol{e}_3 + \boldsymbol{e}_3 \otimes \boldsymbol{e}_r) + \gamma^2\boldsymbol{e}_3 \otimes \boldsymbol{e}_3, \quad \text{where} \quad \boldsymbol{e}_r = \mathbf{1}\hat{\boldsymbol{E}}_R.$$

(b) We have

$$\mu\boldsymbol{m} = \boldsymbol{F}\boldsymbol{M} = \frac{\sqrt{2}}{2}(\boldsymbol{F}\hat{\boldsymbol{E}}_\Theta + \boldsymbol{F}\hat{\boldsymbol{E}}_3) = \frac{\sqrt{2}}{2}(\boldsymbol{e}_\theta + \boldsymbol{e}_3) = \mathbf{1}\boldsymbol{M},$$

$$\text{where} \quad \boldsymbol{e}_\theta = \mathbf{1}\hat{\boldsymbol{E}}_\Theta,$$

yielding $\mu = 1$ and $\boldsymbol{m} = \mathbf{1}\boldsymbol{M}$. For $\boldsymbol{M} = \frac{\sqrt{2}}{2}(\hat{\boldsymbol{E}}_R + \hat{\boldsymbol{E}}_\Theta)$, we have

$$\mu\boldsymbol{m} = \boldsymbol{F}\boldsymbol{M} = \frac{\sqrt{2}}{2}(\boldsymbol{F}\hat{\boldsymbol{E}}_R + \boldsymbol{F}\hat{\boldsymbol{E}}_\Theta) = \frac{\sqrt{2}}{2}(\boldsymbol{e}_r + \gamma\boldsymbol{e}_3 + \boldsymbol{e}_\theta),$$

$$\text{with} \quad \boldsymbol{e}_r = \mathbf{1}\hat{\boldsymbol{E}}_R,$$

so $\mu = \sqrt{1 + \gamma^2/2}$ and $\boldsymbol{m} = \frac{1}{\sqrt{2+\gamma^2}}(\boldsymbol{e}_r + \gamma\boldsymbol{e}_3 + \boldsymbol{e}_\theta)$.

6.

(a) From $r = f(X_1)$ and $\theta = g(X_2)$ it is evident that a line $X_1 = const.$ is mapped to a cylindrical surface $r = const.$ and a line $X_2 = const.$ is mapped to a plane $\theta = const.$

(b) We have

$$\begin{aligned}\boldsymbol{F}d\boldsymbol{X} &= d\boldsymbol{x} = dr\boldsymbol{e}_r(\theta) + r\boldsymbol{e}_\theta(\theta)d\theta + dz\boldsymbol{e}_3 \\ &= f'(X_1)dX_1\boldsymbol{e}_r(\theta) + f(X_1)g'(X_2)dX_2\boldsymbol{e}_\theta(\theta) + dX_3\boldsymbol{e}_3 \\ &= f'(X_1)\boldsymbol{e}_r(\hat{\boldsymbol{E}}_1 \cdot d\boldsymbol{X}) + f(X_1)g'(X_2)\boldsymbol{e}_\theta(\theta)(\hat{\boldsymbol{E}}_2 \cdot d\boldsymbol{X}) \\ &\quad + \boldsymbol{e}_3(\hat{\boldsymbol{E}}_3 \cdot d\boldsymbol{X}).\end{aligned}$$

Therefore,

$$\boldsymbol{F} = f'(X_1)\boldsymbol{e}_r \otimes \hat{\boldsymbol{E}}_1 + f(X_1)g'(X_2)\boldsymbol{e}_\theta \otimes \hat{\boldsymbol{E}}_2 + \boldsymbol{e}_3 \otimes \hat{\boldsymbol{E}}_3.$$

(c) We have

$$\begin{aligned}0 < \det \boldsymbol{F} &= [\boldsymbol{F}\hat{\boldsymbol{E}}_1, \boldsymbol{F}\hat{\boldsymbol{E}}_2, \boldsymbol{F}\hat{\boldsymbol{E}}_3]/[\hat{\boldsymbol{E}}_1, \hat{\boldsymbol{E}}_2, \hat{\boldsymbol{E}}_3] = [\boldsymbol{F}\hat{\boldsymbol{E}}_1, \boldsymbol{F}\hat{\boldsymbol{E}}_2, \boldsymbol{F}\hat{\boldsymbol{E}}_3] \\ &= [f'(X_1)\boldsymbol{e}_r, f(X_1)g'(X_2)\boldsymbol{e}_\theta, \boldsymbol{e}_3] = f(X_1)f'(X_1)g'(X_2)[\boldsymbol{e}_r, \boldsymbol{e}_\theta, \boldsymbol{e}_3] \\ &= f(X_1)f'(X_1)g'(X_2).\end{aligned}$$

(d) Since $r > 0$, we have $f(X_1) > 0$, and, assuming that $g'(X_2) > 0$ we have $f'(X_1) > 0$ from part (c). Define

$$\boldsymbol{Q} = \boldsymbol{e}_r(\theta) \otimes \hat{\boldsymbol{E}}_1 + \boldsymbol{e}_\theta(\theta) \otimes \hat{\boldsymbol{E}}_2 + \boldsymbol{e}_3 \otimes \hat{\boldsymbol{E}}_3.$$

This is a rotation. We then have

$$\boldsymbol{F} = \boldsymbol{Q}[f'(X_1)\hat{\boldsymbol{E}}_1 \otimes \hat{\boldsymbol{E}}_1 + f(X_1)g'(X_2)\hat{\boldsymbol{E}}_2 \otimes \hat{\boldsymbol{E}}_2 + \hat{\boldsymbol{E}}_3 \otimes \hat{\boldsymbol{E}}_3].$$

The term in brackets is symmetric and positive definite. The uniqueness of the polar decomposition then yields

$$\boldsymbol{R} = \boldsymbol{Q} \quad \text{and} \quad \boldsymbol{U} = f'(X_1)\hat{\boldsymbol{E}}_1 \otimes \hat{\boldsymbol{E}}_1 + f(X_1)g'(X_2)\hat{\boldsymbol{E}}_2 \otimes \hat{\boldsymbol{E}}_2 + \hat{\boldsymbol{E}}_3 \otimes \hat{\boldsymbol{E}}_3.$$

Finally, using $\boldsymbol{V} = \boldsymbol{R}\boldsymbol{U}\boldsymbol{R}^t$ we derive

$$\boldsymbol{V} = f'(X_1)\boldsymbol{e}_r \otimes \boldsymbol{e}_r + f(X_1)g'(X_2)\boldsymbol{e}_\theta \otimes \boldsymbol{e}_\theta + \boldsymbol{e}_3 \otimes \boldsymbol{e}_3.$$

7.

(a) Use the spectral representation $\boldsymbol{A} = \sum \lambda_i \boldsymbol{u}_i \otimes \boldsymbol{u}_i$. Then, $I_1(\boldsymbol{A}) = \lambda_1 + \lambda_2 + \lambda_3$ and $I_3(\boldsymbol{A}) = \lambda_1\lambda_2\lambda_3$. From Problem 22(b) of Chapter 1, we have that $I_2(\boldsymbol{A}) = \frac{1}{2}[I_1^2 - tr(\boldsymbol{A}^2)]$, where, in the present case, $\boldsymbol{A}^2 = \sum \lambda_i^2 \boldsymbol{u}_i \otimes \boldsymbol{u}_i$ and $tr(\boldsymbol{A}^2) = \lambda_1^2 + \lambda_2^2 + \lambda_3^2$. Thus,

$$I_2(\boldsymbol{A}) = \frac{1}{2}[(\lambda_1+\lambda_2+\lambda_3)^2 - (\lambda_1^2+\lambda_2^2+\lambda_3^2)] = \lambda_1\lambda_2+\lambda_1\lambda_3+\lambda_2\lambda_3.$$

(b) Use $\boldsymbol{A}^3 = \sum \lambda_i^3 \boldsymbol{u}_i \otimes \boldsymbol{u}_i$ to get

$$\boldsymbol{A}^3 - I_1\boldsymbol{A}^2 + I_2\boldsymbol{A} - I_3\boldsymbol{I} = \sum(\lambda_i^3 - I_1\lambda_i^2 + I_2\lambda_i - I_3)\boldsymbol{u}_i \otimes \boldsymbol{u}_i = \boldsymbol{O},$$

because the characteristic equation for principal values requires that the parenthesis vanishes for each $i \in \{1, 2, 3\}$. Inspection of the derivation indicates that this result is valid for tensors that map a vector space (in this case, E^3) to itself.

8.

The principal values are easily seen to be the roots of the quadratic equation

$$0 = \lambda^2 - (\lambda_1 + \lambda_2)\lambda + \lambda_1\lambda_2 = \lambda^2 - \lambda tr\boldsymbol{A} + \det \boldsymbol{A}.$$

Proceed as in Problem 7.

9.

The first result is a consequence of the result of Problem 8. The remainder is straightforward.

Chapter 4

1.

(a) $(\boldsymbol{AB})^{\cdot} = (A_{ij}B_{jk}\boldsymbol{e}_i \otimes \boldsymbol{e}_k)^{\cdot} = (A_{ij}B_{jk})^{\cdot}\boldsymbol{e}_i \otimes \boldsymbol{e}_k = (\dot{A}_{ij}B_{jk} + A_{ij}\dot{B}_{jk})\boldsymbol{e}_i \otimes \boldsymbol{e}_k = \dot{\boldsymbol{A}}\boldsymbol{B} + \boldsymbol{A}\dot{\boldsymbol{B}}$.

(b) $(\boldsymbol{A}^t)^{\cdot} = \dot{A}_{ji}\boldsymbol{e}_i \otimes \boldsymbol{e}_j = (\dot{A}_{ji}\boldsymbol{e}_j \otimes \boldsymbol{e}_i)^t = (\dot{\boldsymbol{A}})^t$.

(c) Differentiate $\boldsymbol{I} = \boldsymbol{RR}^t$ to get $\boldsymbol{O} = \dot{\boldsymbol{R}}\boldsymbol{R}^t + \boldsymbol{R}(\boldsymbol{R}^t)^{\cdot} = \dot{\boldsymbol{R}}\boldsymbol{R}^t + \boldsymbol{R}(\dot{\boldsymbol{R}})^t = \dot{\boldsymbol{R}}\boldsymbol{R}^t + (\dot{\boldsymbol{R}}\boldsymbol{R}^t)^t$.

(d) Differentiate $\boldsymbol{I} = \boldsymbol{AA}^{-1}$ to get $\boldsymbol{O} = \dot{\boldsymbol{A}}\boldsymbol{A}^{-1} + \boldsymbol{A}(\boldsymbol{A}^{-1})^{\cdot}$. Thus, $(\boldsymbol{A}^{-1})^{\cdot} = -\boldsymbol{A}^{-1}\dot{\boldsymbol{A}}\boldsymbol{A}^{-1}$.

2.

We have $J\boldsymbol{u}\cdot\boldsymbol{v}\times\boldsymbol{w} = \boldsymbol{F}\boldsymbol{u}\cdot\boldsymbol{F}\boldsymbol{v}\times\boldsymbol{F}\boldsymbol{w}$. Then, for fixed $\boldsymbol{u},\boldsymbol{v},\boldsymbol{w}$, and with $\dot{\boldsymbol{F}} = \boldsymbol{L}\boldsymbol{F}$, we get

$$\begin{aligned}\dot{J}\boldsymbol{u}\cdot\boldsymbol{v}\times\boldsymbol{w} &= \dot{\boldsymbol{F}}\boldsymbol{u}\cdot\boldsymbol{F}\boldsymbol{v}\times\boldsymbol{F}\boldsymbol{w} + \boldsymbol{F}\boldsymbol{u}\cdot\dot{\boldsymbol{F}}\boldsymbol{v}\times\boldsymbol{F}\boldsymbol{w} + \boldsymbol{F}\boldsymbol{u}\cdot\boldsymbol{F}\boldsymbol{v}\times\dot{\boldsymbol{F}}\boldsymbol{w}\\ &= \boldsymbol{L}\boldsymbol{F}\boldsymbol{u}\cdot\boldsymbol{F}\boldsymbol{v}\times\boldsymbol{F}\boldsymbol{w} + \boldsymbol{F}\boldsymbol{u}\cdot\boldsymbol{L}\boldsymbol{F}\boldsymbol{v}\times\boldsymbol{F}\boldsymbol{w}\\ &\quad + \boldsymbol{F}\boldsymbol{u}\cdot\boldsymbol{F}\boldsymbol{v}\times\boldsymbol{L}\boldsymbol{F}\boldsymbol{w}\\ &= (tr\boldsymbol{L})\boldsymbol{F}\boldsymbol{u}\cdot\boldsymbol{F}\boldsymbol{v}\times\boldsymbol{F}\boldsymbol{w}\\ &= J(tr\boldsymbol{L})\boldsymbol{u}\cdot\boldsymbol{v}\times\boldsymbol{w}.\end{aligned}$$

Thus,

$$\dot{J} = Jtr\boldsymbol{L} = Jtr(\dot{\boldsymbol{F}}\boldsymbol{F}^{-1}) = J\boldsymbol{F}^{-t}\cdot\dot{\boldsymbol{F}} = \boldsymbol{F}^{*}\cdot\dot{\boldsymbol{F}} = F^{*}_{iA}\dot{F}_{iA},$$

and therefore $J_{\boldsymbol{F}} = \boldsymbol{F}^{*}$, i.e., $\partial J/\partial F_{iA} = F^{*}_{iA}$.

3.

We have $\theta = (A/r)\cos t$ and $r^2 = x_1^2 + x_2^2$. Thus, $\dot{\theta} = \theta' + \boldsymbol{v}\cdot grad\theta$, where $\theta' = \partial\theta/\partial t$ at fixed $\boldsymbol{x}$ (hence, fixed r). Thus, $\theta' = -(A/r)\sin t$. Also, $grad\theta = (A\cos t)grad(r^{-1})$, where $grad(r^{-1}) = -(r^{-2})gradr = -(r^{-3})\boldsymbol{x}_2$, where $\boldsymbol{x}_2 = x_1\boldsymbol{e}_1 + x_2\boldsymbol{e}_2$. Then, $\boldsymbol{v}\cdot grad\theta = -(Ar^{-3})\cos t\boldsymbol{v}\cdot\boldsymbol{x}_2$, where $\boldsymbol{v}\cdot\boldsymbol{x}_2 = x_1v_1 + x_2v_2$. Use $v_1 = x_1^2x_2 + x_3^2$ and $v_2 = -x_1^3 - x_1x_2^2$ to get $\boldsymbol{v}\cdot\boldsymbol{x}_2 = x_1(x_3^2 - x_2^3)$. Finally,

$$\dot{\theta} = -(A/r)\sin t - (A/r^3)(\cos t)x_1(x_3^2 - x_2^3).$$

4.

$\dot{\boldsymbol{v}} - \boldsymbol{v}' = \boldsymbol{L}\boldsymbol{v} = (\boldsymbol{L} - \boldsymbol{L}^t)\boldsymbol{v} + \boldsymbol{L}^t\boldsymbol{v} = 2\boldsymbol{W}\boldsymbol{v} + \boldsymbol{L}^t\boldsymbol{v}$, where $\boldsymbol{L}^t\boldsymbol{v}\cdot d\boldsymbol{x} = \boldsymbol{v}\cdot\boldsymbol{L}d\boldsymbol{x} = \boldsymbol{v}\cdot d\boldsymbol{v} = d(\frac{1}{2}|\boldsymbol{v}|^2) = grad(\frac{1}{2}|\boldsymbol{v}|^2)\cdot d\boldsymbol{x}$, i.e., $\boldsymbol{L}^t\boldsymbol{v} = grad(\frac{1}{2}|\boldsymbol{v}|^2)$. Then, with $\boldsymbol{W}\boldsymbol{v} = \boldsymbol{w}\times\boldsymbol{v}$ we obtain

$$\dot{\boldsymbol{v}} = \boldsymbol{v}' + 2\boldsymbol{w}\times\boldsymbol{v} + \frac{1}{2}grad(|\boldsymbol{v}|^2).$$

5.

(a) $\boldsymbol{L} = v_{i,j}\boldsymbol{e}_i\otimes\boldsymbol{e}_j = ax_3(\boldsymbol{e}_1\otimes\boldsymbol{e}_2 - \boldsymbol{e}_2\otimes\boldsymbol{e}_1) + ax_2\boldsymbol{e}_1\otimes\boldsymbol{e}_3 - ax_1\boldsymbol{e}_2\otimes\boldsymbol{e}_3 + b\boldsymbol{e}_3\otimes\boldsymbol{e}_3$.

(b) $\boldsymbol{D} = \frac{1}{2}(\boldsymbol{L} + \boldsymbol{L}^t) = \frac{1}{2}ax_2(\boldsymbol{e}_1 \otimes \boldsymbol{e}_3 + \boldsymbol{e}_3 \otimes \boldsymbol{e}_1) - \frac{1}{2}ax_1(\boldsymbol{e}_2 \otimes \boldsymbol{e}_3 + \boldsymbol{e}_3 \otimes \boldsymbol{e}_2) + b\boldsymbol{e}_3 \otimes \boldsymbol{e}_3$, $\boldsymbol{W} = \frac{1}{2}(\boldsymbol{L} - \boldsymbol{L}^t) = ax_3(\boldsymbol{e}_1 \otimes \boldsymbol{e}_2 - \boldsymbol{e}_2 \otimes \boldsymbol{e}_1) + \frac{1}{2}ax_2(\boldsymbol{e}_1 \otimes \boldsymbol{e}_3 - \boldsymbol{e}_3 \otimes \boldsymbol{e}_1) - \frac{1}{2}ax_1(\boldsymbol{e}_2 \otimes \boldsymbol{e}_3 - \boldsymbol{e}_3 \otimes \boldsymbol{e}_2)$. Use $\boldsymbol{w} \times \boldsymbol{u} = \boldsymbol{W}\boldsymbol{u}$ for all $\boldsymbol{u}$.

(d) (i) $J = 1$, so $\dot{J} = 0$. Thus, $0 = tr\boldsymbol{L} = tr\boldsymbol{D} = b$.
(ii) $\boldsymbol{w} = \boldsymbol{0}$ if and only if $\boldsymbol{W} = \boldsymbol{O}$; thus, $a = 0$.

6.

$\dot{\boldsymbol{m}} = \boldsymbol{L}\boldsymbol{m} - (\boldsymbol{m} \cdot \boldsymbol{L}\boldsymbol{m})\boldsymbol{m}$, where $\boldsymbol{L} = \dot{\boldsymbol{F}}\boldsymbol{F}^{-1}$, with

$$\dot{\boldsymbol{F}} = \dot{\lambda}\boldsymbol{e}_1 \otimes \hat{\boldsymbol{E}}_1 - \frac{1}{2}\lambda^{-3/2}\dot{\lambda}(\boldsymbol{e}_2 \otimes \hat{\boldsymbol{E}}_2 + \boldsymbol{e}_3 \otimes \hat{\boldsymbol{E}}_3)$$

$$\text{and} \quad \boldsymbol{F}^{-1} = \lambda^{-1}\hat{\boldsymbol{E}}_1 \otimes \boldsymbol{e}_1 + \lambda^{1/2}(\hat{\boldsymbol{E}}_2 \otimes \boldsymbol{e}_2 + \hat{\boldsymbol{E}}_3 \otimes \boldsymbol{e}_3).$$

Then,

$$\boldsymbol{L} = \frac{3\dot{\lambda}}{2\lambda}\boldsymbol{e}_1 \otimes \boldsymbol{e}_1 - \frac{\dot{\lambda}}{2\lambda}\boldsymbol{I} = \boldsymbol{D}; \quad \boldsymbol{W} = \boldsymbol{O}.$$

Also,

$$\boldsymbol{m} = \cos\phi\boldsymbol{e}_r(\theta) + \sin\phi\boldsymbol{e}_3, \quad \text{and} \quad \dot{\boldsymbol{m}} = \dot{\phi}(\cos\phi\boldsymbol{e}_3 - \sin\phi\boldsymbol{e}_r) + \dot{\theta}\cos\phi\boldsymbol{e}_\theta. \tag{1}$$

We also have

$$\boldsymbol{L}\boldsymbol{m} = \frac{\dot{\lambda}}{2\lambda}[3(\boldsymbol{e}_1 \cdot \boldsymbol{m})\boldsymbol{e}_1 - \boldsymbol{m}]; \quad \text{hence,}$$

$$\boldsymbol{m} \cdot \boldsymbol{L}\boldsymbol{m} = \frac{\dot{\lambda}}{2\lambda}[3(\boldsymbol{e}_1 \cdot \boldsymbol{m})^2 - 1] \text{ and } (\boldsymbol{m} \cdot \boldsymbol{L}\boldsymbol{m})\boldsymbol{m} = \frac{\dot{\lambda}}{2\lambda}[3(\boldsymbol{e}_1 \cdot \boldsymbol{m})^2 - 1]\boldsymbol{m}.$$

Then,

$$\boldsymbol{L}\boldsymbol{m} - (\boldsymbol{m} \cdot \boldsymbol{L}\boldsymbol{m})\boldsymbol{m} = \dot{\boldsymbol{m}} = \frac{3\dot{\lambda}}{2\lambda}(\boldsymbol{e}_1 \cdot \boldsymbol{m})[\boldsymbol{e}_1 - (\boldsymbol{e}_1 \cdot \boldsymbol{m})\boldsymbol{m}]. \tag{2}$$

Use (1) and (2) to get

$$\dot{\phi} = (\cos\phi\boldsymbol{e}_3 - \sin\phi\boldsymbol{e}_r) \cdot \dot{\boldsymbol{m}} = \cdots = -\frac{3\dot{\lambda}}{2\lambda}\sin\phi\cos\phi\cos^2\theta$$

and

$$\dot{\theta}\cos\phi = \boldsymbol{e}_\theta \cdot \dot{\boldsymbol{m}} \quad \rightarrow \quad \dot{\theta} = -\frac{3\dot{\lambda}}{2\lambda}\sin\theta\cos\theta.$$

Note that $\dot{\boldsymbol{m}} = \boldsymbol{0}$ if $\phi = 0, \pm\pi/2$ and $\theta = 0, \pi, 3\pi/2$. These correspond to the principal axes of $\boldsymbol{D}$. Recall that $\dot{\boldsymbol{m}} = \boldsymbol{W}\boldsymbol{m}$ ($= \boldsymbol{0}$, in this case), whenever $\boldsymbol{m}$ is a principal vector of $\boldsymbol{D}$. Here, we have an example that illustrates this fact.

7.

(a) From the Piola–Nanson formula we have $\alpha\boldsymbol{n} = \boldsymbol{F}^*\boldsymbol{N}$. Thus, $\alpha = |\boldsymbol{F}^*\boldsymbol{N}|$. To prove that $\alpha > 0$, we use $(3.32)_1$ to obtain $\boldsymbol{F}^{-t} = \boldsymbol{R}\boldsymbol{U}^{-1}$. Then,

$$\boldsymbol{F}^* = J\boldsymbol{R}\boldsymbol{U}^{-1} = \sum \lambda_i^* \boldsymbol{v}_i \otimes \boldsymbol{u}_i, \quad \text{where} \quad \lambda_i^* = J/\lambda_i,$$

(i.e., $\lambda_1^* = \lambda_2\lambda_3$, $\lambda_2^* = \lambda_1\lambda_3$ and $\lambda_3^* = \lambda_1\lambda_2$) and the results of Problems 21(a), (b) in Chapter 1 give $\det \boldsymbol{F}^* = (\det \boldsymbol{R})[\det(J\boldsymbol{U}^{-1})] = J^3/\det \boldsymbol{U} = J^2 (> 0)$. Then, $\alpha > 0$ because $\boldsymbol{N} \neq \boldsymbol{0}$ and hence $\boldsymbol{F}^*\boldsymbol{N} \neq \boldsymbol{0}$. Note that, for fixed $i \in \{1, 2, 3\}$, we have $\alpha = \lambda_i^*$ if $\boldsymbol{N} = \pm\boldsymbol{u}_i$.

(b) We have $\dot{\alpha}\boldsymbol{n} + \alpha\dot{\boldsymbol{n}} = (\boldsymbol{F}^*)^{\cdot}\boldsymbol{N} = \alpha(\boldsymbol{F}^*)^{\cdot}(\boldsymbol{F}^*)^{-1}\boldsymbol{n}$. Use $(\boldsymbol{F}^{-t})^{\cdot} = -\boldsymbol{L}^t\boldsymbol{F}^{-t}$ (obtained by differentiating $\boldsymbol{F}^{-t}\boldsymbol{F}^t = \boldsymbol{I}$) to get

$$\begin{aligned}(\boldsymbol{F}^*)^{\cdot} &= \dot{J}\boldsymbol{F}^{-t} + J(\boldsymbol{F}^{-t})^{\cdot} = J(tr\boldsymbol{D})\boldsymbol{F}^{-t} - J\boldsymbol{L}^t\boldsymbol{F}^{-t} \\ &= [(tr\boldsymbol{D})\boldsymbol{I} - \boldsymbol{L}^t]\boldsymbol{F}^*.\end{aligned}$$

Then,

$$\dot{\alpha}\boldsymbol{n} + \alpha\dot{\boldsymbol{n}} = \alpha[(tr\boldsymbol{D})\boldsymbol{I} - \boldsymbol{L}^t]\boldsymbol{n}.$$

Because $\boldsymbol{n}$ is a unit vector we have $\boldsymbol{n} \cdot \dot{\boldsymbol{n}} = 0$. Thus,

$$\dot{\alpha}/\alpha = tr\boldsymbol{D} - \boldsymbol{n} \cdot \boldsymbol{L}^t\boldsymbol{n} = tr\boldsymbol{D} - \boldsymbol{n} \cdot \boldsymbol{D}\boldsymbol{n},$$

and

$$\dot{\boldsymbol{n}} = [(tr\boldsymbol{D})\boldsymbol{I} - \boldsymbol{L}^t]\boldsymbol{n} - (\dot{\alpha}/\alpha)\boldsymbol{n} = (\boldsymbol{n} \cdot \boldsymbol{D}\boldsymbol{n})\boldsymbol{n} - \boldsymbol{L}^t\boldsymbol{n}.$$

8.

Use $(4.5)_1$ to obtain the general relations

$$\begin{aligned}\boldsymbol{D} &= \frac{1}{2}\boldsymbol{R}(\dot{\boldsymbol{U}}\boldsymbol{U}^{-1} + \boldsymbol{U}^{-1}\dot{\boldsymbol{U}})\boldsymbol{R}^t \quad \text{and} \\ \boldsymbol{W} &= \dot{\boldsymbol{R}}\boldsymbol{R}^t + \frac{1}{2}\boldsymbol{R}(\dot{\boldsymbol{U}}\boldsymbol{U}^{-1} - \boldsymbol{U}^{-1}\dot{\boldsymbol{U}})\boldsymbol{R}^t.\end{aligned}$$

Use the spectral representation of $\boldsymbol{U}$, in the special case for which $\dot{\boldsymbol{u}}_i = \boldsymbol{0}$, to get $\dot{\boldsymbol{U}} = \sum \dot{\lambda}_i \boldsymbol{u}_i \otimes \boldsymbol{u}_i$. Then, $\dot{\boldsymbol{U}}\boldsymbol{U}^{-1} = \boldsymbol{U}^{-1}\dot{\boldsymbol{U}} = \sum(\dot{\lambda}_i/\lambda_i)\boldsymbol{u}_i \otimes \boldsymbol{u}_i = \sum(\ln \lambda_i)^{\cdot}\boldsymbol{u}_i \otimes \boldsymbol{u}_i$. Thus,

$$\boldsymbol{D} = \sum (\ln \lambda_i)^{\cdot} \boldsymbol{R}\boldsymbol{u}_i \otimes \boldsymbol{R}\boldsymbol{u}_i = \sum (\ln \lambda_i)^{\cdot} \boldsymbol{v}_i \otimes \boldsymbol{v}_i \quad \text{and}$$

$$\boldsymbol{W} = \dot{\boldsymbol{R}}\boldsymbol{R}^t = \dot{\boldsymbol{v}}_i \otimes \boldsymbol{R}\boldsymbol{u}_i = \dot{\boldsymbol{v}}_i \otimes \boldsymbol{v}_i.$$

Accordingly, $\dot{\boldsymbol{v}}_i = \boldsymbol{W}\boldsymbol{v}_i = \boldsymbol{w} \times \boldsymbol{v}_i$ when the principal axes are fixed in the body.

9.

(a) $\boldsymbol{L}d\boldsymbol{x} = d\boldsymbol{v} = (\frac{W}{\theta_0})\boldsymbol{e}_3 d\theta = (\frac{W}{r\theta_0})\boldsymbol{e}_3(\boldsymbol{e}_\theta \cdot d\boldsymbol{x}) = [(\frac{W}{r\theta_0})\boldsymbol{e}_3 \otimes \boldsymbol{e}_\theta]d\boldsymbol{x}$. Thus, $\boldsymbol{L} = (\frac{W}{r\theta_0})\boldsymbol{e}_3 \otimes \boldsymbol{e}_\theta$, $\boldsymbol{D} = \frac{1}{2}(\frac{W}{r\theta_0})(\boldsymbol{e}_3 \otimes \boldsymbol{e}_\theta + \boldsymbol{e}_\theta \otimes \boldsymbol{e}_3)$ and $\boldsymbol{W} = \frac{1}{2}(\frac{W}{r\theta_0})(\boldsymbol{e}_3 \otimes \boldsymbol{e}_\theta - \boldsymbol{e}_\theta \otimes \boldsymbol{e}_3)$.

(b) Let

$$\boldsymbol{\nu}_1 = \frac{\sqrt{2}}{2}(\boldsymbol{e}_3 + \boldsymbol{e}_\theta) \quad \text{and} \quad \boldsymbol{\nu}_2 = \frac{\sqrt{2}}{2}(\boldsymbol{e}_3 - \boldsymbol{e}_\theta).$$

Solving these for $\boldsymbol{e}_3$ and $\boldsymbol{e}_\theta$, after a little algebra we obtain

$$\boldsymbol{D} = \frac{1}{2}\left(\frac{W}{r\theta_0}\right)(\boldsymbol{\nu}_2 \otimes \boldsymbol{\nu}_2 - \boldsymbol{\nu}_1 \otimes \boldsymbol{\nu}_1).$$

This is in spectral form, and therefore the principal vectors of $\boldsymbol{D}$ are $\boldsymbol{\nu}_1$, $\boldsymbol{\nu}_2$ and $\boldsymbol{\nu}_3 = \boldsymbol{e}_r$. The motion is isochoric because $tr\boldsymbol{D} = 0$.

(c) To obtain the vorticity $\boldsymbol{w}$, use

$$\boldsymbol{w} \times \boldsymbol{u} = \boldsymbol{W}\boldsymbol{u} = \frac{1}{2}\left(\frac{W}{r\theta_0}\right)(u_\theta \boldsymbol{e}_3 - u_z \boldsymbol{e}_\theta) = \frac{1}{2}\left(\frac{W}{r\theta_0}\right)\boldsymbol{e}_r \times \boldsymbol{u}$$

for arbitrary $\boldsymbol{u} = u_r\boldsymbol{e}_r + u_\theta\boldsymbol{e}_\theta + u_z\boldsymbol{e}_3$. Thus, $\boldsymbol{w} = \frac{1}{2}(\frac{W}{r\theta_0})\boldsymbol{e}_r$.

10.

We have $\boldsymbol{v} = rf(r)\boldsymbol{e}_\theta$ and therefore $\boldsymbol{L}d\boldsymbol{x} = d\boldsymbol{v} = d(rf)\boldsymbol{e}_\theta + rfd\boldsymbol{e}_\theta = (rf)'\boldsymbol{e}_\theta dr - rf\boldsymbol{e}_r d\theta = (rf)'\boldsymbol{e}_\theta(\boldsymbol{e}_r \cdot d\boldsymbol{x}) - f\boldsymbol{e}_r(\boldsymbol{e}_\theta \cdot d\boldsymbol{x})$, and so

$$\boldsymbol{L} = (rf)'\boldsymbol{e}_\theta \otimes \boldsymbol{e}_r - f\boldsymbol{e}_r \otimes \boldsymbol{e}_\theta,$$

as claimed. Note that $tr\boldsymbol{L} = 0$, as it must be in any flow of an incompressible fluid. We obtain

$$\begin{aligned}2\boldsymbol{D} = \boldsymbol{L} + \boldsymbol{L}^t &= (rf)'(\boldsymbol{e}_\theta \otimes \boldsymbol{e}_r + \boldsymbol{e}_r \otimes \boldsymbol{e}_\theta) - f(\boldsymbol{e}_r \otimes \boldsymbol{e}_\theta + \boldsymbol{e}_\theta \otimes \boldsymbol{e}_r)\\ &= rf'(\boldsymbol{e}_r \otimes \boldsymbol{e}_\theta + \boldsymbol{e}_\theta \otimes \boldsymbol{e}_r),\end{aligned}$$

and

$$\begin{aligned}2\boldsymbol{W} = \boldsymbol{L} - \boldsymbol{L}^t &= (rf)'(\boldsymbol{e}_\theta \otimes \boldsymbol{e}_r - \boldsymbol{e}_r \otimes \boldsymbol{e}_\theta) - f(\boldsymbol{e}_r \otimes \boldsymbol{e}_\theta - \boldsymbol{e}_\theta \otimes \boldsymbol{e}_r)\\ &= (rf' + 2f)(\boldsymbol{e}_\theta \otimes \boldsymbol{e}_r - \boldsymbol{e}_r \otimes \boldsymbol{e}_\theta).\end{aligned}$$

Chapter 5

1.

Let the fixed curve C be the image of C_t in κ. We have

$$\begin{aligned}\frac{d}{dt}\int_{C_t} \boldsymbol{a}\cdot d\boldsymbol{x} &= \frac{d}{dt}\int_{C} \boldsymbol{a}\cdot \boldsymbol{F}d\boldsymbol{X}\\ &= \frac{d}{dt}\int_{C} \boldsymbol{F}^t\boldsymbol{a}\cdot d\boldsymbol{X}\\ &= \int_{C} (\boldsymbol{F}^t\boldsymbol{a})^{\cdot}\cdot d\boldsymbol{X}\\ &= \int_{C} (\dot{\boldsymbol{F}}^t\boldsymbol{a} + \boldsymbol{F}^t\dot{\boldsymbol{a}})\cdot d\boldsymbol{X}\\ &= \int_{C} \boldsymbol{F}^t(\boldsymbol{L}^t\boldsymbol{a} + \dot{\boldsymbol{a}})\cdot d\boldsymbol{X}\\ &= \int_{C} (\dot{\boldsymbol{a}} + \boldsymbol{L}^t\boldsymbol{a})\cdot \boldsymbol{F}d\boldsymbol{X}\\ &= \int_{C_t} (\dot{\boldsymbol{a}} + \boldsymbol{L}^t\boldsymbol{a})\cdot d\boldsymbol{x}\\ &= \int_{C_t} [\boldsymbol{a}' + (grad\boldsymbol{a})\boldsymbol{v} + \boldsymbol{L}^t\boldsymbol{a}]\cdot d\boldsymbol{x}.\end{aligned}$$

2.

The problem is to establish that

$$div(\boldsymbol{u} \otimes \boldsymbol{v}) - (grad\boldsymbol{v})\boldsymbol{u} = (div\boldsymbol{u})\boldsymbol{v} - curl(\boldsymbol{v} \times \boldsymbol{u}).$$

From the result of Problem 24(g) in Chapter 1 we have that $div(\boldsymbol{u} \otimes \boldsymbol{v}) = (grad\boldsymbol{u})\boldsymbol{v} + (div\boldsymbol{v})\boldsymbol{u}$. Thus, our problem is equivalent to proving that

$$curl(\boldsymbol{v} \times \boldsymbol{u}) = (grad\boldsymbol{v})\boldsymbol{u} - (grad\boldsymbol{u})\boldsymbol{v} + (div\boldsymbol{u})\boldsymbol{v} - (div\boldsymbol{v})\boldsymbol{u}. \qquad (*)$$

To verify this we use (1.164) to obtain

$$\begin{aligned} \boldsymbol{e}_j \cdot curl(\boldsymbol{v} \times \boldsymbol{u}) &= e_{ijk} e_{ilm} (v_l u_m)_{,k} \\ &= (\delta_{jl}\delta_{km} - \delta_{jm}\delta_{kl})(v_l u_m)_{,k} \\ &= (v_j u_k)_{,k} - (v_k u_j)_{,k} \\ &= v_{j,k} u_k - u_{j,k} v_k + u_{k,k} v_j - v_{k,k} u_j, \end{aligned}$$

which is precisely the inner product of $\boldsymbol{e}_j$ with the right-hand side of (*).

Chapter 6

2.

Multiply (6.52) by ρ_κ and integrate over $P \subset \kappa$ to get

$$\begin{aligned} m(S)\bar{\boldsymbol{x}} &= \int_{P_t} \rho \boldsymbol{x} dv = \int_P \rho_\kappa \boldsymbol{x} dV = \boldsymbol{Q}(t) \int_P \rho_\kappa \boldsymbol{X} dV + m(S)\boldsymbol{c}(t) \\ &= m(S)\boldsymbol{Q}(t)\bar{\boldsymbol{X}} + m(S)\boldsymbol{c}(t), \end{aligned}$$

where

$$\bar{\boldsymbol{X}} = [m(S)]^{-1} \int_P \rho_\kappa \boldsymbol{X} dV$$

is the center of mass of P. Thus,

$$\bar{\boldsymbol{x}} = \boldsymbol{Q}(t)\bar{\boldsymbol{X}} + \boldsymbol{c}(t),$$

and (6.53) follows from (6.41) and (6.52).

3.

We decompose $\boldsymbol{J}$ in the basis $\{\boldsymbol{e}_i^* \otimes \boldsymbol{e}_j^*\}$, i.e, $\boldsymbol{J} = J_{ij}\boldsymbol{e}_i^* \otimes \boldsymbol{e}_j^*$. Then, $\dot{\boldsymbol{J}} = \dot{J}_{ij}\boldsymbol{e}_i^* \otimes \boldsymbol{e}_j^* + J_{ij}(\dot{\boldsymbol{e}}_i^* \otimes \boldsymbol{e}_j^* + \boldsymbol{e}_i^* \otimes \dot{\boldsymbol{e}}_j^*)$. Using $\dot{\boldsymbol{e}}_i^* = \boldsymbol{\omega} \times \boldsymbol{e}_i^* = \boldsymbol{\Omega}\boldsymbol{e}_i^*$, we reduce this to

$$\begin{aligned}\dot{\boldsymbol{J}} &= \dot{J}_{ij}\boldsymbol{e}_i^* \otimes \boldsymbol{e}_j^* + J_{ij}(\boldsymbol{\Omega}\boldsymbol{e}_i^* \otimes \boldsymbol{e}_j^* + \boldsymbol{e}_i^* \otimes \boldsymbol{\Omega}\boldsymbol{e}_j^*)\\ &= \dot{J}_{ij}\boldsymbol{e}_i^* \otimes \boldsymbol{e}_j^* + \boldsymbol{\Omega}\boldsymbol{J} + \boldsymbol{J}\boldsymbol{\Omega}^t = \dot{J}_{ij}\boldsymbol{e}_i^* \otimes \boldsymbol{e}_j^* + \boldsymbol{\Omega}\boldsymbol{J} - \boldsymbol{J}\boldsymbol{\Omega},\end{aligned}$$

and the result follows from (6.68).

4.

We have $\boldsymbol{a} \cdot \boldsymbol{J}\boldsymbol{a} = \int_{P_t} \boldsymbol{a} \cdot \boldsymbol{A}\boldsymbol{a} dv$, where $\boldsymbol{A} = \rho[(\boldsymbol{\pi} \cdot \boldsymbol{\pi})\boldsymbol{I} - \boldsymbol{\pi} \otimes \boldsymbol{\pi}]$. Then, $\boldsymbol{a} \cdot \boldsymbol{A}\boldsymbol{a} = \rho[|\boldsymbol{\pi}|^2 |\boldsymbol{a}|^2 - (\boldsymbol{\pi} \cdot \boldsymbol{a})^2] \geq 0$ for all $\boldsymbol{a}$, by the Cauchy–Schwarz inequality (see Problem 1 in Chapter 1). Note that $\boldsymbol{a} \cdot \boldsymbol{A}\boldsymbol{a} = 0$ if and only if

$$(\boldsymbol{\pi} \cdot \boldsymbol{a})^2 = |\boldsymbol{\pi}|^2 |\boldsymbol{a}|^2 . \tag{*}$$

Using a result from Problem 11(a) in Chapter 1, we have $|\boldsymbol{\pi}|^2 \boldsymbol{a} = (\boldsymbol{a} \cdot \boldsymbol{\pi})\boldsymbol{\pi} + \boldsymbol{\pi} \times (\boldsymbol{a} \times \boldsymbol{\pi})$, and therefore

$$\begin{aligned}|\boldsymbol{\pi}|^2 |\boldsymbol{a}|^2 &= (\boldsymbol{\pi} \cdot \boldsymbol{a})^2 + \boldsymbol{a} \cdot \boldsymbol{\pi} \times (\boldsymbol{a} \times \boldsymbol{\pi})\\ &= (\boldsymbol{\pi} \cdot \boldsymbol{a})^2 + \boldsymbol{a} \times \boldsymbol{\pi} \cdot \boldsymbol{a} \times \boldsymbol{\pi} = (\boldsymbol{\pi} \cdot \boldsymbol{a})^2 + |\boldsymbol{a} \times \boldsymbol{\pi}|^2 .\end{aligned}$$

Thus, (*) is true if and only if $\boldsymbol{a} \times \boldsymbol{\pi} = \boldsymbol{0}$, i.e., if and only if $\boldsymbol{\pi} = \beta\boldsymbol{a}$ for some $\beta \in \mathbb{R}$. Thus, $\boldsymbol{a} \cdot \boldsymbol{A}\boldsymbol{a} \geq 0$ for all $\boldsymbol{a}$, with equality if and only if $\boldsymbol{\pi} = \beta\boldsymbol{a}$ for some $\beta \in \mathbb{R}$. But there are infinitely many $\boldsymbol{\pi} \in P_t$ that do not have this property (unless P_t is a one-dimensional body aligned with $\boldsymbol{a}$, not the three-dimensional body we are considering here). For these, $\boldsymbol{a} \cdot \boldsymbol{A}\boldsymbol{a} > 0$ for all $\boldsymbol{a} \neq \boldsymbol{0}$, and therefore

$$\boldsymbol{a} \cdot \boldsymbol{J}\boldsymbol{a} = \int_{P_t} \boldsymbol{a} \cdot \boldsymbol{A}\boldsymbol{a} dv > 0 \quad \text{for all} \quad \boldsymbol{a} \neq \boldsymbol{0}.$$

5.

We have $\boldsymbol{T} = \sum \sigma_i \boldsymbol{n}_i \otimes \boldsymbol{n}_i$. Let $\boldsymbol{n} = (\boldsymbol{n} \cdot \boldsymbol{n}_i)\boldsymbol{n}_i = n_i\boldsymbol{n}_i$, with $n_1 = \cos\phi\cos\theta$, $n_2 = \cos\phi\sin\theta$, $n_3 = \sin\phi$. Then, $\boldsymbol{t} = \boldsymbol{T}\boldsymbol{n} = \sum \sigma_i \boldsymbol{n}_i (\boldsymbol{n}_i \cdot \boldsymbol{n}) = \sum \sigma_i n_i \boldsymbol{n}_i$, and $\tau\boldsymbol{s} = \boldsymbol{t} - (\boldsymbol{n} \cdot \boldsymbol{t})\boldsymbol{n}$, where $\boldsymbol{n} \cdot \boldsymbol{t} = \sum \sigma_i n_i^2$.

(a) Thus, $\tau \boldsymbol{s} = \sum_i [\sigma_i n_i - (\sum_j \sigma_j n_j^2) n_i] \boldsymbol{n}_i$.
(b) In the first octant, we have $n_1 = n_2 = n_3 = n(> 0)$, so $\boldsymbol{n} = n(\boldsymbol{n}_1 + \boldsymbol{n}_2 + \boldsymbol{n}_3)$. Then $|\boldsymbol{n}| = 1$ implies that $n = 1/\sqrt{3}$, yielding

$$\begin{aligned}\tau \boldsymbol{s} &= \frac{1}{\sqrt{3}} \sum_i [\sigma_i - \frac{1}{3}(\sum_j \sigma_j)] \boldsymbol{n}_i \\ &= \frac{1}{3\sqrt{3}}[2\sigma_1 - (\sigma_2 + \sigma_3)]\boldsymbol{n}_1 + \frac{1}{3\sqrt{3}}[2\sigma_2 - (\sigma_1 + \sigma_3)]\boldsymbol{n}_2 \\ &\quad + \frac{1}{3\sqrt{3}}[2\sigma_3 - (\sigma_1 + \sigma_2)]\boldsymbol{n}_3,\end{aligned}$$

and therefore

$$\begin{aligned}\tau_{oct}^2 = \frac{1}{27}\{&[2\sigma_1 - (\sigma_2 + \sigma_3)]^2 + [2\sigma_2 - (\sigma_1 + \sigma_3)]^2 \\ &+ [2\sigma_3 - (\sigma_1 + \sigma_2)]^2\}.\end{aligned}$$

6.

(a) $\boldsymbol{P} = \boldsymbol{P}\hat{\boldsymbol{I}} = \boldsymbol{P}(\hat{\boldsymbol{E}}_A \otimes \hat{\boldsymbol{E}}_A) = \boldsymbol{P}\hat{\boldsymbol{E}}_A \otimes \hat{\boldsymbol{E}}_A = \boldsymbol{p}_A \otimes \hat{\boldsymbol{E}}_A$.
(b) $P_{iA} = \boldsymbol{e}_i \cdot \boldsymbol{P}\hat{\boldsymbol{E}}_A = \boldsymbol{e}_i \cdot \boldsymbol{p}_A = \boldsymbol{e}_i \cdot \boldsymbol{p}(\hat{\boldsymbol{E}}_A)$. Thus,

$$(P_{iA}) = \begin{pmatrix} a & 0 & 0 \\ 10 & 3 & c \\ 0 & b-6 & 3 \end{pmatrix}.$$

(c) Use $J\boldsymbol{T} = \boldsymbol{P}\boldsymbol{F}^t$ to get

$$(JT_{ij}) = (P_{iA}F_{jA}) = \begin{pmatrix} 2a & a & 0 \\ 29 & 16 & c \\ 3b-18 & 2b-12 & 3 \end{pmatrix}.$$

This must be symmetric. Therefore, $a = 29$, $b = 6$, $c = 0$, and

$$(JT_{ij}) = \begin{pmatrix} 58 & 29 & 0 \\ 29 & 16 & 0 \\ 0 & 0 & 3 \end{pmatrix}.$$

(d) $J = \det \boldsymbol{F} = \boldsymbol{F}\hat{\boldsymbol{E}}_1 \times \boldsymbol{F}\hat{\boldsymbol{E}}_2 \cdot \boldsymbol{F}\hat{\boldsymbol{E}}_3 = F_{i1}F_{j2}F_{k3}e_{ijk} = 1$, so the above matrix is (T_{ij}).

7.

We have $\boldsymbol{F} = F_{iA}\boldsymbol{e}_i \otimes \hat{\boldsymbol{E}}_A$, where

$$(F_{iA}) = \begin{pmatrix} a_1 & ba_1 & 0 \\ 0 & a_2 & 0 \\ 0 & 0 & a_3 \end{pmatrix}.$$

Then, $0 < J = \det \boldsymbol{F} = \boldsymbol{F}\hat{\boldsymbol{E}}_1 \times \boldsymbol{F}\hat{\boldsymbol{E}}_2 \cdot \boldsymbol{F}\hat{\boldsymbol{E}}_3 = a_1\boldsymbol{e}_1 \times (ba_1\boldsymbol{e}_1 + a_2\boldsymbol{e}_2) \cdot a_3\boldsymbol{e}_3 = a_1a_2a_3$, so $a_i \neq 0$. It is easy to verify that

$$\boldsymbol{F}^{-1} = a_1^{-1}\hat{\boldsymbol{E}}_1 \otimes \boldsymbol{e}_1 - ba_2^{-1}\hat{\boldsymbol{E}}_1 \otimes \boldsymbol{e}_2 + a_2^{-1}\hat{\boldsymbol{E}}_2 \otimes \boldsymbol{e}_2 + a_3^{-1}\hat{\boldsymbol{E}}_3 \otimes \boldsymbol{e}_3,$$

i.e., that $\boldsymbol{F}\boldsymbol{F}^{-1} = \boldsymbol{e}_i \otimes \boldsymbol{e}_i = \boldsymbol{I}$.

Next, $\boldsymbol{T} = \boldsymbol{t}(\boldsymbol{e}_i) \otimes \boldsymbol{e}_i = \boldsymbol{t}(\boldsymbol{e}_1) \otimes \boldsymbol{e}_1 + \boldsymbol{t}(\boldsymbol{e}_2) \otimes \boldsymbol{e}_2 + \boldsymbol{t}(\boldsymbol{e}_3) \otimes \boldsymbol{e}_3 = \tau(\boldsymbol{e}_1 \otimes \boldsymbol{e}_2 + \boldsymbol{e}_2 \otimes \boldsymbol{e}_1)$. This is symmetric, as required. We easily verify that $\boldsymbol{T} = \tau(\boldsymbol{m} \otimes \boldsymbol{m} - \boldsymbol{n} \otimes \boldsymbol{n})$, where $\boldsymbol{m} = \pm\frac{1}{\sqrt{2}}(\boldsymbol{e}_1 + \boldsymbol{e}_2)$ and $\boldsymbol{n} = \pm\frac{1}{\sqrt{2}}(\boldsymbol{e}_1 - \boldsymbol{e}_2)$. This is in spectral form, and the principal stresses are $\pm\tau, 0$, corresponding to tension (of amount τ, if $\tau > 0$) along $\boldsymbol{m} \otimes \boldsymbol{m}$ and compression along $\boldsymbol{n} \otimes \boldsymbol{n}$.

Next, $\boldsymbol{P} = \boldsymbol{T}\boldsymbol{F}^* = J\boldsymbol{T}\boldsymbol{F}^{-t}$. Thus,

$$\boldsymbol{P} = a_1a_2a_3\tau(\boldsymbol{e}_1 \otimes \boldsymbol{F}^{-1}\boldsymbol{e}_2 + \boldsymbol{e}_2 \otimes \boldsymbol{F}^{-1}\boldsymbol{e}_1),$$

where

$$\boldsymbol{F}^{-1}\boldsymbol{e}_1 = a_1^{-1}\hat{\boldsymbol{E}}_1 \quad \text{and} \quad \boldsymbol{F}^{-1}\boldsymbol{e}_2 = a_2^{-1}(\hat{\boldsymbol{E}}_2 - b\hat{\boldsymbol{E}}_1).$$

Then,

$$\boldsymbol{P} = \tau[a_2a_3\boldsymbol{e}_2 \otimes \hat{\boldsymbol{E}}_1 + a_1a_3\boldsymbol{e}_1 \otimes (\hat{\boldsymbol{E}}_2 - b\hat{\boldsymbol{E}}_1)].$$

Finally,

$$\begin{aligned} \boldsymbol{S} = \boldsymbol{F}^{-1}\boldsymbol{P} &= \tau[a_2a_3\boldsymbol{F}^{-1}\boldsymbol{e}_2 \otimes \hat{\boldsymbol{E}}_1 + a_1a_3\boldsymbol{F}^{-1}\boldsymbol{e}_1 \otimes (\hat{\boldsymbol{E}}_2 - b\hat{\boldsymbol{E}}_1)] \\ &= \tau a_3(\hat{\boldsymbol{E}}_1 \otimes \hat{\boldsymbol{E}}_2 + \hat{\boldsymbol{E}}_2 \otimes \hat{\boldsymbol{E}}_1 - 2b\hat{\boldsymbol{E}}_1 \otimes \hat{\boldsymbol{E}}_1). \end{aligned}$$

This is also symmetric, as required.

8.

To establish (6.138) we use $Div\boldsymbol{P} = P_{iA,A}\boldsymbol{e}_i$ with $P_{iA} = T_{ij}F^*_{jA}$. Then, $P_{iA,A} = T_{ij}F^*_{jA,A} + T_{ij,A}F^*_{jA} = T_{ij,A}F^*_{jA}$ because $F^*_{jA,A} = 0$. Continuing, $P_{iA,A} = T_{ij,k}x_{k,A}F^*_{jA} = T_{ij,k}F^*_{jA}F_{kA}$, where $F^*_{jA}F_{kA}$ is the j,k component of $\boldsymbol{F}^*\boldsymbol{F}^t = J\boldsymbol{I}$, namely $J\delta_{jk}$. Thus, $Div\boldsymbol{P} = JT_{ij,k}\delta_{jk}\boldsymbol{e}_i = JT_{ij,j}\boldsymbol{e}_i = Jdiv\boldsymbol{T}$.

Equation (6.175) follows similarly: $Div(J\boldsymbol{F}^{-1}\boldsymbol{q}) = Div\boldsymbol{q}_\kappa = q_{(\kappa)A,A} = (F^*_{iA}q_i)_{,A} = F^*_{iA,A}q_i + F^*_{iA}x_{j,A}q_{i,j} = F^*_{iA}F_{jA}q_{i,j} = J\delta_{ij}q_{i,j} = Jq_{i,i} = Jdiv\boldsymbol{q}$.

9.

In equilibrium, we have $\rho b_i = -T_{ij,j}$, and so $\rho b_i u_i = -T_{ij,j}u_i = -(T_{ij}u_i)_{,j} + T_{ij}u_{i,j}$. Thus,

$$\int_{P_t} \rho b_i u_i dv = \int_{P_t} T_{ij}u_{i,j}dv - \int_{\partial P_t} T_{ij}u_i n_j da = \int_{P_t} T_{ij}u_{i,j}dv - \int_{\partial P_t} t_i u_i da.$$

Note that this applies to dynamics too, if we make the substitution $b_i \to b_i - \dot{v}_i$.

10.

We have $V\bar{T}_{ij} = \int_{\kappa_t} T_{ij}dv$.

(a) In equilibrium we have $\rho b_i x_j = -T_{ik,k}x_j = -(T_{ik}x_j)_{,k} + T_{ik}x_{j,k} = -(T_{ik}x_j)_{,k} + T_{ij}$, where we've used $x_{j,k} = \delta_{jk}$. Therefore,

$$\begin{aligned} V\bar{T}_{ij} &= \int_{\kappa_t} [(T_{ik}x_j)_{,k} + \rho b_i x_j]dv \\ &= \int_{\partial\kappa_t} T_{ik}n_k x_j da + \int_{\kappa_t} \rho b_i x_j dv \\ &= \int_{\partial\kappa_t} t_i x_j da + \int_{\kappa_t} \rho b_i x_j dv. \end{aligned}$$

(b) Thus,

$$Ve_{kji}\bar{T}_{ij} = e_{kji}\left(\int_{\partial\kappa_t} x_j t_i da + \int_{\kappa_t} \rho x_j b_i dv\right) = 0,$$

by Euler's second postulate, and therefore

$$\bar{T}_{ij} = \bar{T}_{ji}.$$

Note that these results are also valid for dynamics. Simply replace b_i by $b_i - \dot{v}_i$, as in the previous problem.

(c) We have

$$V\bar{T}_{33} = \int_{\partial\kappa_t} t_3 x_3 da + \int_{\kappa_t} \rho b_3 x_3 dv, \tag{1}$$

where

$$\int_{\kappa_t} \rho b_3 x_3 dv = -g \int_{\kappa_t} \rho x_3 dv = -gM\bar{x}_3,$$

in which

$$M = \int_{\kappa_t} \rho dv = V\bar{\rho},$$

where $\bar{\rho}$ is the average mass density, and so

$$\int_{\kappa_t} \rho b_3 x_3 dv = -(\bar{\rho} g \bar{x}_3)V. \tag{2}$$

Consider $\int_{\partial\kappa_t} t_3 x_3 da$. The only place where t_3 is non-zero is on the plane, where $x_3 = 0$. Therefore, this integral is zero. Then (1) and (2) combine to give

$$\bar{T}_{33} = -\bar{\rho} g \bar{x}_3.$$

If we replace x_3 by $x_3 + c$, i.e., if we shift the origin up or down by $|c|$, we get

$$V\bar{T}_{33} = \int_{\partial\kappa_t} t_3 x_3 da + \int_{\kappa_t} \rho b_3 x_3 dv + c\left(\int_{\partial\kappa_t} t_3 da + \int_{\kappa_t} \rho b_3 dv\right).$$

The first two integrals add up to $-(\bar{\rho} g \bar{x}_3)V$, as before, and the integrals in parentheses add up to zero (by equilibrium), so the result is independent of the origin.

(d) We have $V\bar{T}_{ij} = \int_{\partial\kappa_t} t_i x_j da$ in the absence of body force, where, for the block, $V = whl_2$. Thus,

$$V\bar{T}_{21} = \int_{\partial\kappa_t} t_2 x_1 da = -p\int_{-w}^{0}\int_{l_1}^{l_2} x_1 dx_1 dx_3 = -\frac{pw}{2}(l_2^2 - l_1^2).$$

We observe that the built-in wall makes no contribution because $x_1 = 0$ there. Finally,

$$\bar{T}_{21} = -\frac{p}{2hl_2}(l_2^2 - l_1^2),$$

and the result of part (b) gives $\bar{T}_{12} = \bar{T}_{21}$. The formula $V\bar{T}_{12} = \int_{\partial\kappa_t} t_1 x_2 da$ is also valid, of course, but it is not useful here because we do not know the distribution of t_1 at the wall.

11.

We wrote the energy balance in the form

$$\frac{d}{dt}\int_{P_t}\rho\left(\frac{1}{2}|\boldsymbol{v}|^2 + \varepsilon\right)dv = \int_{P_t}\rho(\boldsymbol{b}\cdot\boldsymbol{v} + r)dv + \int_{\partial P_t}(\boldsymbol{t}\cdot\boldsymbol{v} + h)da.$$

Thus, with $\phi = \frac{1}{2}|\boldsymbol{v}|^2 + \varepsilon$ we have $\sigma = \boldsymbol{b}\cdot\boldsymbol{v} + r$, and

$$\boldsymbol{\pi}\cdot\boldsymbol{n} = \boldsymbol{t}\cdot\boldsymbol{v} + h = \boldsymbol{T}\boldsymbol{n}\cdot\boldsymbol{v} - \boldsymbol{q}\cdot\boldsymbol{n} = (\boldsymbol{T}^t\boldsymbol{v} - \boldsymbol{q})\cdot\boldsymbol{n},$$

yielding the energy flux

$$\boldsymbol{\pi} = \boldsymbol{T}^t\boldsymbol{v} - \boldsymbol{q} = \boldsymbol{T}\boldsymbol{v} - \boldsymbol{q}.$$

12.

(a) $(\boldsymbol{AB})^* = \det(\boldsymbol{AB})(\boldsymbol{AB})^{-t} = (\det\boldsymbol{A})(\det\boldsymbol{B})(\boldsymbol{B}^{-1}\boldsymbol{A}^{-1})^t = [(\det\boldsymbol{A})\boldsymbol{A}^{-t}][(\det\boldsymbol{B})\boldsymbol{B}^{-t}] = \boldsymbol{A}^*\boldsymbol{B}^*$.

(b) If $\boldsymbol{Q}$ is a rotation, then $\det\boldsymbol{Q} = 1$, $\boldsymbol{Q}^{-1} = \boldsymbol{Q}^t$ and $\boldsymbol{Q}^{-t} = \boldsymbol{Q}$. Thus, $\boldsymbol{Q}^* = (\det\boldsymbol{Q})\boldsymbol{Q}^{-t} = \boldsymbol{Q}$. The proof of the converse statement is left to the reader.

(c) Using the polar decomposition $\boldsymbol{F} = \boldsymbol{RU}$, we write the stress power as

$$\begin{aligned}\mathcal{S}(S,t) &= \int_P \boldsymbol{P}\cdot\dot{\boldsymbol{F}}dV = \int_P \boldsymbol{P}\cdot(\dot{\boldsymbol{R}}\boldsymbol{U} + \boldsymbol{R}\dot{\boldsymbol{U}})dV \\ &= \int_P tr(\boldsymbol{P}\boldsymbol{U}\dot{\boldsymbol{R}}^t + \boldsymbol{P}\dot{\boldsymbol{U}}\boldsymbol{R}^t)dV.\end{aligned}$$

We have $\boldsymbol{P} = \boldsymbol{TF}^* = \boldsymbol{T}(\boldsymbol{RU})^* = \boldsymbol{TR}^*\boldsymbol{U}^* = \boldsymbol{TRU}^* = J\boldsymbol{TRU}^{-1}$. Thus, $tr(\boldsymbol{PU}\dot{\boldsymbol{R}}^t) = Jtr(\boldsymbol{TR}\dot{\boldsymbol{R}}^t) = J\boldsymbol{T}\cdot\dot{\boldsymbol{R}}\boldsymbol{R}^t$, which vanishes because $\boldsymbol{T}$ is symmetric whereas $\dot{\boldsymbol{R}}\boldsymbol{R}^t$ is skew. Then the stress power simplifies to

$$\mathcal{S}(S,t) = \int_P tr(\boldsymbol{P}\dot{\boldsymbol{U}}\boldsymbol{R}^t)dV = \int_P tr(\boldsymbol{R}^t\boldsymbol{P}\dot{\boldsymbol{U}})dV = \int_P \boldsymbol{R}^t\boldsymbol{P}\cdot\dot{\boldsymbol{U}}dV.$$

Because $\dot{\boldsymbol{U}}$ is symmetric, the inner product in the final integral picks up only the symmetric part of the first factor. Thus,

$$\mathcal{S}(S,t) = \int_P \boldsymbol{\sigma}\cdot\dot{\boldsymbol{U}}dV, \quad \text{where} \quad \boldsymbol{\sigma} = Sym(\boldsymbol{R}^t\boldsymbol{P}) = \frac{1}{2}(\boldsymbol{R}^t\boldsymbol{P}+\boldsymbol{P}^t\boldsymbol{R}).$$

(d) $\boldsymbol{\Sigma} = Sym(\boldsymbol{PR}^t) = \frac{1}{2}(\boldsymbol{PR}^t + \boldsymbol{RP}^t)$. This yields $\boldsymbol{R}^t\boldsymbol{\Sigma}\boldsymbol{R} = \frac{1}{2}(\boldsymbol{R}^t\boldsymbol{P} + \boldsymbol{P}^t\boldsymbol{R}) = \boldsymbol{\sigma}$. Thus,

$$\begin{aligned}\mathcal{S}(S,t) &= \int_P \boldsymbol{R}^t\boldsymbol{\Sigma}\boldsymbol{R}\cdot\dot{\boldsymbol{U}}dV = \int_P tr(\boldsymbol{R}^t\boldsymbol{\Sigma}\boldsymbol{R}\dot{\boldsymbol{U}})dV \\ &= \int_P tr(\boldsymbol{\Sigma}\boldsymbol{R}\dot{\boldsymbol{U}}\boldsymbol{R}^t)dV = \int_P \boldsymbol{\Sigma}\cdot\boldsymbol{A}dV,\end{aligned}$$

where $\boldsymbol{A} = \boldsymbol{R}\dot{\boldsymbol{U}}\boldsymbol{R}^t$.

Chapter 7

1.

Differentiate (7.8) to obtain

$$\boldsymbol{a}^+ = \dot{\boldsymbol{v}}^+ = \boldsymbol{Q}\dot{\boldsymbol{v}} + \dot{\boldsymbol{Q}}\boldsymbol{v} + \dot{\boldsymbol{\Omega}}(\boldsymbol{x}^+ - \boldsymbol{c}) + \boldsymbol{\Omega}(\dot{\boldsymbol{x}}^+ - \dot{\boldsymbol{c}}) + \ddot{\boldsymbol{c}}.$$

Use $\dot{\boldsymbol{x}}^+ = \boldsymbol{v}^+$ and $\dot{\boldsymbol{v}} = \boldsymbol{a}$, and solve for $\boldsymbol{v}$ from (7.8):

$$\begin{aligned}\boldsymbol{a}^+ - \boldsymbol{Q}\boldsymbol{a} &= \dot{\boldsymbol{Q}}\boldsymbol{Q}^t[\boldsymbol{v}^+ - \dot{\boldsymbol{c}} - \boldsymbol{\Omega}(\boldsymbol{x}^+ - \boldsymbol{c})] + \dot{\boldsymbol{\Omega}}(\boldsymbol{x}^+ - \boldsymbol{c}) + \boldsymbol{\Omega}(\boldsymbol{v}^+ - \dot{\boldsymbol{c}}) + \ddot{\boldsymbol{c}} \\ &= \boldsymbol{\Omega}[\boldsymbol{v}^+ - \dot{\boldsymbol{c}} - \boldsymbol{\Omega}(\boldsymbol{x}^+ - \boldsymbol{c})] + \boldsymbol{\Omega}(\boldsymbol{v}^+ - \dot{\boldsymbol{c}}) + \dot{\boldsymbol{\Omega}}(\boldsymbol{x}^+ - \boldsymbol{c}) + \ddot{\boldsymbol{c}} \\ &= 2\boldsymbol{\Omega}(\boldsymbol{v}^+ - \dot{\boldsymbol{c}}) + (\dot{\boldsymbol{\Omega}} - \boldsymbol{\Omega}^2)(\boldsymbol{x}^+ - \boldsymbol{c}) + \ddot{\boldsymbol{c}}.\end{aligned}$$

2.

The result follows immediately upon differentiating $\mathbf{1}^+ = \boldsymbol{Q}(t)\mathbf{1}\boldsymbol{K}^t$. Thus,

$$(\mathbf{1}^+)^{\cdot} = \dot{\boldsymbol{Q}}\mathbf{1}\boldsymbol{K}^t = \dot{\boldsymbol{Q}}\boldsymbol{Q}^t\mathbf{1}^+ = \boldsymbol{\Omega}\mathbf{1}^+.$$

To aid in the interpretation of this result, we use (2.68)–(2.70) to obtain

$$\mathbf{1}^+ = \delta_{iA}\boldsymbol{Q}(t)\boldsymbol{e}_i \otimes \boldsymbol{K}\hat{\boldsymbol{E}}_A, \quad \text{where} \quad \delta_{iA} = \boldsymbol{e}_i \cdot \mathbf{1}\hat{\boldsymbol{E}}_A = \boldsymbol{e}_i \cdot \boldsymbol{E}_A.$$

With reference to (7.35), let $\boldsymbol{n} \in \{\boldsymbol{e}_i\}$. Then, $\boldsymbol{n}^+$ is the corresponding element of $\{\boldsymbol{e}_i^+\}$, where $\boldsymbol{e}_i^+ = \boldsymbol{Q}(t)\boldsymbol{e}_i$. Similarly, from (7.36) it follows that if $\boldsymbol{N} \in \{\hat{\boldsymbol{E}}_A\}$, then $\boldsymbol{N}^+$ is the corresponding element of $\{\hat{\boldsymbol{E}}_A^+\}$, where $\hat{\boldsymbol{E}}_A^+ = \boldsymbol{K}\hat{\boldsymbol{E}}_A$. Thus,

$$\begin{aligned}
\mathbf{1}^+ &= \delta_{iA}^+\boldsymbol{e}_i^+ \otimes \hat{\boldsymbol{E}}_A^+, \quad \text{where} \\
\delta_{iA}^+ &= \boldsymbol{e}_i^+ \cdot \mathbf{1}^+\hat{\boldsymbol{E}}_A^+ = \boldsymbol{Q}(t)\boldsymbol{e}_i \cdot \boldsymbol{Q}(t)\mathbf{1}\boldsymbol{K}^t\boldsymbol{K}\hat{\boldsymbol{E}}_A = \boldsymbol{Q}(t)\boldsymbol{e}_i \cdot \boldsymbol{Q}(t)\mathbf{1}\hat{\boldsymbol{I}}\hat{\boldsymbol{E}}_A \\
&= \boldsymbol{Q}(t)\boldsymbol{e}_i \cdot \boldsymbol{Q}(t)\mathbf{1}\hat{\boldsymbol{E}}_A = \boldsymbol{e}_i \cdot \mathbf{1}\hat{\boldsymbol{E}}_A \\
&= \delta_{iA}.
\end{aligned}$$

Because $\{\boldsymbol{e}_i\}$ is fixed in the frame of the (non-spinning) inertial frame E^3 of $\mathcal{O}$, we have $\dot{\boldsymbol{e}}_i = \mathbf{0}$, and therefore

$$\begin{aligned}
(\mathbf{1}^+)^{\cdot} &= \delta_{iA}\dot{\boldsymbol{e}}_i^+ \otimes \hat{\boldsymbol{E}}_A^+ = \delta_{iA}^+\dot{\boldsymbol{Q}}\boldsymbol{e}_i \otimes \hat{\boldsymbol{E}}_A^+ \\
&= \delta_{iA}^+\dot{\boldsymbol{Q}}\boldsymbol{Q}^t\boldsymbol{e}_i^+ \otimes \hat{\boldsymbol{E}}_A^+ \\
&= \boldsymbol{\Omega}\mathbf{1}^+.
\end{aligned}$$

3.

The relevant equations are Eqs. (7.62) and (7.63). Near the surface of the planet, the body force due to gravity is given approximately by $\boldsymbol{b}^+ = -g\boldsymbol{e}_\rho$, where $\boldsymbol{e}_\rho$ is the radial unit vector of Problem 25 in Chapter 1.

The problem is to derive the appropriate expression for the inertial body force per unit mass

$$\boldsymbol{i}^+ = \ddot{\boldsymbol{c}} + 2\boldsymbol{\Omega}(\boldsymbol{v}^+ - \dot{\boldsymbol{c}}) + (\dot{\boldsymbol{\Omega}} - \boldsymbol{\Omega}^2)(\boldsymbol{x}^+ - \boldsymbol{c}). \qquad (*)$$

To this end, let $\{\boldsymbol{e}_i^+\} = \{\boldsymbol{e}_\rho, \boldsymbol{e}_\theta, \boldsymbol{e}_\phi\}$, as defined in that problem, with the latitude ϕ fixed and longitude given by $\theta = \omega t + \theta_0$, constitute the frame used by observer $\mathcal{O}^+$, centered on the surface of the planet. Let $\{\boldsymbol{e}_i\}$ be a fixed orthonormal basis in the frame of an inertial observer $\mathcal{O}$, with origin located at the center of the planet (see the solution to the preceding problem). Then, $\boldsymbol{e}_i^+ = \boldsymbol{Q}\boldsymbol{e}_i$ and $\boldsymbol{Q} = \boldsymbol{Q}\boldsymbol{I} = \boldsymbol{Q}\boldsymbol{e}_i \otimes \boldsymbol{e}_i = \boldsymbol{e}_i^+ \otimes \boldsymbol{e}_i$. Thus, $\boldsymbol{\Omega} = \dot{\boldsymbol{Q}}\boldsymbol{Q}^T = \dot{\boldsymbol{e}}_i^+ \otimes \boldsymbol{Q}\boldsymbol{e}_i = \dot{\boldsymbol{e}}_i^+ \otimes \boldsymbol{e}_i^+$. Using

$$\boldsymbol{e}_1^+ = \boldsymbol{e}_\rho = \cos\phi \boldsymbol{e}_r(\theta) + \sin\phi \boldsymbol{e}_3, \quad \boldsymbol{e}_2^+ = \boldsymbol{e}_\theta(\theta) \quad \text{and}$$

$$\boldsymbol{e}_3^+ = \boldsymbol{e}_\phi = \cos\phi \boldsymbol{e}_3 - \sin\phi \boldsymbol{e}_r(\theta),$$

with $\dot{\theta} = \omega = const.$, we get

$$\dot{\boldsymbol{e}}_1^+ = \omega\cos\phi \boldsymbol{e}_\theta, \quad \dot{\boldsymbol{e}}_2^+ = -\omega \boldsymbol{e}_r \quad \text{and} \quad \dot{\boldsymbol{e}}_3^+ = -\omega\sin\phi \boldsymbol{e}_\theta,$$

and therefore

$$\boldsymbol{\Omega} = \omega(\boldsymbol{e}_\theta \otimes \boldsymbol{e}_r - \boldsymbol{e}_r \otimes \boldsymbol{e}_\theta), \quad \text{where}$$

$$\boldsymbol{e}_r = \cos\phi \boldsymbol{e}_\rho - \sin\phi \boldsymbol{e}_\phi = \cos\phi \boldsymbol{e}_1^+ - \sin\phi \boldsymbol{e}_3^+ \quad \text{and} \quad \boldsymbol{e}_\theta = \boldsymbol{e}_2^+.$$

This gives $\dot{\boldsymbol{\Omega}} = \boldsymbol{0}$ and

$$\boldsymbol{\Omega}^2 = -\omega^2(\boldsymbol{e}_\theta \otimes \boldsymbol{e}_\theta + \boldsymbol{e}_r \otimes \boldsymbol{e}_r).$$

It remains to determine $\boldsymbol{c}$ and its time derivatives. Recall that $\boldsymbol{c}$ is the position, relative to $\mathcal{O}^+$, of the origin in the system used by $\mathcal{O}$ (the center of the planet). Since $\mathcal{O}^+$ is located at the position $R\boldsymbol{e}_\rho$ relative to the center of the planet, this means that $\boldsymbol{c} = -R\boldsymbol{e}_\rho = -R\boldsymbol{e}_1^+$. Thus, $\dot{\boldsymbol{c}} = -R\omega\cos\phi \boldsymbol{e}_\theta = -R\omega\cos\phi \boldsymbol{e}_2^+$ and $\ddot{\boldsymbol{c}} = R\omega^2\cos\phi \boldsymbol{e}_r = R\omega^2\cos\phi(\cos\phi \boldsymbol{e}_1^+ - \sin\phi \boldsymbol{e}_3^+)$. Substituting these results into (*) gives the desired equation of motion.

Chapter 9

1.

We derive $(9.13)_1$. Thus, we use (9.8) and (9.11), obtaining

$$\begin{aligned}
\boldsymbol{L}_2 &= \boldsymbol{Q}_2^t\boldsymbol{Q}_1\boldsymbol{L}_1(\boldsymbol{Q}_2^t\boldsymbol{Q}_1)^t + \boldsymbol{Q}_2^t(\dot{\boldsymbol{Q}}_1\boldsymbol{Q}_1^t - \dot{\boldsymbol{Q}}_2\boldsymbol{Q}_2^t)\boldsymbol{Q}_2 \\
&= \boldsymbol{Q}\boldsymbol{L}_1\boldsymbol{Q}^t + \boldsymbol{Q}_2^t\dot{\boldsymbol{Q}}_1\boldsymbol{Q}_1^t\boldsymbol{Q}_2 - \boldsymbol{Q}_2^t\dot{\boldsymbol{Q}}_2 \\
&= \boldsymbol{Q}\boldsymbol{L}_1\boldsymbol{Q}^t + \dot{\boldsymbol{Q}}_2^t\boldsymbol{Q}_2 + \boldsymbol{Q}_2^t\dot{\boldsymbol{Q}}_1\boldsymbol{Q}_1^t\boldsymbol{Q}_2 \\
&= \boldsymbol{Q}\boldsymbol{L}_1\boldsymbol{Q}^t + (\dot{\boldsymbol{Q}}_2^t\boldsymbol{Q}_1 + \boldsymbol{Q}_2^t\dot{\boldsymbol{Q}}_1)\boldsymbol{Q}_1^t\boldsymbol{Q}_2 \\
&= \boldsymbol{Q}\boldsymbol{L}_1\boldsymbol{Q}^t + \dot{\boldsymbol{Q}}\boldsymbol{Q}^t.
\end{aligned}$$

2.

Here, $\boldsymbol{Q}: E^3 \to E^{3+}$. Because $\boldsymbol{I}^+ = \boldsymbol{QIQ}^t$, $\boldsymbol{D}^+ = \boldsymbol{QDQ}^t$, $(\boldsymbol{D}^+)^2 = \boldsymbol{QD}^2\boldsymbol{Q}^t$ we have that $I_k^+ = I_k$. Then, with $\rho^+ = \rho$ and $\theta^+ = \theta$, it follows, from (7.40), that

$$\boldsymbol{T}^+ = \varphi_0(I_1^+, I_2^+, I_3^+, \theta^+, \rho^+; p)\boldsymbol{I}^+ + \varphi_1(I_1^+, I_2^+, I_3^+, \theta^+, \rho^+; p)\boldsymbol{D}^+ + \varphi_2(I_1^+, I_2^+, I_3^+, \theta^+, \rho^+; p)(\boldsymbol{D}^+)^2.$$

Similarly, from (9.42), we have

$$\boldsymbol{K}^+ = \phi_0(I_1^+, \ldots, I_6^+, \theta^+, \rho^+; p)\boldsymbol{I}^+ + \phi_1(I_1^+, \ldots, I_6^+, \theta^+, \rho^+; p)\boldsymbol{D}^+ + \phi_2(I_1^+, \ldots, I_6^+, \theta^+, \rho^+; p)(\boldsymbol{D}^+)^2 = \boldsymbol{QKQ}^t,$$

and (7.44), $(7.55)_1$ give

$$\boldsymbol{q}^+ = \boldsymbol{K}^+ grad^+ \theta^+.$$

3.

This proceeds as in the solution to the previous problem, but now with $\boldsymbol{Q}: E^3 \to E^3$.

4.

Use $\boldsymbol{T}^+ = \boldsymbol{QTQ}^t$, $\boldsymbol{D}^+ = \boldsymbol{QDQ}^t$ and $\boldsymbol{QIQ}^t = \boldsymbol{QQ}^t = \boldsymbol{I}^+$, together with (9.83), to obtain

$$\boldsymbol{T}^+ = -p\boldsymbol{I}^+ + 2\mu^+\boldsymbol{D}^+,$$

where $\mu^+ = \mu$. Thus,

$$\boldsymbol{T}^+ = -p^+\boldsymbol{I}^+ + 2\mu^+\boldsymbol{D}^+,$$

where, from (7.1) and (7.2),

$$p^+(\boldsymbol{x}^+, t^+) = p(\boldsymbol{Q}(t^+ - a)^t(\boldsymbol{x}^+ - \boldsymbol{c}(t^+ - a)), t^+ - a).$$

5.

We have $\boldsymbol{T} = -p\boldsymbol{I} + 2\mu\boldsymbol{D}$ and

$$div\boldsymbol{T} = \rho(\boldsymbol{v}' + \boldsymbol{Lv}) = \rho\boldsymbol{Lv}, \tag{1}$$

where we've used $\boldsymbol{v}' = \mathbf{0}$ because the motion is steady. Now, $\boldsymbol{v} = f(x_1, x_2)\boldsymbol{e}_3$ and $\boldsymbol{L}d\boldsymbol{x} = d\boldsymbol{v} = (f_{,1}dx_1 + f_{,2}dx_2)\boldsymbol{e}_3 = \boldsymbol{e}_3(f_{,1}\boldsymbol{e}_1 \cdot d\boldsymbol{x} + f_{,2}\boldsymbol{e}_2 \cdot d\boldsymbol{x}) = (\boldsymbol{e}_3 \otimes gradf)d\boldsymbol{x}$, where $gradf = f_{,1}\boldsymbol{e}_1 + f_{,2}\boldsymbol{e}_2$. Thus, $\boldsymbol{L} = \boldsymbol{e}_3 \otimes gradf$.

(a) $tr\boldsymbol{L} = \boldsymbol{e}_3 \cdot gradf = f_{,3} = 0$, since f does not depend on x_3. Thus, the motion is isochoric.

(b) Also, $\boldsymbol{L}\boldsymbol{v} = f(\boldsymbol{e}_3 \otimes gradf)\boldsymbol{e}_3 = f\boldsymbol{e}_3(\boldsymbol{e}_3 \cdot gradf) = \mathbf{0}$, and (1) reduces to

$$div\boldsymbol{T} = \mathbf{0}. \tag{2}$$

We need $2\boldsymbol{D} = \boldsymbol{L} + \boldsymbol{L}^t = \boldsymbol{e}_3 \otimes gradf + gradf \otimes \boldsymbol{e}_3$, i.e., $2D_{ij} = \delta_{3i}f_{,j} + f_{,i}\delta_{3j}$. Then, $T_{ij,j} = -p_{,i} + \mu(\delta_{3i}f_{,j} + f_{,i}\delta_{3j})_{,j} = -p_{,i} + \mu\delta_{3i}f_{,jj} + \mu f_{,ij}\delta_{3j} = -p_{,i} + \mu\delta_{3i}\Delta f + \mu f_{,i3}$, where Δ is the Laplacian and $f_{,i3} = (f_{,3})_{,i} = 0$. Thus, $div\boldsymbol{T} = -gradp + \mu(\Delta f)\boldsymbol{e}_3$ and (2) reduces to

$$gradp = \mu(\Delta f)\boldsymbol{e}_3. \tag{3}$$

It follows that $p_{,1} = 0 = p_{,2}$ and that p is a function of x_3 only, with derivative

$$p'(x_3) = \mu\Delta f(x_1, x_2). \tag{4}$$

This implies that $p''(x_3) = 0$ and hence that $p'(x_3) = P$, a constant. We finally arrive at

$$\Delta f(x_1, x_2) = P/\mu = const. \tag{5}$$

(c) To satisfy the boundary condition $\boldsymbol{v} = \mathbf{0}$, i.e., $f = 0$, on the boundary where $\phi = 0$, we try a solution of the form $f = c\phi$, where c is a constant. Since $\Delta\phi = const.$ for the elliptical boundary, this results in $\Delta f = const.$ Substitute into (5) to find the constant c. Fortuitously, this simple solution method works for the elliptical boundary and a few other shapes too, but not for general boundary shapes.

6.

(a) We have $\boldsymbol{v} = v_1(x_2)\boldsymbol{e}_1$. Then, $\boldsymbol{L}d\boldsymbol{x} = d\boldsymbol{v} = v_1'(x_2)\boldsymbol{e}_1 dx_2 = v_1'(x_2)\boldsymbol{e}_1(\boldsymbol{e}_2 \cdot d\boldsymbol{x})$, so $\boldsymbol{L} = v_1'(x_2)\boldsymbol{e}_1 \otimes \boldsymbol{e}_2$. Then $tr\boldsymbol{L} = 0$ and the flow is isochoric. Also, $\boldsymbol{D} = \frac{1}{2}v_1'(x_2)(\boldsymbol{e}_1 \otimes \boldsymbol{e}_2 + \boldsymbol{e}_2 \otimes \boldsymbol{e}_1)$ and $D_{12} = D_{21} = \frac{1}{2}v_1'(x_2)$, with all other D_{ij} equal to zero. Using $\boldsymbol{v}' = \boldsymbol{0}$ and $\boldsymbol{Lv} = \boldsymbol{0}$, we solve

$$-grad p + 2\mu div \boldsymbol{D} = \rho(\boldsymbol{v}' + \boldsymbol{Lv}) = \boldsymbol{0}; \quad \text{thus,} \quad grad p = 2\mu div \boldsymbol{D}$$
$$= 2\mu D_{ij,j}\boldsymbol{e}_i = 2\mu D_{12,2}\boldsymbol{e}_1 = \mu v_1''(x_2)\boldsymbol{e}_1.$$

This gives $p_{,2} = p_{,3} = 0$ and $p'(x_1) = \mu v_1''(x_2)$. Thus, $p''(x_1) = 0$ and $p'(x_1) = P$, a constant. Then, $v_1''(x_2) = P/\mu (= const)$. Integrate and apply the boundary conditions to get

$$v_1(x_2) = \frac{P}{2\mu}x_2(x_2 - d) + Vx_2/d.$$

(b) We have $\boldsymbol{v} = v(r)\boldsymbol{e}_3$. Then, $\boldsymbol{L}d\boldsymbol{x} = d\boldsymbol{v} = v'(r)\boldsymbol{e}_3 dr = v'(r)\boldsymbol{e}_3(\boldsymbol{e}_r \cdot d\boldsymbol{x})$, so $\boldsymbol{L} = v'(r)\boldsymbol{e}_3 \otimes \boldsymbol{e}_r$. Then $tr\boldsymbol{L} = 0$ and the flow is isochoric. Also, $\boldsymbol{D} = \frac{1}{2}v'(r)(\boldsymbol{e}_3 \otimes \boldsymbol{e}_r + \boldsymbol{e}_r \otimes \boldsymbol{e}_3)$. Then, $D_{zr} = D_{rz} = \frac{1}{2}v'(r)$, and all other polar components are zero. We again have $\boldsymbol{v}' = \boldsymbol{0}$ and $\boldsymbol{Lv} = \boldsymbol{0}$, and so we solve

$$grad p = 2\mu div \boldsymbol{D} = 2\mu(D_{zr,r} + \frac{1}{r}D_{zr})\boldsymbol{e}_3 = \frac{2\mu}{r}(rD_{zr})'\boldsymbol{e}_3$$
$$= \frac{\mu}{r}(rv')'\boldsymbol{e}_3.$$

Then, $p_{,r} = p_{,\theta} = 0$ and $p'(z) = \frac{\mu}{r}(rv')'$. This gives $p''(z) = 0$ and $p'(z) = P$, a constant. Then, $(rv')' = rP/\mu$. Integrate and apply the boundary condition to get the bounded solution

$$v(r) = \frac{P}{4\mu}(r^2 - a^2).$$

7.

We have $\boldsymbol{v} = r\omega\boldsymbol{e}_\theta$.

(a) This is an azimuthal velocity field about the $\boldsymbol{e}_3$ axis. Mass conservation for incompressible liquids, i.e., $0 = \dot{\rho} = \rho' + \boldsymbol{v} \cdot grad\rho = 0$, is satisfied because $\rho' = 0$ and $grad\rho = \boldsymbol{0}$.
(b) We have $\boldsymbol{L}d\boldsymbol{x} = d\boldsymbol{v} = \omega\boldsymbol{e}_\theta dr - \omega\boldsymbol{e}_r r d\theta = \omega\boldsymbol{e}_\theta(\boldsymbol{e}_r \cdot d\boldsymbol{x}) - \omega\boldsymbol{e}_r(\boldsymbol{e}_\theta \cdot d\boldsymbol{x})$, so $\boldsymbol{L} = \omega(\boldsymbol{e}_\theta \otimes \boldsymbol{e}_r - \boldsymbol{e}_r \otimes \boldsymbol{e}_\theta)$. Then $tr\boldsymbol{L} = 0$ and the flow is isochoric. Also, $\boldsymbol{D} = \boldsymbol{O}$ and $\boldsymbol{W} = \boldsymbol{L}$ and so the motion is rigid.

(c) With $\boldsymbol{v}' = \mathbf{0}$, $\boldsymbol{Lv} = -r\omega^2\boldsymbol{e}_r$ and $\boldsymbol{D} = \boldsymbol{O}$, we solve

$$grad p + \rho g \boldsymbol{e}_3 = \rho r\omega^2 \boldsymbol{e}_r.$$

Then, $p_{,\theta} = 0$, $p_{,3} = -\rho g$ and $p_{,r} = \rho r\omega^2$. Integrate the last of these to get

$$p(r, z) = \frac{1}{2}\rho r^2\omega^2 + q(z)$$

and substitute back to get $q'(z) = -\rho g$, and hence $q(z) = -\rho gz + c$, where c is a constant. Thus,

$$p(r, z) = \frac{1}{2}\rho r^2\omega^2 - \rho gz + c.$$

The boundary conditions give $-p_a\boldsymbol{n} = \boldsymbol{Tn} = -p\boldsymbol{n}$ on the surface of the liquid with unit normal $\boldsymbol{n}$. Thus, $p = p_a$ on the surface. If we locate the origin $(r, z) = (0, 0)$ on the surface we get $c = p_a$. But $p(r, z(r)) = p_a$ on the entire surface, defined by the function $z(r)$. Thus, the equation of the surface is

$$z(r) = \frac{1}{2g}r^2\omega^2.$$

This describes a paraboloid, obtained by rotating a parabola about the $\boldsymbol{e}_3$ axis.

8.

From the result of Problem 10 of Chapter 4, we have $\boldsymbol{v} = rf(r)\boldsymbol{e}_\theta$,

$$\begin{aligned}
\boldsymbol{L} &= (rf)'\boldsymbol{e}_\theta \otimes \boldsymbol{e}_r - f\boldsymbol{e}_r \otimes \boldsymbol{e}_\theta, \\
2\boldsymbol{D} &= \boldsymbol{L} + \boldsymbol{L}^T = (rf)'(\boldsymbol{e}_\theta \otimes \boldsymbol{e}_r + \boldsymbol{e}_r \otimes \boldsymbol{e}_\theta) - f(\boldsymbol{e}_r \otimes \boldsymbol{e}_\theta + \boldsymbol{e}_\theta \otimes \boldsymbol{e}_r) \\
&= rf'(\boldsymbol{e}_r \otimes \boldsymbol{e}_\theta + \boldsymbol{e}_\theta \otimes \boldsymbol{e}_r),
\end{aligned}$$

and

$$\begin{aligned}
2\boldsymbol{W} &= \boldsymbol{L} - \boldsymbol{L}^T = (rf)'(\boldsymbol{e}_\theta \otimes \boldsymbol{e}_r - \boldsymbol{e}_r \otimes \boldsymbol{e}_\theta) - f(\boldsymbol{e}_r \otimes \boldsymbol{e}_\theta - \boldsymbol{e}_\theta \otimes \boldsymbol{e}_r) \\
&= (rf' + 2f)(\boldsymbol{e}_\theta \otimes \boldsymbol{e}_r - \boldsymbol{e}_r \otimes \boldsymbol{e}_\theta).
\end{aligned}$$

(a) We need to solve

$$-grad p + \mu div(2\boldsymbol{D}) = \rho \boldsymbol{Lv},$$

where $\boldsymbol{Lv} = -rf^2\boldsymbol{e}_r$ and $2D_{r\theta} = 2D_{\theta r} = rf'$, functions of r only, with all other polar components equal to zero. According to (1.153) and (1.162), we then have

$$div(2\boldsymbol{D}) = [(2D_{\theta r})' + \frac{2}{r}(2D_{r\theta})]\boldsymbol{e}_\theta = (g' + \frac{2}{r}g)\boldsymbol{e}_\theta, \quad \text{where} \quad g = rf'.$$

Assuming that p depends only on r, the equation to be solved reduces to

$$-p'(r)\boldsymbol{e}_r + \mu(g' + \frac{2}{r}g)\boldsymbol{e}_\theta = -\rho rf^2\boldsymbol{e}_r.$$

This yields the two equations

$$p'(r) = \rho rf^2 \quad \text{and} \quad 0 = g' + \frac{2}{r}g = \frac{1}{r^2}(r^2 g)'.$$

The second equation gives $g(= rf') = const./r^2$, and therefore $f' = const./r^3$, which integrates to $f(r) = A + B/r^2$, where A and B are constants.

The boundary conditions are $f(a) = \omega_a$ and $f(b) = \omega_b$, and provide two equations to determine A and B, yielding

$$A = (b^2\omega_b - a^2\omega_a)/(b^2 - a^2) \quad \text{and} \quad B = a^2b^2(\omega_a - \omega_b)/(b^2 - a^2).$$

(b) We calculate the torque acting on a unit length of the inner cylinder. This is given by

$$\boldsymbol{m} = a\int_0^1 \left(\int_0^{2\pi} \boldsymbol{x} \times \boldsymbol{t} d\theta\right) dz,$$

where $\boldsymbol{x} = a\boldsymbol{e}_r + z\boldsymbol{e}_3$ and

$$\boldsymbol{t} = \boldsymbol{T}\boldsymbol{e}_r = -p(a)\boldsymbol{e}_r + 2\mu\boldsymbol{D}\boldsymbol{e}_{r|r=a} = -p(a)\boldsymbol{e}_r + \mu af'(a)\boldsymbol{e}_\theta.$$

Then,

$$\boldsymbol{x} \times \boldsymbol{t} = \mu a^2 f'(a)\boldsymbol{e}_3 - zp(a)\boldsymbol{e}_\theta - z\mu af'(a)\boldsymbol{e}_r.$$

Since $\boldsymbol{e}_\theta = \boldsymbol{e}_r'(\theta)$ and $\boldsymbol{e}_r = -\boldsymbol{e}_\theta'(\theta)$, the second and third terms make no contribution to the interior integral. Thus,

$$\boldsymbol{m} = 2\pi\mu a^3 f'(a)\boldsymbol{e}_3 = -4\pi\mu B\boldsymbol{e}_3.$$

(c) Note that $rf' + 2f = 2A$, and so the spin tensor is

$$\boldsymbol{W} = A(\boldsymbol{e}_\theta \otimes \boldsymbol{e}_r - \boldsymbol{e}_r \otimes \boldsymbol{e}_\theta).$$

We therefore have a flow with zero vorticity if and only if $A = 0$, i.e., if and only if $\omega_a/\omega_b = b^2/a^2$.

9.

Assuming the fluid to be at rest, we have $\boldsymbol{T} = -p\boldsymbol{I}$, and since it is in equilibrium we have $\boldsymbol{0} = \rho\boldsymbol{b} + div\boldsymbol{T} = -\rho g\boldsymbol{e}_3 - gradp$. Thus,

$$dp = gradp \cdot d\boldsymbol{x} = -\rho g\boldsymbol{e}_3 \cdot d\boldsymbol{x} = d(-\rho g\boldsymbol{e}_3 \cdot \boldsymbol{x}) = d(-\rho g x_3),$$

where we've assumed that $\rho = const$. Thus,

$$p = c - \rho g x_3,$$

where c is a constant.

(a) The resultant force on κ_t due to the pressure exerted on its boundary is

$$\boldsymbol{f} = \int_{\partial\kappa_t} \boldsymbol{t}da = -\int_{\partial\kappa_t} p\boldsymbol{n}da = \rho g \int_{\partial\kappa_t} x_3\boldsymbol{n}da - c\int_{\partial\kappa_t} \boldsymbol{n}da.$$

Using $\int_{\partial\kappa_t} \boldsymbol{n}da = \boldsymbol{0}$ (because $\partial\kappa_t$ is a closed surface; see (5.9)) and $\int_{\partial\kappa_t} x_3\boldsymbol{n}da = \int_{\kappa_t} gradx_3 dv = vol(\kappa_t)\boldsymbol{e}_3$, we get

$$\boldsymbol{f} = \rho g vol(\kappa_t)\boldsymbol{e}_3,$$

as claimed.

(b) The moment about the centroid is

$$\begin{aligned}
\boldsymbol{m}_c &= \int_{\partial\kappa_t} (\boldsymbol{x} - \boldsymbol{x}_c) \times \boldsymbol{t}da = e_{ijk}\boldsymbol{e}_k \int_{\partial\kappa_t} (x_i - x_i^c)t_j da \\
&= -e_{ijk}\boldsymbol{e}_k \int_{\partial\kappa_t} p(x_i - x_i^c)n_j da \\
&= e_{ijk}\boldsymbol{e}_k[\rho g \int_{\partial\kappa_t} x_3(x_i - x_i^c)n_j da - c\int_{\partial\kappa_t} (x_i - x_i^c)n_j da] \\
&= e_{ijk}\boldsymbol{e}_k[\rho g \int_{\kappa_t} \{x_3(x_i - x_i^c)\}_{,j} dv - c\int_{\kappa_t} (x_i - x_i^c)_{,j} dv] \\
&= e_{ijk}\boldsymbol{e}_k[\rho g\delta_{3j} \int_{\kappa_t} (x_i - x_i^c)dv + \rho g\delta_{ij} \int_{\kappa_t} x_3 dv - c\delta_{ij} vol(\kappa_t)] \\
&= \boldsymbol{0},
\end{aligned}$$

because $\int_{\kappa_t} (x_i - x_i^c)dv = 0$ (by definition of the centroid) and $e_{ijk}\delta_{ij} = 0$.

10.

If the internal heating supply vanishes, i.e., if $r = 0$, then the energy balance reduces to

$$\rho\dot{\varepsilon} = \boldsymbol{T} \cdot \boldsymbol{L} - div\boldsymbol{q},$$

where, in the present case, $\dot{\varepsilon} = \varepsilon'(\theta)\dot{\theta} = 0$, because $\dot{\theta} = 0$. Thus the energy balance reduces to $\boldsymbol{T} \cdot \boldsymbol{L} = div\boldsymbol{q}$. Noting that $\dot{\theta} = \theta' + \boldsymbol{v} \cdot grad\theta$, we see that this situation also obtains if $\theta' = 0$, provided that $\boldsymbol{v} \cdot grad\theta = 0$.

11.

We have $\boldsymbol{v} \cdot grad\theta = 0$, where $grad\theta = \theta'(r)\boldsymbol{e}_r$, and therefore the conditions required for validity of the energy balance equation

$$\boldsymbol{T} \cdot \boldsymbol{L} = div\boldsymbol{q}$$

are satisfied. Here, $div\boldsymbol{q} = Kdiv(grad\theta) = Kr^{-1}(r\theta')'$, after using the formula (1.158) for the divergence. Also,

$$\begin{aligned} \boldsymbol{T} \cdot \boldsymbol{L} &= -ptr\boldsymbol{L} + 2\mu\boldsymbol{D} \cdot \boldsymbol{D} \\ &= 2\mu tr(\boldsymbol{D}^2), \end{aligned}$$

where

$$4\boldsymbol{D}^2 = (rf')^2(\boldsymbol{e}_r \otimes \boldsymbol{e}_r + \boldsymbol{e}_\theta \otimes \boldsymbol{e}_\theta).$$

Thus, $tr(\boldsymbol{D}^2) = \frac{1}{2}(rf')^2$, and the energy balance reduces to

$$Kr^{-1}(r\theta')' = \mu(rf')^2.$$

From the solution to Problem 8, we have $rf' = -2B/r^2$, and we thus arrive at the simple differential equation

$$(r\theta')' = \frac{4\mu}{K}\frac{B^2}{r^3},$$

which integrates to

$$\theta(r) = \frac{\mu}{K}\frac{B^2}{r^2} + C\ln r + D,$$

where C and D are constants. These are easily determined from the linear system obtained by imposing the boundary conditions $\theta(a) = \theta_a$ and $\theta(b) = \theta_b$.

12.

(a) We have $\boldsymbol{v}(\boldsymbol{x},t) = r^{-1}v(r)\boldsymbol{x}$, where $r = |\boldsymbol{x}|$. This is a function of $\boldsymbol{x}$ alone; thus, $\boldsymbol{v}'$ vanishes and the flow is steady. The spatial velocity gradient follows from

$$\boldsymbol{L}d\boldsymbol{x} = d\boldsymbol{v} = r^{-1}v(r)d\boldsymbol{x} + r^{-1}v'(r)\boldsymbol{x}dr - r^{-2}v(r)\boldsymbol{x}dr,$$

where $dr = r^{-1}\boldsymbol{x} \cdot d\boldsymbol{x}$.

Thus,

$$\boldsymbol{L} = (v/r)\boldsymbol{I} + r^{-2}(v' - v/r)\boldsymbol{x} \otimes \boldsymbol{x}.$$

We have $\boldsymbol{D} = \boldsymbol{L}$ and $\boldsymbol{W} = \boldsymbol{O}$, and the motion is isochoric if and only if

$$0 = tr\boldsymbol{L} = 3v/r + v' - v/r = v' + 2v/r = r^{-2}(r^2v)'.$$

Then, $r^2v(r) = const.$ and

$$v(r) = (a/r)^2 v_a.$$

(b) The acceleration is

$$\begin{aligned}
\boldsymbol{a} = \dot{\boldsymbol{v}} &= \boldsymbol{v}' + \boldsymbol{L}\boldsymbol{v} = \boldsymbol{L}\boldsymbol{v} \\
&= (v/r)\boldsymbol{v} + r^{-2}(v' - v/r)(\boldsymbol{x} \cdot \boldsymbol{v})\boldsymbol{x} \\
&= [(v/r)^2 + (v/r)(v' - v/r)]\boldsymbol{x} \\
&= (v/r)v'\boldsymbol{x} \\
&= -2\frac{a^4 v_a^2}{r^6}\boldsymbol{x}.
\end{aligned}$$

With $\boldsymbol{x} = r\boldsymbol{e}_\rho$, we have $v(r)\boldsymbol{e}_\rho = \boldsymbol{v} = \dot{\boldsymbol{x}} = \dot{r}\boldsymbol{e}_\rho$. Thus, $\dot{r} = v(r) = (a/r)^2 v_a$, i.e., $r^2\dot{r}(t) = a^2 v_a$. Integrating and imposing $r(0) = a$ then gives the radius of the material surface as a function of time:

$$r^3 = a^3 + 3a^2 v_a t.$$

Setting $r = 2a$, the time at which the radius doubles is found to be

$$t = \frac{7a}{3v_a}.$$

(c) We solve

$$div\boldsymbol{T} = \rho\boldsymbol{a} = -2\frac{a^4 v_a^2}{r^6}\rho\boldsymbol{x},$$

where, for incompressible fluids, mass conservation reduces to

$$0 = \dot{\rho} = \rho' + \boldsymbol{v}\cdot grad\rho = v\boldsymbol{e}_\rho \cdot grad\rho = v\rho_{,r},$$

i.e., $\rho_{,r} = 0$; and where

$$\boldsymbol{T} = -p\boldsymbol{I} + 2\mu\boldsymbol{D} = -p\boldsymbol{I} + 2\mu\boldsymbol{L}$$

in this case, with

$$\boldsymbol{L} = \frac{a^2 v_a}{r^3}\left(\boldsymbol{I} - \frac{3}{r^2}\boldsymbol{x}\otimes\boldsymbol{x}\right).$$

Thus, we seek p such that

$$grad p = 2\frac{a^4 v_a^2}{r^6}\rho\boldsymbol{x} + 2\mu div\boldsymbol{L}.$$

To evaluate the divergence we use the Cartesian formula $div\boldsymbol{L} = L_{ij,j}\boldsymbol{e}_i$, with

$$L_{ij} = \frac{a^2 v_a}{r^3}(\delta_{ij} - \frac{3}{r^2}x_i x_j).$$

Then,

$$L_{ij,j} = -\frac{3a^2 v_a}{r^4}r_{,j}(\delta_{ij} - \frac{3}{r^2}x_i x_j) + \frac{a^2 v_a}{r^3} \times \left(\frac{6}{r^3}r_{,j}x_i x_j - \frac{3}{r^2}x_{i,j}x_j - \frac{3}{r^2}x_i x_{j,j}\right),$$

which is reduced, with the aid of $r_{,j} = r^{-1}x_j$, $x_{i,j} = \delta_{ij}$ and $x_{j,j} = 3$, to $L_{ij,j} = 0$. Thus, $div\boldsymbol{L} = \boldsymbol{0}$ and the problem is now to solve

$$grad p = 2\frac{a^4 v_a^2}{r^6}\rho\boldsymbol{x}.$$

This, in turn, implies that p is a function of r, with derivative

$$p'(r) = 2\frac{a^4 v_a^2}{r^5}\rho,$$

which further implies that ρ is a function of r. Recalling that $\rho_{,r}$ vanishes by conservation of mass, we then have that $\rho = const.$

The expression for $\boldsymbol{L}$ yields $\boldsymbol{L} \to \boldsymbol{O}$ as $r \to \infty$. Thus, $p(r) \to p_\infty$ as $r \to \infty$, yielding

$$\begin{aligned} p(r) &= p_\infty + \int_\infty^r p'(x)dx \\ &= p_\infty + 2a^4 v_a^2 \rho \int_\infty^r x^{-5} dx \\ &= p_\infty - \frac{1}{2}(a/r)^4 v_a^2 \rho. \end{aligned}$$

Chapter 10

1.

The constraint of incompressibility is $g(\boldsymbol{C}) = \det \boldsymbol{C} - 1$. From (10.10) and (7.26), we have

$$\begin{aligned} g^+(\boldsymbol{C}^+) + 1 = g(\boldsymbol{K}^t\boldsymbol{C}^+\boldsymbol{K}) + 1 &= \det(\boldsymbol{K}^t\boldsymbol{C}^+\boldsymbol{K}) = \det(\boldsymbol{K}\boldsymbol{K}^t\boldsymbol{C}^+) \\ &= \det(\boldsymbol{K}\hat{\boldsymbol{I}}\boldsymbol{K}^t\boldsymbol{C}^+) = \det(\hat{\boldsymbol{I}}^+\boldsymbol{C}^+) \\ &= \det(\boldsymbol{C}^+). \end{aligned}$$

For inextensibility, we have

$$\begin{aligned} g^+(\boldsymbol{C}^+) + 1 = g(\boldsymbol{K}^t\boldsymbol{C}^+\boldsymbol{K}) + 1 &= \boldsymbol{M} \cdot \boldsymbol{K}^t\boldsymbol{C}^+\boldsymbol{K}\boldsymbol{M} \\ &= \boldsymbol{K}\boldsymbol{M} \cdot \boldsymbol{C}^+\boldsymbol{K}\boldsymbol{M} \\ &= \boldsymbol{M}^+ \cdot \boldsymbol{C}^+\boldsymbol{M}^+, \end{aligned}$$

where (7.30) has been used in the last step. Rigidity is equivalent to $\hat{\boldsymbol{I}} = \boldsymbol{C} = \boldsymbol{K}^t\boldsymbol{C}^+\boldsymbol{K}$, which, with (7.26), is seen to be equivalent to $\boldsymbol{C}^+ = \boldsymbol{K}\hat{\boldsymbol{I}}\boldsymbol{K}^t = \hat{\boldsymbol{I}}^+$.

2.

We have $\boldsymbol{T} = \boldsymbol{N} + \boldsymbol{G}$, in which $\boldsymbol{G}$ is the constitutive part of the stress. According to the hypothesis of material frame indifference, which entails observer consensus concerning material response, we should require that $\boldsymbol{G}^+ = \boldsymbol{Q}\boldsymbol{G}\boldsymbol{Q}^t$ (see (8.35)), where $\boldsymbol{Q}$: $E^3 \to E^{3+}$ is the orthogonal tensor characterizing the change of frame. But (7.40) requires that $\boldsymbol{T}^+ = \boldsymbol{QTQ}^t$, whether or not any constraints are in effect. Thus, $\boldsymbol{T}^+ = \boldsymbol{N}^+ + \boldsymbol{G}^+$, where,

$$\begin{aligned}\boldsymbol{N}^+ = \boldsymbol{QNQ}^t &= \boldsymbol{QF}\left[\sum \lambda_i g_{\boldsymbol{C}}^{(i)}\right]\boldsymbol{F}^t\boldsymbol{Q}^t \\ &= \boldsymbol{F}^+\boldsymbol{K}\left[\sum \lambda_i g_{\boldsymbol{C}}^{(i)}\right]\boldsymbol{K}^t(\boldsymbol{F}^+)^t,\end{aligned}$$

if $n(\leq 6)$ independent constraints are operative. From (10.10), we have, for each $i \in \{1, \ldots, n\}$,

$$g_{\boldsymbol{C}^+}^{+(i)} \cdot \dot{\boldsymbol{C}}^+ = g_{\boldsymbol{C}}^{(i)} \cdot \dot{\boldsymbol{C}} = g_{\boldsymbol{C}}^{(i)} \cdot \boldsymbol{K}^t\dot{\boldsymbol{C}}^+\boldsymbol{K} = \boldsymbol{K}g_{\boldsymbol{C}}^{(i)}\boldsymbol{K}^t \cdot \dot{\boldsymbol{C}}^+.$$

Thus, $\boldsymbol{K}g_{\boldsymbol{C}}^{(i)}\boldsymbol{K}^t = g_{\boldsymbol{C}^+}^{+(i)}$ and

$$\boldsymbol{N}^+ = \boldsymbol{F}^+\left[\sum \lambda_i^+ g_{\boldsymbol{C}^+}^{+(i)}\right](\boldsymbol{F}^+)^t, \quad \text{where} \quad \lambda_i^+ = \lambda_i.$$

3.

The constraints are $g_{\alpha\beta}(\boldsymbol{C}) = 0$, with $\alpha, \beta \in \{1, 2\}$, where

$$g_{\alpha\beta}(\boldsymbol{C}) = C_{\alpha\beta} - \delta_{\alpha\beta}, \quad \text{with} \quad C_{\alpha\beta} = \hat{\boldsymbol{E}}_\alpha \otimes \hat{\boldsymbol{E}}_\beta \cdot \boldsymbol{C}.$$

Then,

$$(g_{\alpha\beta})_{\boldsymbol{C}} \cdot \dot{\boldsymbol{C}} = \dot{g}_{\alpha\beta}(\boldsymbol{C}) = Sym(\hat{\boldsymbol{E}}_\alpha \otimes \hat{\boldsymbol{E}}_\beta) \cdot \dot{\boldsymbol{C}},$$

and, thus, $(g_{\alpha\beta})_{\boldsymbol{C}} = Sym(\hat{\boldsymbol{E}}_\alpha \otimes \hat{\boldsymbol{E}}_\beta)$. This yields the constraint stress in the form

$$\begin{aligned}\boldsymbol{N} = \lambda_{\alpha\beta}\boldsymbol{F}[Sym(\hat{\boldsymbol{E}}_\alpha \otimes \hat{\boldsymbol{E}}_\beta)]\boldsymbol{F}^t &= \boldsymbol{F}[\lambda_{(\alpha\beta)}\hat{\boldsymbol{E}}_\alpha \otimes \hat{\boldsymbol{E}}_\beta]\boldsymbol{F}^t \\ &= \lambda_{(\alpha\beta)}\boldsymbol{m}_\alpha \otimes \boldsymbol{m}_\beta,\end{aligned}$$

where a double sum over α and β is implied, $\lambda_{\alpha\beta}$ are functions of $\boldsymbol{x}$ and t, $\lambda_{(\alpha\beta)} = \frac{1}{2}(\lambda_{\alpha\beta} + \lambda_{\beta\alpha})$ and $\boldsymbol{m}_\alpha = \boldsymbol{F}\hat{\boldsymbol{E}}_\alpha$. The constraints require that the vectors $\boldsymbol{m}_\alpha$ be orthonormal. They span the tangent plane of the deformed image of a sheet at $\boldsymbol{x}$.

4.

We can express the constraint $\alpha = 1$ (where α is the local areal stretch) in terms of $\boldsymbol{C}$ and then proceed as explained in the text. However, it is much more convenient to use the result of Problem 7(b) in Chapter 4 directly, which we write in the form

$$\dot{\alpha}/\alpha = (\boldsymbol{I} - \boldsymbol{n} \otimes \boldsymbol{n}) \cdot \boldsymbol{D}.$$

Thus, we seek $\boldsymbol{N}$ such that

$$\boldsymbol{N} \cdot \boldsymbol{D} = 0 \quad \text{for all} \quad \boldsymbol{D} \quad \text{such that} \quad (\boldsymbol{I} - \boldsymbol{n} \otimes \boldsymbol{n}) \cdot \boldsymbol{D} = 0,$$

and it follows immediately that

$$\boldsymbol{N} = \lambda(\boldsymbol{I} - \boldsymbol{n} \otimes \boldsymbol{n}).$$

Recall that the tensor in parentheses is the projection onto a plane with unit normal $\boldsymbol{n}$. This is the tangent plane, at $\boldsymbol{x}$, of the deformed image of a constrained sheet. We have $\boldsymbol{N}\boldsymbol{\nu} = \lambda\boldsymbol{\nu}$ for any unit vector $\boldsymbol{\nu}$ lying in such a tangent plane. Thus furnishes the mechanical interpretation of $\lambda(\boldsymbol{x}, t)$ — if positive — as a surface tension in the deformed sheet.

Chapter 11

1.

$\boldsymbol{A}_1 \cdot \boldsymbol{A}_2\boldsymbol{A}_3 = tr(\boldsymbol{A}_1\boldsymbol{A}_3^t\boldsymbol{A}_2^t) = tr(\boldsymbol{A}_2^t\boldsymbol{A}_1\boldsymbol{A}_3^t) = \boldsymbol{A}_2^t\boldsymbol{A}_1 \cdot \boldsymbol{A}_3 = tr(\boldsymbol{A}_3\boldsymbol{A}_1^t\boldsymbol{A}_2) = \boldsymbol{A}_3\boldsymbol{A}_1^t \cdot \boldsymbol{A}_2^t$.

5.

For Problem 3 in Chapter 10, we have $g_{\alpha\beta}(\boldsymbol{R}^t\boldsymbol{C}\boldsymbol{R}) = \hat{\boldsymbol{E}}_\alpha \cdot \boldsymbol{R}^t\boldsymbol{C}\boldsymbol{R}\hat{\boldsymbol{E}}_\beta - \delta_{\alpha\beta}$ for $\alpha, \beta \in \{1, 2\}$. This is equal to $g_{\alpha\beta}(\boldsymbol{C})$ for all positive definite symmetric $\boldsymbol{C}$ if and only if $\hat{\boldsymbol{E}}_\alpha \cdot \boldsymbol{C}\hat{\boldsymbol{E}}_\beta = \hat{\boldsymbol{E}}_\alpha \cdot \boldsymbol{R}^t\boldsymbol{C}\boldsymbol{R}\hat{\boldsymbol{E}}_\beta = \boldsymbol{R}\hat{\boldsymbol{E}}_\alpha \cdot \boldsymbol{C}\boldsymbol{R}\hat{\boldsymbol{E}}_\beta$, where, in the case of isotropy, $\boldsymbol{R}$ is an arbitrary orthogonal tensor. Isotropy thus requires, in particular, that $C_{11} = \boldsymbol{R}\hat{\boldsymbol{E}}_1 \cdot \boldsymbol{C}\boldsymbol{R}\hat{\boldsymbol{E}}_1$ for all orthogonal $\boldsymbol{R}$. Let $\boldsymbol{R}$ be a 90° rotation about the axis $\hat{\boldsymbol{E}}_2$, so that $\boldsymbol{R}\hat{\boldsymbol{E}}_1 = -\hat{\boldsymbol{E}}_3$. We then require that $C_{11} = C_{33}$, which imposes an undue restriction on $\boldsymbol{C}$. Therefore, a material subject to this constraint is not isotropic.

For Problem 4 in Chapter 10, the constraint may expressed in the form $g(\boldsymbol{F}) = |\boldsymbol{F}^*\hat{\boldsymbol{E}}_3| - 1$. If $\boldsymbol{R}$ is a rotation, then $(\boldsymbol{F}\boldsymbol{R})^* = \boldsymbol{F}^*\boldsymbol{R}$ and $g(\boldsymbol{F}\boldsymbol{R}) = |\boldsymbol{F}^*\boldsymbol{R}\hat{\boldsymbol{E}}_3| - 1$. For an arbitrary admissible $\boldsymbol{F}$ this is not equal to $g(\boldsymbol{F})$ unless $\boldsymbol{R}\hat{\boldsymbol{E}}_3 = \pm\hat{\boldsymbol{E}}_3$. However, there are infinitely many rotations that do not have this property, and thus, since the rotation is arbitrary in the case of isotropy, a material constrained in this manner is not isotropic.

8.

(a) For the special strain-energy function $W^* = \frac{1}{2}G(I_1 - 3)$, we have $W_1^* = \frac{1}{2}G$ and $W_2^* = 0$, and (11.36) yields

$$\boldsymbol{T} = -p\boldsymbol{I} + G\boldsymbol{B},$$

as claimed.
In the case of equilibrium without body force we have $div\boldsymbol{T} = \boldsymbol{0}$, which reduces, in this case, to

$$grad p = G div\boldsymbol{B}. \qquad (*)$$

(b) We have $\boldsymbol{P}\cdot\dot{\boldsymbol{F}} = \dot{W} = \dot{W}^* = \frac{1}{2}G\dot{I}_1 = \frac{1}{2}G(3+\gamma^2)^{\cdot} = G\gamma\dot{\gamma}$. On the other hand, for simple shear (3.51) yields $\dot{\boldsymbol{F}} = \dot{\gamma}\boldsymbol{e}_1 \otimes \hat{\boldsymbol{E}}_2$. Thus, $\boldsymbol{P}\cdot\dot{\boldsymbol{F}} = \dot{\gamma}\boldsymbol{P}\cdot\boldsymbol{e}_1 \otimes \hat{\boldsymbol{E}}_2 = \dot{\gamma}\boldsymbol{e}_1 \cdot \boldsymbol{P}\hat{\boldsymbol{E}}_2 = \tau\dot{\gamma}$, where $\tau = \boldsymbol{e}_1 \cdot \boldsymbol{p}(\hat{\boldsymbol{E}}_2)$ is the shear component of the Piola traction acting on a plane with unit normal $\hat{\boldsymbol{E}}_2$ (see Figure 3.2). Thus, $\tau = G\gamma$.

(c) $\boldsymbol{C} = \lambda^2\hat{\boldsymbol{E}}_1 \otimes \hat{\boldsymbol{E}}_1 + \lambda^{-1}(\hat{\boldsymbol{I}} - \hat{\boldsymbol{E}}_1 \otimes \hat{\boldsymbol{E}}_1)$, $\boldsymbol{B} = \lambda^2\boldsymbol{e}_1 \otimes \boldsymbol{e}_1 + \lambda^{-1}(\boldsymbol{I} - \boldsymbol{e}_1 \otimes \boldsymbol{e}_1)$. Note that $\boldsymbol{B}$ is spatially uniform; thus $div\boldsymbol{B} = \boldsymbol{0}$ and (*) requires $grad p = \boldsymbol{0}$, i.e., $p = const$.

(d) $\boldsymbol{0} = \boldsymbol{T}\boldsymbol{n} = (-p+G\lambda^{-1})\boldsymbol{n}+G(\lambda^2-\lambda^{-1})(\boldsymbol{e}_1\cdot\boldsymbol{n})\boldsymbol{e}_1 = (-p+G\lambda^{-1})\boldsymbol{n}$; hence, $p = G\lambda^{-1}$.

(e) $\boldsymbol{T} = (G\lambda^{-1} - p)\boldsymbol{I} + G(\lambda^2 - \lambda^{-1})\boldsymbol{e}_1 \otimes \boldsymbol{e}_1 = T(\lambda)\boldsymbol{e}_1 \otimes \boldsymbol{e}_1$, where $T(\lambda) = G(\lambda^2 - \lambda^{-1})$.

(f) $E = T'(\lambda)_{|\lambda=1} = G(2\lambda + \lambda^{-2})_{|\lambda=1} = 3G$.

(g) $\boldsymbol{P} = \boldsymbol{T}\boldsymbol{F}^* = J\boldsymbol{T}\boldsymbol{F}^{-t} = T(\lambda)\boldsymbol{e}_1 \otimes \boldsymbol{F}^{-1}\boldsymbol{e}_1 = \lambda^{-1}T(\lambda)\boldsymbol{e}_1 \otimes \hat{\boldsymbol{E}}_1 = P(\lambda)\boldsymbol{e}_1 \otimes \hat{\boldsymbol{E}}_1$, where $P(\lambda) = \lambda^{-1}T(\lambda) = G(\lambda - \lambda^{-2})$. Also, $\boldsymbol{S} = \boldsymbol{F}^{-1}\boldsymbol{P} = P(\lambda)\boldsymbol{F}^{-1}\boldsymbol{e}_1 \otimes \hat{\boldsymbol{E}}_1 = \lambda^{-1}P(\lambda)\hat{\boldsymbol{E}}_1 \otimes \hat{\boldsymbol{E}}_1 = S(\lambda)\hat{\boldsymbol{E}}_1 \otimes \hat{\boldsymbol{E}}_1$, where $S(\lambda) = \lambda^{-1}P(\lambda) = G(1 - \lambda^{-3})$. We get $P'(\lambda)_{|\lambda=1} = G(1 + 2\lambda^{-3})_{|\lambda=1} = 3G = E$ and $S'(\lambda)_{|\lambda=1} = 3G\lambda^{-4}_{|\lambda=1} = 3G = E$.

9.

We have $\boldsymbol{F} = \mathbf{1} + \gamma \boldsymbol{e}_3 \otimes \hat{\boldsymbol{E}}_R$, where $\gamma = w'(r)$ and $r = R$. Thus,

$$\det \boldsymbol{F} = [\boldsymbol{F}\hat{\boldsymbol{E}}_R, \boldsymbol{F}\hat{\boldsymbol{E}}_\Theta, \boldsymbol{F}\hat{\boldsymbol{E}}_3] = [\boldsymbol{E}_R + \gamma \boldsymbol{e}_3, \boldsymbol{E}_\Theta, \boldsymbol{E}_3]$$
$$= [\boldsymbol{E}_R + \gamma \boldsymbol{E}_3, \boldsymbol{E}_\Theta, \boldsymbol{E}_3] = 1,$$

for all γ, where $\boldsymbol{E}_R = \mathbf{1}\hat{\boldsymbol{E}}_R$ and $\boldsymbol{E}_\Theta = \mathbf{1}\hat{\boldsymbol{E}}_\Theta$, and so this deformation is possible in our incompressible material. Here, we have chosen $\{\boldsymbol{e}_i\}$ such that $\boldsymbol{e}_3 = \boldsymbol{E}_3 = \mathbf{1}\hat{\boldsymbol{E}}_3$. Also,

$$\boldsymbol{B} = \boldsymbol{I} + \gamma(\boldsymbol{e}_r \otimes \boldsymbol{e}_3 + \boldsymbol{e}_3 \otimes \boldsymbol{e}_r) + \gamma^2 \boldsymbol{e}_3 \otimes \boldsymbol{e}_3,$$

where $\boldsymbol{e}_r(\theta) = \boldsymbol{E}_R(\theta)$. Thus, $B_{rr} = B_{\theta\theta} = 1$, $B_{zz} = 1 + \gamma^2$, $B_{rz} = B_{zr} = \gamma$, and all other polar components are equal to zero. Then, from (1.162),

$$grad p = G div \boldsymbol{B} = G(\gamma' + \frac{1}{r}\gamma)\boldsymbol{e}_3 = \frac{G}{r}(r\gamma)'\boldsymbol{e}_3.$$

This means that p depends only on z, with derivative

$$p'(z) = \frac{G}{r}(r\gamma)'.$$

Thus, $p''(z) = 0$ and $p'(z) = P$, a constant. Accordingly,

$$(r\gamma)' = \frac{P}{G}r,$$

which integrates to

$$w(r) = \frac{P}{4G}r^2 + C\ln r + D,$$

where C and D are constants. These are determined by imposing $w(A) = W$ and $w(B) = 0$.

Note that the traction on a cross-section is

$$\boldsymbol{t} = \boldsymbol{T}\boldsymbol{e}_3 = [-p(z) + G\gamma^2]\boldsymbol{e}_3 + G\gamma\boldsymbol{e}_r.$$

If this is the same at every cross section, then it is independent of z. In this case, $p'(z)(= P)$ is zero and the problem simplifies accordingly. We obtain

$$w(r) = W\ln(r/B)/\ln(A/B).$$

10.

If $\boldsymbol{F} = \mathbf{1}$, we have $\boldsymbol{B} = \boldsymbol{I}$ and (11.34) or (11.36) give $\boldsymbol{T} = \beta(\boldsymbol{X})\boldsymbol{I}$ for some scalar field β. From (2.76), we have $\boldsymbol{x} = \mathbf{1}\boldsymbol{X}$. Then, $\boldsymbol{T} = \alpha(\boldsymbol{x})\boldsymbol{I}$, where $\alpha(\boldsymbol{x}) = \beta(\mathbf{1}^t\boldsymbol{x})$. This must satisfy $\mathbf{0} = div\boldsymbol{T} = grad\alpha$, yielding $\alpha = const$. If the traction $\boldsymbol{T}\boldsymbol{n} = \alpha\boldsymbol{n}$ vanishes on a part of the boundary, then $\alpha = 0$ on that part. But $\alpha = const.$; hence, $\alpha = 0$ everywhere in the body, and $\boldsymbol{T} = \boldsymbol{O}$.

11.

(a) We have $r\boldsymbol{e}_r(\theta)+z\boldsymbol{e}_3 = \boldsymbol{x} = \mathbf{1}\boldsymbol{X}+w(\Theta)\boldsymbol{e}_3 = R\mathbf{1}\hat{\boldsymbol{E}}_R(\Theta)+Z\mathbf{1}\hat{\boldsymbol{E}}_3+w(\Theta)\boldsymbol{e}_3$. We select $\boldsymbol{e}_r(\theta) = \boldsymbol{E}_R(\theta)$, $\boldsymbol{e}_\theta(\theta) = \boldsymbol{E}_\Theta(\theta)$ and $\boldsymbol{e}_3 = \boldsymbol{E}_3$, where $\boldsymbol{E}_R(\Theta) = \mathbf{1}\hat{\boldsymbol{E}}_R(\Theta)$, $\boldsymbol{E}_\Theta(\Theta) = \mathbf{1}\hat{\boldsymbol{E}}_\Theta(\Theta)$ and $\boldsymbol{E}_3 = \mathbf{1}\hat{\boldsymbol{E}}_3$. Then, $r\boldsymbol{e}_r(\theta) + z\boldsymbol{e}_3 = R\boldsymbol{e}_r(\Theta) + [Z + w(\Theta)]\boldsymbol{e}_3$. This implies that $r\boldsymbol{e}_r(\theta) = R\boldsymbol{e}_r(\Theta)$, and hence that $R = r$ and $\Theta = \theta$ (or $\theta + 2n\pi$, for arbitrary integer n.). Then, $Z + w(\theta) = z$.

(b) We have

$$\begin{aligned}\boldsymbol{F}d\boldsymbol{X} &= d\boldsymbol{x} = \mathbf{1}d\boldsymbol{X} + w'(\Theta)\boldsymbol{e}_3 d\Theta \\ &= \mathbf{1}d\boldsymbol{X} + R^{-1}w'(\Theta)\boldsymbol{e}_3[\hat{\boldsymbol{E}}_\Theta(\Theta)\cdot d\boldsymbol{X}] \\ &= [\mathbf{1} + R^{-1}w'(\Theta)\boldsymbol{e}_3 \otimes \hat{\boldsymbol{E}}_\Theta(\Theta)]d\boldsymbol{X},\end{aligned}$$

and, thus,

$$\boldsymbol{F} = \mathbf{1} + R^{-1}w'(\Theta)\boldsymbol{e}_3 \otimes \hat{\boldsymbol{E}}_\Theta(\Theta) = \mathbf{1} + r^{-1}w'(\theta)\boldsymbol{e}_3 \otimes \hat{\boldsymbol{E}}_\Theta(\theta).$$

Using $\det \boldsymbol{F} = [\boldsymbol{F}\hat{\boldsymbol{E}}_R, \boldsymbol{F}\hat{\boldsymbol{E}}_\Theta, \boldsymbol{F}\hat{\boldsymbol{E}}_3]$, we easily verify that $\det \boldsymbol{F} = 1$, and hence that the deformation is isochoric.

(c) The stress is

$$\boldsymbol{T} = -p\boldsymbol{I} + G\boldsymbol{B},$$

where

$$\boldsymbol{B} = \boldsymbol{I} + r^{-1}w'(\theta)(\boldsymbol{e}_3 \otimes \boldsymbol{e}_\theta + \boldsymbol{e}_\theta \otimes \boldsymbol{e}_3) + (r^{-1}w')^2\boldsymbol{e}_3 \otimes \boldsymbol{e}_3.$$

Then, $B_{rr} = B_{\theta\theta} = 1$, $B_{zz} = 1 + (r^{-1}w')^2$, $B_{\theta z} = B_{z\theta} = r^{-1}w'$, and all other polar components vanish. Thus, with reference to (1.162), we solve

$$grad p = G div\boldsymbol{B} = G(\frac{1}{r}B_{z\theta,\theta})\boldsymbol{e}_3 = G[r^{-2}w''(\theta)]\boldsymbol{e}_3.$$

This implies that p depends only on z, with $p'(z) = Gr^{-2}w''(\theta)$. Then, $p''(z) = 0$ and $p'(z) = P$, a constant. Thus, $r^2P/G = w''(\theta)$. Differentiate with respect to r to get $2rP/G = 0$. Thus, $P = 0$ and

$$w(\theta) = C\theta + D,$$

where C and D are constants. Let $B = w(2\pi) - w(0)$. Then, $C = B/2\pi$. If $B \neq 0$, then this deformation represents a *screw dislocation* with Burgers displacement B, whereas D represents a rigid-body displacement along the axis of the cylinder.
Using $\boldsymbol{B}\boldsymbol{e}_r = \boldsymbol{e}_r$, the tractions at the outer and inner cylindrical surfaces are found to be $\boldsymbol{t} = \pm(G - p)\boldsymbol{e}_r$, respectively. If these vanish, then $p(z) = G$, and this, in turn, is consistent with the finding that $P(= p'(z))$ vanishes.

12.

From (3.91), (3.96) and (3.97), we have

$$\boldsymbol{B} = \boldsymbol{G}\boldsymbol{G}^t = \boldsymbol{I} + r\tau(\boldsymbol{e}_\theta \otimes \boldsymbol{e}_3 + \boldsymbol{e}_3 \otimes \boldsymbol{e}_\theta) + r^2\tau^2\boldsymbol{e}_\theta \otimes \boldsymbol{e}_\theta.$$

Then, $B_{rr} = 1$, $B_{\theta\theta} = 1 + r^2\tau^2$, $B_{\theta z} = B_{z\theta} = r\tau$ and all other polar components vanish. We solve

$$grad p = G div \boldsymbol{B},$$

where, from (1.162),

$$div\boldsymbol{B} = \left[B_{rr,r} + \frac{1}{r}(B_{rr} - B_{\theta\theta})\right]\boldsymbol{e}_r = -r\tau^2\boldsymbol{e}_r,$$

yielding $p_{,\theta} = 0 = p_{,z}$ and $p'(r) = -G\tau^2 r$. Thus, $p(r) = p_0 - \frac{1}{2}G\tau^2r^2$, where $p_0 = p(0)$. With $\boldsymbol{T}\boldsymbol{e}_r = (G - p)\boldsymbol{e}_r$, the zero-traction condition at $r = a$ yields $p(a) = G$ and hence

$$p(r) = G + \frac{1}{2}G\tau^2(a^2 - r^2).$$

The force on a cross section of radius a is

$$\boldsymbol{f} = \int_0^a \left(\int_0^{2\pi} \boldsymbol{T}\boldsymbol{e}_3 d\theta\right) r dr,$$

where

$$\boldsymbol{T}\boldsymbol{e}_3 = [G - p(r)]\boldsymbol{e}_3 + G\tau\boldsymbol{e}_\theta.$$

Since $\boldsymbol{e}_\theta = -\boldsymbol{e}_r'(\theta)$, the interior integral reduces to

$$\int_0^{2\pi} \boldsymbol{T}\boldsymbol{e}_3 d\theta = 2\pi[G - p(r)]\boldsymbol{e}_3.$$

Thus,

$$\boldsymbol{f} = -\frac{\pi}{4}Ga^4\tau^2\boldsymbol{e}_3.$$

This is the compressive force required to maintain the end-to-end length of the cylinder. The length would increase in its absence. This prediction is corroborated by experiments.

The torque on a cross-section is

$$\boldsymbol{m} = \int_0^a \left(\int_0^{2\pi} \boldsymbol{x} \times \boldsymbol{T}\boldsymbol{e}_3 d\theta\right) r dr.$$

With $\boldsymbol{x} = r\boldsymbol{e}_r(\theta) + z\boldsymbol{e}_3$, we find that

$$\boldsymbol{m} = \frac{\pi}{2}Ga^4\tau\boldsymbol{e}_3,$$

in agreement with the classical strength-of-materials result.

13.

(a) We have

$$\begin{aligned}\boldsymbol{F}d\boldsymbol{X} = d\boldsymbol{x} &= f'(R)\boldsymbol{e}_r(\theta)dR + f(R)\boldsymbol{e}_r'(\theta)g'(\Theta)d\Theta + \boldsymbol{e}_3 dZ\\ &= f'(R)\boldsymbol{e}_r(\theta)[\hat{\boldsymbol{E}}_R(\Theta)\cdot d\boldsymbol{X}] + R^{-1}f(R)g'(\Theta)\boldsymbol{e}_\theta(\theta)\\ &\quad [\hat{\boldsymbol{E}}_\Theta(\Theta)\cdot d\boldsymbol{X}] + \boldsymbol{e}_3(\hat{\boldsymbol{E}}_3\cdot d\boldsymbol{X}).\end{aligned}$$

Thus,

$$\boldsymbol{F} = f'(R)\boldsymbol{e}_r(\theta)\otimes\hat{\boldsymbol{E}}_R(\Theta) + R^{-1}f(R)g'(\Theta)\boldsymbol{e}_\theta(\theta)\otimes\hat{\boldsymbol{E}}_\Theta(\Theta) + \boldsymbol{e}_3\otimes\hat{\boldsymbol{E}}_3.$$

(b) We require

$$\begin{aligned}1 = \det\boldsymbol{F} &= [\boldsymbol{F}\hat{\boldsymbol{E}}_R, \boldsymbol{F}\hat{\boldsymbol{E}}_\Theta, \boldsymbol{F}\hat{\boldsymbol{E}}_3]\\ &= [f'\boldsymbol{e}_r, R^{-1}fg'\boldsymbol{e}_\theta, \boldsymbol{e}_3]\\ &= R^{-1}f(R)f'(R)g'(\Theta).\end{aligned}$$

Thus,

$$R^{-1}f(R)f'(R) = C \quad \text{and} \quad g'(\Theta) = 1/C,$$

where C is a constant, and $g(\Theta) = \frac{\Theta}{C} + D$, where D is another constant. From the figure, $g(\pm\frac{\gamma}{2}) = \pm\pi$. Thus, $C = \frac{\gamma}{2\pi}$, $D = 0$ and

$$\theta = g(\Theta) = \frac{2\pi\Theta}{\gamma},$$

and the equation for $f(R)$ reduces to $(f^2)' = \frac{\gamma}{\pi}R$, which combines with $f(0) = 0$ to give

$$r = f(R) = \sqrt{\frac{\gamma}{2\pi}}R.$$

Thus, $a = \sqrt{\frac{\gamma}{2\pi}}A$.

(c) These results imply that

$$\boldsymbol{F} = \sqrt{\frac{\gamma}{2\pi}}\boldsymbol{e}_r(\theta) \otimes \hat{\boldsymbol{E}}_R(\Theta) + \sqrt{\frac{2\pi}{\gamma}}\boldsymbol{e}_\theta(\theta) \otimes \hat{\boldsymbol{E}}_\Theta(\Theta) + \boldsymbol{e}_3 \otimes \hat{\boldsymbol{E}}_3.$$

Let

$$\boldsymbol{Q} = \boldsymbol{e}_r(\theta) \otimes \hat{\boldsymbol{E}}_R(\Theta) + \boldsymbol{e}_\theta(\theta) \otimes \hat{\boldsymbol{E}}_\Theta(\Theta) + \boldsymbol{e}_3 \otimes \hat{\boldsymbol{E}}_3.$$

This is a rotation that rotates $\{\hat{\boldsymbol{E}}_R, \hat{\boldsymbol{E}}_\Theta, \hat{\boldsymbol{E}}_3\}$ to $\{\boldsymbol{e}_r, \boldsymbol{e}_\theta, \boldsymbol{e}_3\}$. Then,

$$\boldsymbol{F} = \boldsymbol{Q}\left[\sqrt{\frac{\gamma}{2\pi}}\hat{\boldsymbol{E}}_R(\Theta)\otimes\hat{\boldsymbol{E}}_R(\Theta) + \sqrt{\frac{2\pi}{\gamma}}\hat{\boldsymbol{E}}_\Theta(\Theta)\otimes\hat{\boldsymbol{E}}_\Theta(\Theta) + \hat{\boldsymbol{E}}_3\otimes\hat{\boldsymbol{E}}_3\right],$$

where the term in brackets is symmetric and positive definite. The uniqueness of the polar decomposition then yields

$$\boldsymbol{R} = \boldsymbol{Q}, \quad \boldsymbol{U} = \sqrt{\frac{\gamma}{2\pi}}\hat{\boldsymbol{E}}_R(\Theta) \otimes \hat{\boldsymbol{E}}_R(\Theta)$$
$$+ \sqrt{\frac{2\pi}{\gamma}}\hat{\boldsymbol{E}}_\Theta(\Theta) \otimes \hat{\boldsymbol{E}}_\Theta(\Theta) + \hat{\boldsymbol{E}}_3 \otimes \hat{\boldsymbol{E}}_3$$

and

$$\boldsymbol{V} = \boldsymbol{R}\boldsymbol{U}\boldsymbol{R}^t = \sqrt{\frac{\gamma}{2\pi}}\boldsymbol{e}_r(\theta) \otimes \boldsymbol{e}_r(\theta) + \sqrt{\frac{2\pi}{\gamma}}\boldsymbol{e}_\theta(\theta) \otimes \boldsymbol{e}_\theta(\theta) + \boldsymbol{e}_3 \otimes \boldsymbol{e}_3.$$

(d) The expressions for $\boldsymbol{U}$ and $\boldsymbol{V}$ are in spectral form. Thus, $\{\lambda_i\} = \{\sqrt{\frac{\gamma}{2\pi}}, \sqrt{\frac{2\pi}{\gamma}}, 1\}$.

(e) Again we solve

$$grad p = G div \boldsymbol{B},$$

this time with

$$\boldsymbol{B}(=\boldsymbol{V}^2) = \frac{\gamma}{2\pi}\boldsymbol{e}_r(\theta) \otimes \boldsymbol{e}_r(\theta) + \frac{2\pi}{\gamma}\boldsymbol{e}_\theta(\theta) \otimes \boldsymbol{e}_\theta(\theta) + \boldsymbol{e}_3 \otimes \boldsymbol{e}_3.$$

Thus,

$$div\boldsymbol{B} = \frac{1}{r}(B_{rr} - B_{\theta\theta})\boldsymbol{e}_r = \frac{1}{r}\left(\frac{\gamma}{2\pi} - \frac{2\pi}{\gamma}\right)\boldsymbol{e}_r.$$

Accordingly, p is a function of r, with derivative

$$p'(r) = \frac{G}{r}(\frac{\gamma}{2\pi} - \frac{2\pi}{\gamma}).$$

Thus,

$$p(r) = G\left(\frac{\gamma}{2\pi} - \frac{2\pi}{\gamma}\right)\ln r + E,$$

where E is a constant.

The traction furnished by this solution at $r = a$ is

$$\boldsymbol{T}\boldsymbol{e}_{r|r=a} = [G\frac{\gamma}{2\pi} - p(a)]\boldsymbol{e}_r = -P\boldsymbol{e}_r,$$

where P is a uniform pressure. This furnishes the strongly singular reaction pressure field

$$p(r) = G\left(\frac{\gamma}{2\pi} - \frac{2\pi}{\gamma}\right)\ln(r/a) + G\frac{\gamma}{2\pi} + P,$$

in terms of an arbitrary applied pressure P.

The traction on the upper plane surface of the closed gap is

$$\boldsymbol{T}\boldsymbol{e}_\theta(\pi) = \left[G\frac{2\pi}{\gamma} - p(r)\right]\boldsymbol{e}_\theta(\pi).$$

If the external traction vanishes, then $P = 0$ and this reduces to

$$\boldsymbol{T}\boldsymbol{e}_\theta(\pi) = \left\{G\left(\frac{2\pi}{\gamma} - \frac{\gamma}{2\pi}\right)[1 - \ln(r/a)]\right\}\boldsymbol{e}_\theta(\pi).$$

14.

The referential energy balance is given by (see (6.173))

$$\rho_\kappa \dot{\varepsilon} = \boldsymbol{P} \cdot \dot{\boldsymbol{F}} - Div\boldsymbol{q}_\kappa + \rho_\kappa r,$$

where $\varepsilon = \psi + \theta\eta$. Thus, $\dot{\varepsilon} = \dot{\psi} + \dot{\theta}\eta + \theta\dot{\eta} = \dot{\Psi} - \dot{\theta}\Psi_\theta + \theta\dot{\eta}$, yielding $\rho_\kappa \dot{\varepsilon} = \dot{W} - \dot{\theta} W_\theta + \rho_\kappa \theta\dot{\eta}$, with $\dot{W} = W_{\boldsymbol{F}} \cdot \dot{\boldsymbol{F}} + W_\theta \dot{\theta}$. This gives $\rho_\kappa \dot{\varepsilon} = \boldsymbol{P} \cdot \dot{\boldsymbol{F}} + \rho_\kappa \theta\dot{\eta}$, which reduces the energy balance to

$$\rho_\kappa \theta\dot{\eta} = \rho_\kappa r - Div\boldsymbol{q}_\kappa, \quad \text{where } \rho_\kappa \theta\dot{\eta} = -\theta(W_\theta)^{\cdot} = -\theta(W_{\theta\theta}\dot{\theta} + W_{\theta\boldsymbol{F}} \cdot \dot{\boldsymbol{F}}),$$

and the stated result follows.

15.

From (11.49), we have

$$\begin{aligned}
\boldsymbol{q}_\kappa &= J\boldsymbol{F}^{-1}\boldsymbol{q} = J(\phi_0 \boldsymbol{F}^{-1} + \phi_1 \boldsymbol{F}^t + \phi_2 \boldsymbol{C}\boldsymbol{F}^t) grad\theta \\
&= J(\phi_0 \boldsymbol{F}^{-1} + \phi_1 \boldsymbol{F}^t + \phi_2 \boldsymbol{C}\boldsymbol{F}^t)\boldsymbol{F}^{-t} Grad\theta \\
&= J(\phi_0 \boldsymbol{C}^{-1} + \phi_1 \hat{\boldsymbol{I}} + \phi_2 \boldsymbol{C}) Grad\theta,
\end{aligned}$$

where $\phi_{0,1,2}$ are functions of θ; of $I_k(\boldsymbol{B}) = I_k(\boldsymbol{C})$ for $k = 1, 2, 3$; and, of

$$\begin{aligned}
grad\theta \cdot grad\theta &= \boldsymbol{F}^{-t} Grad\theta \cdot \boldsymbol{F}^{-t} Grad\theta = Grad\theta \cdot \boldsymbol{F}^{-1}\boldsymbol{F}^{-t} Grad\theta \\
&= Grad\theta \cdot \boldsymbol{C}^{-1} Grad\theta, \\
grad\theta \cdot \boldsymbol{B} grad\theta &= grad\theta \cdot \boldsymbol{F}\boldsymbol{F}^t grad\theta = \boldsymbol{F}^t grad\theta \cdot \boldsymbol{F}^t grad\theta \\
&= Grad\theta \cdot Grad\theta, \quad \text{and} \\
grad\theta \cdot \boldsymbol{B}^2 grad\theta &= grad\theta \cdot \boldsymbol{F}(\boldsymbol{F}^t\boldsymbol{F})\boldsymbol{F}^t grad\theta \\
&= \boldsymbol{F}^t grad\theta \cdot (\boldsymbol{F}^t\boldsymbol{F})\boldsymbol{F}^t grad\theta = Grad\theta \cdot \boldsymbol{C} Grad\theta.
\end{aligned}$$

We can use the Cayley–Hamilton formula (see Problem 7 in Chapter 3) in the form

$$\boldsymbol{C}^{-1} = I_3^{-1}(\boldsymbol{C}^2 - I_1\boldsymbol{C} + I_2\hat{\boldsymbol{I}}),$$

together with $J = \sqrt{I_3}$, to write

$$\boldsymbol{q}_\kappa = (\psi_0 \hat{\boldsymbol{I}} + \psi_1 \boldsymbol{C} + \psi_2 \boldsymbol{C}^2) Grad\theta,$$

where $\psi_{0,1,2}$ are functions of the variables mentioned in the problem statement.

16.

Recall that for the torsion problem, the left Cauchy–Green tensor is given by

$$\boldsymbol{B} = \boldsymbol{I} + r\tau(\boldsymbol{e}_\theta \otimes \boldsymbol{e}_3 + \boldsymbol{e}_3 \otimes \boldsymbol{e}_\theta) + r^2\tau^2 \boldsymbol{e}_\theta \otimes \boldsymbol{e}_\theta.$$

Then,

$$\boldsymbol{B}^2 = \boldsymbol{B} + r\tau(\boldsymbol{B}\boldsymbol{e}_\theta \otimes \boldsymbol{e}_3 + \boldsymbol{B}\boldsymbol{e}_3 \otimes \boldsymbol{e}_\theta) + r^2\tau^2 \boldsymbol{B}\boldsymbol{e}_\theta \otimes \boldsymbol{e}_\theta.$$

With $grad\theta = \theta'(z)\boldsymbol{e}_3$, we have

$$\boldsymbol{B}grad\theta = \theta'(z)(\boldsymbol{e}_3 + r\tau\boldsymbol{e}_\theta) \quad \text{and} \quad \boldsymbol{B}^2 grad\theta = \theta'(z)[\boldsymbol{B}\boldsymbol{e}_3 + r\tau\boldsymbol{B}\boldsymbol{e}_\theta]$$
$$= \theta'(z)[(1 + r^2\tau^2)\boldsymbol{e}_3 + r\tau(2 + r^2\tau^2)\boldsymbol{e}_\theta],$$

and

$$\boldsymbol{q} = \theta'(z)\{[(\phi_0 + \phi_1) + \phi_2(1 + r^2\tau^2)]\boldsymbol{e}_3 + [\phi_1 r\tau + \phi_2 r\tau(2 + r^2\tau^2)]\boldsymbol{e}_\theta\}.$$

The flux across any cylindrical surface, including the exterior surface, is $\boldsymbol{q} \cdot \boldsymbol{e}_r = 0$.

We have

$$I_1 = I_2 = 3 + r^2\tau^2, \quad I_3 = 1, \quad I_4 = (\theta')^2, \quad I_5 = (\theta')^2(1 + r^2\tau^2)$$

and

$$I_6 = |\theta'|.$$

Then, with $\theta'(z) = (\theta_l - \theta_0)/l$, we conclude that

$$\boldsymbol{q} = q_\theta(r)\boldsymbol{e}_\theta + q_z(r)\boldsymbol{e}_3,$$

for some functions q_θ and q_z, and hence, with (1.158), that $div\boldsymbol{q} = 0$. This, in turn, implies that $Div\boldsymbol{q}_\kappa$ vanishes, and thus, because the deformation and temperature fields are static (i.e., $\dot{\theta} = 0$ and $\dot{\boldsymbol{F}} = \boldsymbol{O}$), that the energy balance with zero internal heating is identically satisfied.

Chapter 12

1.

Recall that $\boldsymbol{H}^+ = \boldsymbol{QHK}^t$ and $\mathbf{1}^+ = \boldsymbol{Q}\mathbf{1}\boldsymbol{K}^t$. Then, $\boldsymbol{G}^+ = (\mathbf{1}^+)^t \boldsymbol{H}^+ = \boldsymbol{K}\mathbf{1}^t\boldsymbol{Q}^t\boldsymbol{QHK}^t = \boldsymbol{K}\mathbf{1}^t\boldsymbol{IHK}^t = \boldsymbol{K}\mathbf{1}^t\boldsymbol{HK}^t = \boldsymbol{KGK}^t$. This gives $|\boldsymbol{G}^+|^2 = tr[\boldsymbol{G}^+(\boldsymbol{G}^+)^t] = tr(\boldsymbol{KGK}^t\boldsymbol{KG}^t\boldsymbol{K}^t) = tr(\boldsymbol{KG}\hat{\boldsymbol{I}}\boldsymbol{G}^t\boldsymbol{K}^t) = tr(\boldsymbol{K}^t\boldsymbol{KGG}^t) = tr(\hat{\boldsymbol{I}}\boldsymbol{GG}^t) = tr(\boldsymbol{GG}^t) = |\boldsymbol{G}|^2$. Further, $2\epsilon^+ = \boldsymbol{G}^+ + (\boldsymbol{G}^+)^t = \boldsymbol{K}(\boldsymbol{G} + \boldsymbol{G}^t)\boldsymbol{K}^t = 2\boldsymbol{K}\epsilon\boldsymbol{K}^t$.

2.

The problem is to establish that the stress power is given by

$$\mathcal{S}(S,t) = \frac{d}{dt}\int_P U dV.$$

From (12.32), we have

$$\frac{d}{dt}\int_P U dV = \int_P \dot{U} dV = \int_P U_{\boldsymbol{E}} \cdot \dot{\boldsymbol{E}} dV = \int_P \boldsymbol{S} \cdot \dot{\boldsymbol{E}} dV,$$

and the result then follows from (6.160).

3.

From $(12.50)_1$ and (12.71), we have $\boldsymbol{0} = Div(\boldsymbol{S}_0) = Div(\alpha\hat{\boldsymbol{I}}) = Grad\alpha$. Thus, α is constant. From (12.64) we have $\alpha\boldsymbol{N} = \boldsymbol{0}$ on a portion of the boundary. Thus, $\alpha = 0$ there, and hence everywhere in the body. Thus, $\boldsymbol{S}_0 = \boldsymbol{O}$. (See also Problem 10 in Chapter 11).

4.

We have $\boldsymbol{F}^* = (\boldsymbol{RU})^* = \boldsymbol{R}^*\boldsymbol{U}^* = \boldsymbol{RU}^* = J\boldsymbol{RU}^{-1}$. Use $(12.20)_{2,3}$ and (12.25) to estimate this as

$$\begin{aligned}\boldsymbol{F}^* &\simeq (1 + tr\boldsymbol{G})\mathbf{1}(\hat{\boldsymbol{I}} + \hat{\boldsymbol{\omega}})(\hat{\boldsymbol{I}} - \epsilon)\\ &\simeq (1 + tr\boldsymbol{G})\mathbf{1}(\hat{\boldsymbol{I}} + \hat{\boldsymbol{\omega}} - \epsilon)\\ &= (1 + tr\boldsymbol{G})\mathbf{1}(\hat{\boldsymbol{I}} - \boldsymbol{G}^t)\\ &\simeq \mathbf{1} + \mathbf{1}[(tr\boldsymbol{G})\hat{\boldsymbol{I}} - \boldsymbol{G}^t].\end{aligned}$$

Thus, $\boldsymbol{u} \cdot \boldsymbol{n}da = \boldsymbol{u} \cdot \boldsymbol{F}^*\boldsymbol{N}dA \simeq \boldsymbol{u} \cdot \mathbf{1}\boldsymbol{N}dA = \mathbf{1}^t\boldsymbol{u} \cdot \boldsymbol{N}dA = \boldsymbol{w} \cdot \boldsymbol{N}dA$.

5.

We have

$$\begin{aligned}\mathcal{C}_{ijkl} &= \delta_{iA}\delta_{jB}\delta_{kC}\delta_{lD}\mathcal{D}_{ABCD} \\ &= \delta_{kC}\delta_{lD}\delta_{iA}\delta_{jB}\mathcal{D}_{CDAB} \quad \text{from} \quad (12.38) \\ &= \mathcal{C}_{klij}.\end{aligned}$$

Also,

$$\begin{aligned}\mathcal{C}_{ijkl} &= \delta_{jB}\delta_{iA}\delta_{kC}\delta_{lD}\mathcal{D}_{BACD} \quad \text{from} \quad (12.40) \\ &= \mathcal{C}_{jikl},\end{aligned}$$

and, finally,

$$\begin{aligned}\mathcal{C}_{ijkl} &= \delta_{iA}\delta_{jB}\delta_{lD}\delta_{kC}\mathcal{D}_{ABDC} \quad \text{from} \quad (12.41) \\ &= \mathcal{C}_{ijlk}.\end{aligned}$$

6.

We have $U = \frac{1}{2}\boldsymbol{E}\cdot\mathcal{D}[\boldsymbol{E}] \simeq \frac{1}{2}\boldsymbol{\epsilon}\cdot\mathcal{D}[\boldsymbol{\epsilon}]$. Thus, $U_{\boldsymbol{\epsilon}}\cdot\dot{\boldsymbol{\epsilon}} \simeq \dot{U} \simeq \mathcal{D}[\boldsymbol{\epsilon}]\cdot\dot{\boldsymbol{\epsilon}} \simeq \boldsymbol{S}\cdot\dot{\boldsymbol{\epsilon}}$, and hence $\boldsymbol{S} \simeq U_{\boldsymbol{\epsilon}}$. Using $\boldsymbol{\epsilon} \simeq \mathbf{1}^t\boldsymbol{\varepsilon}\mathbf{1}$, we also have $U \simeq \frac{1}{2}\mathbf{1}^t\boldsymbol{\varepsilon}\mathbf{1}\cdot\mathcal{D}[\mathbf{1}^t\boldsymbol{\varepsilon}\mathbf{1}] = \frac{1}{2}\boldsymbol{\varepsilon}\cdot\mathbf{1}(\mathcal{D}[\mathbf{1}^t\boldsymbol{\varepsilon}\mathbf{1}])\mathbf{1}^t = \frac{1}{2}\boldsymbol{\varepsilon}\cdot\mathcal{C}[\boldsymbol{\varepsilon}]$. Thus, $U_{\boldsymbol{\varepsilon}}\cdot\dot{\boldsymbol{\varepsilon}} \simeq \dot{U} \simeq \mathcal{C}[\boldsymbol{\varepsilon}]\cdot\dot{\boldsymbol{\varepsilon}} \simeq \boldsymbol{T}\cdot\dot{\boldsymbol{\varepsilon}}$, and hence $\boldsymbol{T} \simeq U_{\boldsymbol{\varepsilon}}$.

7.

We have $S_{AB} \simeq \mathcal{D}_{ABCD}\epsilon_{CD}$. It is easily verified that

$$\mathcal{D}_{ABCD} = \lambda\delta_{AB}\delta_{CD} + \mu(\delta_{AC}\delta_{BD} + \delta_{AD}\delta_{BC})$$

has all the required symmetries, and delivers

$$S_{AB} \simeq \lambda\epsilon_{CC}\delta_{AB} + \mu(\epsilon_{AB} + \epsilon_{BA}) = \lambda\epsilon_{CC}\delta_{AB} + 2\mu\epsilon_{AB},$$

as required by (12.53). Then,

$$\begin{aligned}\mathcal{C}_{ijkl} &= \delta_{iA}\delta_{jB}\delta_{kC}\delta_{lD}\mathcal{D}_{ABCD} \\ &= \lambda\delta_{iA}\delta_{jA}\delta_{kC}\delta_{lC} + \mu(\delta_{iC}\delta_{kC}\delta_{jD}\delta_{lD} + \delta_{iD}\delta_{lD}\delta_{jC}\delta_{kC}) \\ &= \lambda\delta_{ij}\delta_{kl} + \mu(\delta_{ik}\delta_{jl} + \delta_{il}\delta_{jk}).\end{aligned}$$

To prove that the strain-energy function is positive definite if and only if $\mu > 0$ and $\kappa > 0$, we decompose $\boldsymbol{\epsilon}$ or $\boldsymbol{\varepsilon}$ as the sum of spherical and deviatoric parts and proceed exactly as in the passage from (9.75) to (9.79), with obvious notational adjustments.

8.

The result follows easily from $\rho \simeq \rho_\kappa$, together with (6.109) and (12.68).

9.

To prove (12.73), we follow the steps leading to (7.52), with $\boldsymbol{T}$ replaced by $\boldsymbol{S}$ and $\boldsymbol{Q}$ replaced by $\boldsymbol{K}$. From (7.18) and (7.21), we have

$$\boldsymbol{w}^+ = (\boldsymbol{1}^+)^t \boldsymbol{u}^+ = \boldsymbol{K}\boldsymbol{1}^t\boldsymbol{Q}^t\boldsymbol{Q}\boldsymbol{1}\boldsymbol{w} + \boldsymbol{K}\boldsymbol{1}^t\boldsymbol{Q}^t\boldsymbol{d},$$

with $\boldsymbol{d} = \boldsymbol{c} - \boldsymbol{Q}\boldsymbol{1}\boldsymbol{K}^t\boldsymbol{c}^+$ and $\boldsymbol{c}^+ = (\boldsymbol{1}_0^+)^t\boldsymbol{c}_0$, where $\boldsymbol{c}_0 = \boldsymbol{c}(t_0)$ and $\boldsymbol{1}_0^+ = \boldsymbol{1}^+(t_0^+) = \boldsymbol{Q}(t_0)\boldsymbol{1}\boldsymbol{K}^t$. Thus, with $\boldsymbol{K}^t\boldsymbol{K} = \hat{\boldsymbol{I}}$, $\boldsymbol{1}\hat{\boldsymbol{I}} = \boldsymbol{1}$ and $\boldsymbol{Q}\boldsymbol{I} = \boldsymbol{Q}$ we obtain

$$\begin{aligned}\boldsymbol{d}(t) &= \boldsymbol{c}(t) - \boldsymbol{Q}(t)\boldsymbol{1}\boldsymbol{1}^t\boldsymbol{Q}(t_0)^t\boldsymbol{c}_0 = \boldsymbol{c}(t) - \boldsymbol{Q}(t)\boldsymbol{I}\boldsymbol{Q}(t_0)^t\boldsymbol{c}_0 \\ &= \boldsymbol{c}(t) - \boldsymbol{Q}(t)\boldsymbol{Q}(t_0)^t\boldsymbol{c}_0,\end{aligned}$$

which combines with $\boldsymbol{Q}^t\boldsymbol{Q} = \boldsymbol{I}$ and $\boldsymbol{1}^t\boldsymbol{I} = (\boldsymbol{I}\boldsymbol{1})^t = \boldsymbol{1}^t$ to give

$$\boldsymbol{w}^+ = \boldsymbol{K}\boldsymbol{1}^t[\boldsymbol{1}\boldsymbol{w} + \boldsymbol{Q}(t)^t\boldsymbol{c}(t) - \boldsymbol{Q}(t_0)^t\boldsymbol{c}(t_0)].$$

Finally, with $\boldsymbol{K}\boldsymbol{1}^t\boldsymbol{1}\boldsymbol{w} = \boldsymbol{K}\hat{\boldsymbol{I}}\boldsymbol{w} = \boldsymbol{K}\boldsymbol{w}$ we have

$$\boldsymbol{w}^+ = \boldsymbol{K}\boldsymbol{w} + \boldsymbol{K}\boldsymbol{1}^t[\boldsymbol{Q}(t)^t\boldsymbol{c}(t) - \boldsymbol{Q}(t_0)^t\boldsymbol{c}(t_0)]. \qquad (*)$$

Under a Galilean transformation we have, from (7.10), that $\boldsymbol{Q}(t) = \boldsymbol{Q}(t_0)$ and $\boldsymbol{c}(t) = \boldsymbol{V}t + \boldsymbol{c}(t_0)$. Then,

$$\boldsymbol{w}^+ = \boldsymbol{K}\boldsymbol{w} + (\boldsymbol{1}_0^+)^t\boldsymbol{V}t \quad \text{and} \quad \ddot{\boldsymbol{w}}^+ = \boldsymbol{K}\ddot{\boldsymbol{w}}.$$

10.

Continuing from the general formula (*), we have

$$\begin{aligned}\dot{\boldsymbol{w}}^+ &= \boldsymbol{K}\dot{\boldsymbol{w}} + \boldsymbol{K}\mathbf{1}^t(\dot{\boldsymbol{Q}}^t\boldsymbol{c} + \boldsymbol{Q}^t\dot{\boldsymbol{c}}) \\ &= \boldsymbol{K}[\dot{\boldsymbol{w}} + \mathbf{1}^t\boldsymbol{Q}^t(\dot{\boldsymbol{c}} - \boldsymbol{\Omega}\boldsymbol{c})], \quad \text{from (7.8)}.\end{aligned}$$

Then,

$$\begin{aligned}\ddot{\boldsymbol{w}}^+ &= \boldsymbol{K}[\ddot{\boldsymbol{w}} + \mathbf{1}^t\dot{\boldsymbol{Q}}^t(\dot{\boldsymbol{c}} - \boldsymbol{\Omega}\boldsymbol{c}) + \mathbf{1}^t\boldsymbol{Q}^t(\ddot{\boldsymbol{c}} - \dot{\boldsymbol{\Omega}}\boldsymbol{c} - \boldsymbol{\Omega}\dot{\boldsymbol{c}})] \\ &= \boldsymbol{K}[\ddot{\boldsymbol{w}} + \mathbf{1}^t\boldsymbol{Q}^t(\ddot{\boldsymbol{c}} - \dot{\boldsymbol{\Omega}}\boldsymbol{c} - \boldsymbol{\Omega}\dot{\boldsymbol{c}}) - \mathbf{1}^t\boldsymbol{Q}^t\boldsymbol{\Omega}(\dot{\boldsymbol{c}} - \boldsymbol{\Omega}\boldsymbol{c})] \\ &= \boldsymbol{K}[\ddot{\boldsymbol{w}} + \mathbf{1}^t\boldsymbol{Q}^t[\ddot{\boldsymbol{c}} - 2\boldsymbol{\Omega}\dot{\boldsymbol{c}} - (\dot{\boldsymbol{\Omega}} - \boldsymbol{\Omega}^2)\boldsymbol{c}]] \\ &= \boldsymbol{K}\ddot{\boldsymbol{w}} + (\mathbf{1}^+)^t[\ddot{\boldsymbol{c}} - 2\boldsymbol{\Omega}\dot{\boldsymbol{c}} - (\dot{\boldsymbol{\Omega}} - \boldsymbol{\Omega}^2)\boldsymbol{c}],\end{aligned}$$

and substitution into (12.74) gives the equation of motion

$$Div^+\boldsymbol{S}^+ + \rho^+_{\kappa^+}(\boldsymbol{f}^+ + \boldsymbol{j}^+) = \rho^+_{\kappa^+}\ddot{\boldsymbol{w}}^+$$

in the frame of $\mathcal{O}^+$, where

$$\boldsymbol{j}^+ = (\mathbf{1}^+)^t[\ddot{\boldsymbol{c}} - 2\boldsymbol{\Omega}\dot{\boldsymbol{c}} - (\dot{\boldsymbol{\Omega}} - \boldsymbol{\Omega}^2)\boldsymbol{c}].$$

References

[1] R. M. Bowen and C. C. Wang. *Introduction to Vectors and Tensors, Volumes 1 & 2*. Plenum Press, 1976.

[2] W. Fleming. *Functions of Several Variables*. Springer, N.Y., 1977.

[3] M. E. Gurtin. *An Introduction to Continuum Mechanics*. Academic Press, Orlando, 1981.

[4] R. W. Ogden. *Non-linear Elastic Deformations*. Dover, N.Y., 1997.

[5] R. A. Stephenson. "On the uniqueness of the square-root of a symmetric, positive-definite tensor". *J. Elasticity* 10(2) (1980), 213–214.

[6] P. Chadwick. *Continuum Mechanics: Concise Theory and Problems*. Dover, N.Y., 1999.

[7] I-Shih Liu. *Continuum Mechanics*. Springer, 2002.

[8] M. E. Gurtin, E. Fried and L. Anand. *The Mechanics and Thermodynamics of Continua*. Cambridge University Press, 2010.

[9] M. Shirani and D. J. Steigmann. "A Cosserat model of elastic solids reinforced by a family of curved and twisted fibers". *Symmetry* 12(7) (2020), 1133.

[10] W. Noll. "The foundations of classical mechanics in the light of recent advances in continuum mechanics". In *The Foundations of Mechanics and Thermodynamics: Selected Papers*: Berlin, Heidelberg: Springer Berlin Heidelberg, 1974, pp. 31–47.

[11] C. A. Truesdell. *A First Course in Rational Continuum Mechanics V1*. Academic Press, Boston, 1992.

[12] C. A. Truesdell. *Rational Thermodynamics*. New York, McGraw-Hill, 1969.

[13] A. I. Murdoch. "Objectivity in classical continuum physics: a rationale for discarding the 'principle of invariance under superposed rigid body motions' in favour of purely objective considerations". *Cont. Mech. & Thermodynam.* 15(3) (2003), 309–320.

[14] A. Wineman. "Nonlinear viscoelastic solids: a review". *Math. Mech. Solids* 14(3) (2009), 300–366.

[15] R. I. Tanner. *Engineering Rheology.* Oxford University Press, 2000.

[16] W. Noll. "A mathematical theory of the mechanical behavior of continuous media". In *The Foundations of Mechanics and Thermodynamics: Selected Papers*: Berlin, Heidelberg: Springer Berlin Heidelberg, 1974, pp. 1–30.

[17] M. Epstein and M. Elzanowski. *Material Inhomogeneities and Their Evolution: A Geometric Approach.* Springer, Berlin, 2007.

[18] D. J. Steigmann. *A Course on Plasticity Theory.* Oxford University Press, 2022.

[19] W. Noll. "Materially Uniform Simple Bodies with Inhomogeneities". In *The Foundations of Mechanics and Thermodynamics: Selected Papers*: Berlin, Heidelberg: Springer Berlin Heidelberg, 1974, pp. 211–242.

[20] B. D. Coleman and W. Noll. "The thermodynamics of elastic materials with heat conduction and viscosity". In *The Foundations of Mechanics and Thermodynamics*: Selected Papers: Berlin, Heidelberg: Springer Berlin Heidelberg, 1974, pp. 145–156.

[21] R. M. Bowen. *Introduction to Continuum Mechanics for Engineers.* New York, Dover, 2009.

[22] M. E. Gurtin. *Configurational Forces as Basic Concepts of Continuum Physics.* Springer, N.Y., 2000.

[23] C. D. Coman. *Continuum Mechanics and Linear Elasticity.* Springer, 2020.

[24] D. E. Carlson. "On the range of applicability of linearized elasticity". *Math. Mech. Solids* 16(5) (2011), 467–481.

[25] D. J. Steigmann. "On the frame invariance of linear elasticity theory". *Zeitschrift für angewandte Mathematik und Physik.* 58 (2007), 121–136.

Index

G

H

I

J

K

L

O

P

R

S

W

Y

Z

www.ingramcontent.com/pod-product-compliance
Lightning Source LLC
Chambersburg PA
CBHW061624220326
41598CB00026BA/3865
9789819807543